Chief Editors

Wu Wen-tsun   Cheng Min-de

Chinese Mathematics into the 21st Century

Peking University Press
Beijing, China

Springer-Verlag
Berlin Heidelberg New York London Paris
Tokyo Hong Kong Barcelona Budapest

Responsible Editor Liu Yong   Wang Mingzhou

Distribution rights throughout the world excluding the People's Republic of China granted to Springer-Verlag

ISBN 7-301-01622-1/O · 260 Peking University Press

ISBN 3-540-54378-3 Springer-Verlag Berlin Heidelberg New York
ISBN 0-387-54378-3 Springer-Verlag New York Berlin Heidelberg

Published by Peking University Press, Beijing, China

Printed by A & U Publication HK Ltd., Hong Kong
First edition 1991

Preface

This is a collection of papers presented to the symposium on "Chinese Mathematics into the 21st Century". Its aim was to bring together the young mathematicians in and outside of China for a discussion of the prospects and issues on the development of mathematics in China. The meeting took place on Aug. 20–24,1988 at the Nankai Institute of Mathematics, Nankai University, Tianjin, China. Participants include 122 from China and 45 from abroad. We wish to thank the Chinese National Science Foundation for the support which made the meeting possible. We earnestly hope that further meetings for the same purpose could be organized.

S. S. Chern

Contents

Obstruction Sets, Minimal Rambling Sets and Their Applications

Liao Shantao
Institute of Mathematics, Peking University
Beijing, 100871, P.R.China

Abstract

In this article, we survey the so-called obstruction sets, minimal rambling sets and some topics related to them in the theory of differentiable dynamical systems. In particular, we are concerned with the stability problems of vector fields.

The study of differentiable dynamical systems began to become alive in mathematical world in the early part of the sixties of this century, starting with the pioneering work [1] of M. Peixoto. The developments in many aspects by S. Smale and other mathematicians have made this field fruitful and far-reaching. A good reference for this history is the book [2]; in particular, the 1967 Bulletin AMS survey article of Smale is reprinted there.

In the present article, we are mainly concerned with the so-called obstruction sets and minimal rambling sets. These are concepts which appear useful but unfortunately not so well-known as many other important things such as hyperbolicity, Axiom A and strong transversality etc.. In the last section of this article, we describe the use of obstruction sets and minimal rambling sets in the study of stability problems of differential systems.

1 Obstruction sets

Let M^n be a compact n-dimensional C^∞ Riemann manifold ($n \geq 2$), and $\mathcal{X}(M^n)$ the linear space of all C^1 differential systems (tangent vector fields) on M^n, endowed with the C^1 norm $\| \cdot \|_1$. A system $S \in \mathcal{X}(M^n)$

induces a C^1 one-parameter transformation group

$$\phi_t : M^n \to M^n \quad (-\infty < t < \infty)$$

and induces therefore an one-parameter transformation group

$$\Phi_t = d\phi_t : C \to C(-\infty < t < \infty)$$

over the tangent bundle

$$C = \bigcup_{x \in M^n} C_x$$

(C_x=the tangent space at x of M^n).

Denote by M the set of all ordinary points of S. Consider the conjugate bundle

$$D = \bigcup_{x \in M} D_x$$

of S, which is the bundle with base M and with fibre

$$D_x = \{u \in C_x \mid \langle u, S(x) \rangle = 0\}$$

over $x \in M$. For $u \in D_x$ ($x \in M$), take $\psi_t(u)$ as the projection of $\Phi_t(u)$ on $D_{\phi_t(x)}$. This gives an one-parameter transformation group

$$\psi_t : D \to D(-\infty < t < \infty).$$

For $u \in D_x$ ($x \in M$), denote by $\otimes u$ the linear subspace $\{v \in D_x \mid \langle u, v \rangle = 0\}$. Then, for each $t \in (-\infty, \infty)$ we can take a unique $\Psi_t(u) \in D_{\phi_t(x)}$ orthogonal to $\psi_t(\otimes u)$ and such that

$$\psi_t(u) - \Psi_t(u) \in \psi_t(\otimes u).$$

This gives an one-parameter transformation group

$$\Psi_t : D \to D(-\infty < t < \infty).$$

This transformation group can extend uniquely to an one-parameter transformation group (still denoted by)

$$\Psi_t : D^* \to D^*(-\infty < t < \infty)$$

over the closure D^* of D in C. The obstruction set $Ob(S)$ of S is defined as

$$Ob(S) = \left\{ x \in M^n \mid \exists u \in D^* \bigcap C_x \right.$$

$$\left. \text{such that } \|u\| = 1 = \inf_{t \in (-\infty, \infty)} \|\Psi_t(u)\| \right\}$$

([3], [4]). This is a closed subset of M^n.

2 Relations between obstruction sets and transversality conditions

For any $x \in M^n$, write

$$
\begin{aligned}
F_-(x) &= \left\{ u \in C_x \mid \lim_{t \to \infty} \|\Phi_t(u)\| = 0 \right\}, \\
F_+(x) &= \left\{ u \in C_x \mid \lim_{t \to -\infty} \|\Phi_t(u)\| = 0 \right\}.
\end{aligned}
$$

These are linear subspace of C_x.

Let Λ be a closed subset of M^n, invariant under $\phi_t(-\infty < t < \infty)$. We say that S satisfies the linear transversality condition over Λ, if for every $x \in \Lambda$,

$$
C_x = F_-(x) + F_+(x) + \{S(x)\}.
$$

This is a concept which generalizes an analogous one for discrete systems introduced previously in Robbin [5]. When S has hyperbolic structure over Λ, S satisfies obviously the linear transversality condition over Λ.

Denote by $I(S)$ the interior of the set of all singularities of S.

Theorem 2.1 ([3],[4]). *S satisfies the linear transversality condition over a closed subset Λ of M^n, invariant under $\phi_t(-\infty < t < \infty)$, if and only if*

$$
\Lambda \bigcap \left(I(S) \bigcup Ob(S) \right) = 0.
$$

Corollary 2.2. *A singularity a of S is hyperbolic if and only if*

$$
a \overline{\in} I(S) \bigcup Ob(S).
$$

A periodic orbit P of S is hyperbolic if and only if

$$
P \bigcap Ob(S) = 0.
$$

Remark. In [6,p.278], a linear system of differential equations under consideration may be regarded as the linear part of a system of standard equations for S, and a lemma there describes a property of the linear system which suggested the present writer to find a certain subset of M^n associated with S, which becomes vacuous over every hyperbolic set of S. This is the obstruction set which we obtained later in [3].

Theorem 2.3 ([7]). *S satisfies Axiom A and the strong transversality condition if and only if*

$$I(S) \bigcup Ob(S) = 0.$$

The following theorem, which is analogous to Theorem 2.3, can be also proved. Note that the chain recurrent set $\check{R}(S)$ of S contains the nonwandering set $\Omega(s)$ of S.

Theorem 2.4. *S satisfies Axiom A and the no-cycle condition if and only if*

$$\check{R}(S) \bigcap \left(I(S) \bigcup Ob(S) \right) = 0.$$

For definitions of hyperbolicity, Axiom A and strong transversality etc., one can consult with [2, pp.1-82] and some references cited there.

In view of the above theorems, a closed subset Λ of M^n, invariant under $\phi_t(-\infty < t < \infty)$ and satisfying $\Lambda \bigcap (I(S) \bigcup Ob(S)) = 0$, is of special interest. Such a set Λ will be called a normal set of S.

3   A filtration for normal sets

For a p-dimensional linear subspace L of C_x, $x \in M^n$, let $*L$ denote the $(n - p)$-dimensional linear subspace of C_x, orthogonal to L. Write

$$\begin{aligned}
B_-(x) &= *(F_+(x) + \{S(x)\}), \\
B_+(x) &= *(F_-(x) + \{S(x)\})
\end{aligned}$$

for any $x \in M^n$.

Theorem 3.1 ([3], [4]). *For any normal set Λ of S there are numbers $\eta_\Lambda < 0$ and $d_\Lambda > 0$ such that*

$$\|\Psi_t(u)\| \le \|u\| \exp(\eta_\Lambda t) \qquad \textit{for all } u \in B_-(x), \; x \in \Lambda, \textit{ and } t \ge d_\Lambda,$$
$$\|\Psi_t(u)\| \ge \|u\| \exp(-\eta_\Lambda t) \qquad \textit{for all } u \in B_+(x), \; x \in \Lambda, \textit{ and } t \ge d_\Lambda.$$

Let Λ be a normal set of S. For each $x \in \Lambda$, write

$$\mathrm{gr}(x) = n - \dim B_-(x) - \dim B_+(x)$$

and call the triple

$$(\mathrm{gr}(x), \dim B_-(x), \dim B_+(x))$$

the grade type of x. We see easily that $\mathrm{gr}(x) = 0$ if and only if x is a singular point of S. For any integers $p \geq 0$, $q \geq 0$ and $r \geq 0$ with $p + q + r = n$, let

$$\begin{aligned}
\Lambda_{q,r}^p &= \{x \in \Lambda \mid x \text{ has grade type } (p,q,r)\}, \\
\Lambda^p &= \bigcup_{(p',q',r'),\, p' \leq p} \Lambda_{q',r'}^{p'}.
\end{aligned}$$

Denote by Ω_Λ^* the set consisting of all $x \in \Lambda^1$ such that both the ω-limit set and the α-limit set of the orbit of S through x contain no singularities of S. Denote also by Ω_Λ the nonwandering set of the dynamical system $\phi_t(-\infty < t < \infty)$ restricted on Λ.

Theorem 3.2 ([3],[4]). *Let Λ be a normal set of S. Then:*

(1) *In the filtration*

$$0 = \Lambda^{-1} \subset \Lambda^0 \subset \Lambda^1 \subset \Lambda^2 \subset \cdots \subset \Lambda^n = \Lambda,$$

each Λ^p is closed in Λ, each $\Lambda_{q,r}^p$ is open in $\Lambda^p - \Lambda^{p-1}$ and they are all invariant under $\phi_t(-\infty < t < \infty)$.

(2) *For each $x \in \Lambda_{q,r}^p$ with $p > 0$, the ω-limit set Γ_x of the orbit P_x of S through x is contained in $\Lambda_{p+q,r}^0$ or in $\Lambda_{p+q-1,r}^1$ according as Γ_x contains singularities of S or not, and the α-limit set Γ_x' of P_x is contained in $\Lambda_{p,q+r}^0$ or $\Lambda_{q,p+r-1}^1$ according as Γ_x' contains singularities of S or not.*

(3) *Ω_Λ^* is closed in Λ and invariant under $\phi_t(-\infty < t < \infty)$, and S has hyperbolic structure over Ω_Λ^* and hence over $\Omega_\Lambda \subset \Lambda^0 \bigcup \Omega_\Lambda^*$.*

(4) *Each singularity x_0 in Λ of S has a neighborhood V in M^n such that for any $x \in \Lambda \bigcap (M^n - V)$, V cannot meet both the positive orbit $\{\phi_t(x) \mid t \in \langle 0, \infty)\}$ and the negative orbit $\{\phi_t(x) \mid t \in (-\infty, 0\rangle\}$.*

The following can be deduced by virtue of the above theorem.

Theorem 3.3 ([7]). *If Λ is a normal set of S, Then Ω_Λ lies in the closure of the set of all singularities and points in periodic orbits of S.*

4   Minimal rambling sets

A subset Λ of M^n is called a rambling set of S if it is closed in M^n, invariant under $\phi_t(-\infty < t < \infty)$, and $\Lambda \bigcap Ob(S) \neq 0$. A rambling set is called minimal if it contains no proper rambling subset.

It is clear by using the Brouwer reduction theorem that every rambling set of S contains at least one minimal rambling subset.

A minimal rambling set of S is called simple if either (1) Λ contains no ordinary points of S or (2) $\Lambda \bigcup Ob(S)$ contains at least one ordinary point a of S such that both the ω-limit set and the α-limit set of the orbit of S through a are proper subset of Λ. Otherwise, a minimal rambling set will be called non-simple. (See [8].)

It is an important procedure to analyze the structure of minimal rambling sets in order to understand the dynamical behavior of a system. The structure of a simple minimal rambling set has a rather nice picture in phase as shown in a theorem in [8,§3], from which the following can be easily deduced.

Theorem 4.1. *If $S \in \mathcal{X}(M^n)$ has a simple minimal rambling set Λ containing an ordinary point, then for any given $\varepsilon > 0$, any given neighborhood of Λ in M^n contains hyperbolic singularities or periodic orbits Q_1 and Q_2 of some $X \in \mathcal{X}(M^n)$ with $\|X - S\|_1 < \varepsilon$ such that, the stable manifold of Q_1 and the unstable manifold of Q_2 intersect but they do not intersect transversally.*

The structure of a non-simple minimal rambling set is more complicated in general. However, for the family $\mathcal{X}^*(M^n)$ of differential systems in $\mathcal{X}(M^n)$ we can say something which reveals significance and appears useful.

Here, by $\mathcal{X}^*(M^n)$ we denote the family of all $X \in \mathcal{X}(M^n)$ such that there is a neighborhood U of X such that all singularities and all periodic orbits of each $Y \in U$ are hyperbolic. It is easily seen that all Ω-stable and also structurally stable differential systems on M^n are in $\mathcal{X}^*(M^n)$.

We use the following notations. For any orbit Q of S through a point q of M, write $\mathrm{Ind}_S Q$ for the dimension of the linear subspace $F_-(q)$ of C_q. For a non-empty connected hyperbolic set K of S, write

$$\mathrm{Ind}_S K = \mathrm{Ind}_S Q$$

with $Q =$ an arbitrary orbit of S, contained in K. Then, $\text{Ind}_S Q$ and $\text{Ind}_S K$ are defined independently of the choice of the point $q \in Q$ and the orbit $Q \subset K$.

By a basic sequence of S with $p =$ an integer $\in \langle 0, n-1 \rangle$, we mean a sequence $\{X_i, P_i\}$ where $X_i \in \mathcal{X}(M^n)$ and P_i is a periodic orbit of X, such that

$$\lim_{i \to \infty} \|X_i - S\|_1 = 0,$$
$$\text{Ind}_{X_i} P_i = p, \quad i = 1, 2, 3, \cdots.$$

Theorem 4.2 ([8,§5]). *Let $S \in \mathcal{X}^*(M^n)$. Let Λ be a non-simple minimal rambling set of S, containing no singularities of S. Then for a certain integer $p \in \langle 0, n-1 \rangle$, there is a basic p-sequence $\{X_i, P_i\}$ of S such that the sequence $\{P_i\}$ converges to Λ. Moreover, if p is any such integer, then Λ contains a non-empty connected hyperbolic set K of S with*

$$\text{Ind}_S K \neq p.$$

5   The families $\mathcal{X}^{**}(M^n)$ and $\mathcal{X}^{\otimes}(M^n)$

Denote by $\mathcal{X}^{**}(M^n)$ the family of all $X \in \mathcal{X}^*(M^n)$ possessing the following property: There is a neighborhood U of X in $\mathcal{X}^*(M^n)$ and there are mutually exclusive open subsets $G, W_1, W_2, \cdots, W_{n-1}$ of M^n such that all singularities of any $Y \in U$ are in G and all periodic orbits P of any $Y \in U$ with $\text{Ind}_Y P = k$ are contained in W_k, $k = 1, 2, \cdots, n-1$.

Denote also by $\mathcal{X}^{\otimes}(M^n)$ the family of all $X \in \mathcal{X}(M^n)$ possessing the following property: X has a neighborhood V such that, whenever $\lambda \mapsto X_\lambda$ gives a mapping $\langle 0, 1 \rangle \to V$ with $X_0 = X$, we have for each singularities or each periodic orbit Q of X a homotopy

$$h_{Q,\lambda} : Q \to M^n, \, 0 \leq \lambda \leq 1$$

such that for each λ, $h_{Q,\lambda}$ carries Q topologically onto a singularity or a periodic orbit of X_λ and that

$$Q \mapsto h_{Q,\lambda}(Q)$$

yields a (1,1)-correspondence between the set of all singularities and periodic orbits of X and that of X_λ. By the theorem of Kupka-Smale

[2,p.58] we see easily that

$$\mathcal{X}^{\otimes}(M^n) \subset \mathcal{X}^*(M^n).$$

For any singularity or periodic orbit Q of $X \in \mathcal{X}(M^n)$, let us write

$$J_X(Q) = \text{Ind}_{(-X)}(Q) - \text{Ind}_X(Q).$$

A nonwandering point c of S will be called chaotic [8] if for any given $\varepsilon > 0$ and any given neighborhood W of c in M^n there are X_1 and $X_2 \in \mathcal{X}(M^n)$ such that X_i has a singularity or a periodic orbit Q_i satisfying

(1) $\|X_i - S\|_1 \leq \varepsilon$ and $Q_i \cap W \neq 0$, $i = 1, 2$,

(2) $J_{X_1}(Q_1) \neq J_{X_2}(Q_2)$.

Theorem 5.1 ([8,§6]). *For $S \in \mathcal{X}(M^n)$ the following three statements (1)–(3) are equivalent:*

(1) $S \in \mathcal{X}^{**}(M^n)$.

(2) *S does not possess chaotic nonwandering points.*

(3) *S satisfies Axiom A and the no-cycle condition.*

In the proof of this theorem, we used in [8] properties of minimal rambling sets.

For discrete dynamics, by using an ergodic closing lemma, Mañé obtained a theorem which takes a nice and simpler form [9] . Let $\mathcal{F}^*(M^n)$ be the subset of $\text{Diff}(M^n)$ (= the space of all C^1 diffeomorphisms of M^n) consisting of all such g possessing the property: g has a neighborhood W in $\text{Diff}(M^n)$ such that all periodic points of each $h \in W$ are hyperbolic. For any $f \in \text{Diff}(M^n)$, denote by $\text{per}(f)$ the set of all periodic points of f, and by $\text{Ind}_f(x)$ for $x \in \text{per}(f)$ the dimension of the linear space

$$\{u \in C_x \mid \lim_{j \to \infty} \|d\, f^j(u)\| = 0\}.$$

For $k = 0, 1, \cdots, n$, let $\Lambda_k(f)$ be the set of all points $y \in M^n$ possessing the property : There is a sequence $\{y_i\}$ of points in $\text{per}(f)$ with $\text{Ind}_f(y_i) = k$ which converges to y.

Theorem 5.2 (Mañé [9]). *$f \in \mathrm{Diff}(M^n)$ satisfies Axiom A and the no-cycle condition if and only if*

$$f \in \mathcal{F}^*(M^n) \quad and \quad \Lambda_i(f) \bigcap \Lambda_j(f) = 0, \quad for \ 0 \le i < j \le n.$$

However, an analogous theorem for a differential system $\in \mathcal{X}(M^n)$ does not hold in general at least in the case when the system admits singularities, as shown by an example in [10].

It would be interesting to deduce Theorem 5.2 from Theorem 5.1. For this purpose, let $M_0(S)$ be the set of all singularities of S, and $\Delta_k(S)$ the set of all points x in M^n possessing the following property: There is a sequence $\{x_i\}$ of points which converges to x with x_i belonging to a periodic orbit P_i of S and $\mathrm{Ind}_S P_i = k$, $k = 0, 1, \cdots, n - 1$.

Theorem 5.3 ([11]). *Let $S \in \mathcal{X}^{\otimes}(M^n)$. Then S satisfies Axiom A and the no-cycle condition when and only when $M_0(S), \Delta_0(S), \ldots, \Delta_{n-1}(S)$ mutually exclusive subsets of M^n.*

This theorem can be derived directly from Theorem 5.1. Consider the suspended manifold M_f^{n+1} for $f \in \mathcal{F}^*(M^n)$. As the suspended differential system of f is in $\mathcal{X}^{\otimes}(M_f^{n+1})$, we can deduce easily Theorem 5.2 from Theorem 5.3.

6 On stability of differential systems

A basic problem in the theory of differentiable dynamical systems is to characterize the stability of differential systems as well as that of discrete systems. There is a long history for this problem. In [12] Smale proved that $f \in \mathrm{Diff}(M^n)$ is Ω-stable if it satisfies Axiom A and the no-cycle condition. The corresponding result for differential systems is due to Pugh-Shub [13]. According to Robbin [5] and Robinson [14], $f \in \mathrm{Diff}(M^n)$ as well as $S \in \mathcal{X}(M^n)$ is structurally stable, if the system satisfies Axiom A and the strong transversality condition. The question whether the converse of these results are also true, known as the stability conjecture, kept open until quite recently, Mañé and Palis proved this conjecture for discrete systems, namely:

Theorem 6.1 (Mañé [15]). *A structurally stable $f \in \mathrm{Diff}(M^n)$ satisfies Axiom A and the strong transversality condition.*

Theorem 6.2 (Palis [16]). *An Ω-stable $f \in \mathrm{Diff}(M^n)$ satisfies Axiom A and the no-cycle condition.*

However, the stability conjecture for differential systems remains unproved, except in a few cases in dimension ≤ 4 for which Theorems as stated in the following 6.3–6.5 were published before. Of course, the first one is extremely well-known.

Theorem 6.3 (Peixoto [17]). *For $S \in \mathcal{X}(M^2)$ to be structurally stable it is necessary and sufficient that (1) $\Omega(S)$ is the union of finitely many hyperbolic singularities and finitely many hyperbolic periodic orbits, and (2) there is no saddle connection.*

Thus, the structural stability conjecture holds for $S \in \mathcal{X}(M^2)$. Combining this and the density theorem in [17], we see easily that the Ω-stability conjecture holds also for $S \in \mathcal{X}(M^2)$.

Theorem 6.4 ([18],[8]). *Suppose that $S \in \mathcal{X}(M^3)$ has no singularities. Then: (1) S is structurally stable if and only if S satisfies Axiom A and the strong transversality condition. (2) S is Ω-stable if and only if S satisfies Axiom A and the no-cycle condition, and also if and only if $S \in \mathcal{X}^*(M^3)$.*

Let $X \in \mathcal{X}(M^n)$ be such that all singularities and all periodic orbits of X are hyperbolic. Denote by $\check{\pi}(X)$ the set of all points $x \in M^n$ such that for some singularity or periodic orbit Q_i of X, $i = 1, 2$, the stable manifold of Q_1 and the unstable manifold of Q_2 with respect to X both contain x but they do not intersect at x transversally. We call X a Kupka-Smale system if $\check{\pi}(X) = 0$. Denote by $\mathcal{X}^{\#}(M^n)$ the interior of the subset of $\mathcal{X}(M^n)$ consisting of all Kupka-Smale systems.

We sketched a proof of the following theorem in [19].

Theorem 6.5. *Suppose that $S \in \mathcal{X}(M^4)$ has no singularities. The following three statements (1)–(3) are equivalent:*

(1) *S is structurally stable.*

(2) *$S \in \mathcal{X}^{\#}(M^4)$.*

(3) *S satisfies Axiom A and the strong transversality condition.*

Based upon the above discussion, the following conjectures seem reasonable.

Conjecture I. $S \in \mathcal{X}(M^n)$ is structurally stable if and only if $S \in \mathcal{X}^{\#}(M^n)$.

Conjecture II. $S \in \mathcal{X}(M^n)$ is Ω-stable if and only if $S \in \mathcal{X}^*(M^n)$ and has a neighborhood V in $\mathcal{X}^*(M^n)$ such that

$$\hat{r}(X) \bigcap \check{R}(X) = 0$$

for every $X \in V$.

Here, the "only if" part in Conjecture I follows easily from the theorem of Kupka-Smale [2,p.58] and the definition of structural stability.

The techniques used in the proofs of the stability conjecture for diffeomorphisms may not be always easily extended to treat the case of differential systems at least for two reasons (see [8,p.35] and also [9,p.508] for this comment): Firstly, the periodic points of $f \in \mathcal{F}^*(M^n)$ are dense in the nonwandering set of f, but an analogous result may not hold for differential systems in $\mathcal{X}^*(M^n)$ in general (see[10]). Secondly, the existence of singularities of a differential system makes the situation of its phase-portrait more complicated than that of a diffeomorphism (note that the suspended differential system $S_f \in \mathcal{X}(M_f^{n+1})$ of $f \in \text{Diff}(M^n)$ possess no singularities).

We have thus propose the use of obstruction sets and minimal rambling sets in the study of differentiable dynamical systems as previously in [8,p.35].

To conclude our report, let us make for example a description of a proof of the following theorem (without the "non-singular" assumption of Theorem 6.4) to show the use of such sets in practice.

Theorem 6.6. *For $S \in \mathcal{X}(M^3)$ the following three statements are equivalent:*

(1) *S is structurally stable.*

(2) *$S \in \mathcal{X}^{\#}(M^3)$.*

(3) *S satisfies Axiom A and the strong transversality condition.*

In fact, (1) $\Rightarrow$ (2) is clear, as explained above. (2) $\Rightarrow$ (1) was given in [14]. It remains to show (2) $\Rightarrow$ (3).

Suppose that this does not hold. Then, since all singularities of $S \in X^*(M^3)$ are hyperbolic, by Theorem 2.3 above we have $Ob(S) \neq 0$, and hence there is a minimal rambling set Δ of S. But Δ cannot be a simple one, for otherwise, using Theorem 4.1 above we could get a contradiction to the hypothesis $S \in X^\#(M^3)$. Therefore, Δ is non-simple.

We assert that Δ must contain at least a singularity x_0 of S, for otherwise, by some arguments similar to those in [8, §6], Δ would be a hyperbolic set of S, contradicting to that Δ is rambling. Next, by the definition of a minimal rambling se, we see easily that x_0 cannot be a sink or a source of S. Thus, x_0 is a saddle point of S. Then, there is a stable or unstable manifold N of S through x_0 with $\dim N = 1$, because $\dim M^3 = 3$. By considering $-S$ instead of S if this is necessary, we may assume that N is the unstable manifold of S through x_0. We can then choose $x_1 \in N, \neq x_0$, such that x_1 is in Δ. Let Γ_- and Γ_+ be respectively the ω-limit set and the α-limit set of the orbit $P = \{\phi_t(x_1) \mid t \in (-\infty, \infty)\}$. Clearly, $\Gamma_+ = \{x_0\}$.

Suppose that $\Lambda = \Gamma_- \bigcup P \bigcup \Gamma_+$ is a proper subset of the minimal rambling set Δ of S. Then Λ is normal, and using Theorem 3.2 above, we see easily that x_1 is of grade type $(1, 2, 0)$ and Γ_- is either a sink or a subset of $\Lambda^1_{2,0}$. In the latter case, Γ_- contains no singularities of S and hence contains a positively contractible periodic orbit of S. Both cases would lead to a contradiction to that Δ is a rambling set of S. This contradiction shows that

$$\Delta = \Gamma_- \bigcup P \bigcup \Gamma_+.$$

It follows that x_1 is an ω-limit point of the orbit of S through x_1.

Now, in order to complete the proof of the theorem, we need analyze in a further step some global behavior of the singularity x_0 and the minimal rambling set Δ of S.

Similarly, the following characterizations for Ω-stability of a differential system in dimension 3 can be also obtained.

Theorem 6.7. *For $S \in X(M^3)$ the following statements (1)–(3) are equivalent:*

(1) *S is Ω-stable.*

(2) $S \in \mathcal{X}^*(M^3)$ *and there is a neighborhood* V *of* S *in* $\mathcal{X}^*(M^3)$ *such that*

$$\not{*}(X) \bigcap \check{R}(X) = 0$$

for every X *in* V.

(3) S *satisfies Axiom A and the no-cycle condition.*

References

[1] M. Peixoto, On structural stability, *Annals of Math.*, **69**(1959), 199–222.

[2] S. Smale, *The Mathematics of Time*, Springer-Verlag, 1980.

[3] Liao Shantao, Obstruction sets and strong transversality, *Acta Math. Sinica*, **19**(1976), 203–209. (in Chinese)

[4] Liao Shantao, Obstruction sets (I), *Acta Math. Sinica*, **23**(1980), 411–453.

[5] J. Robbin, A structural stability theorem, *Annals of Math.*, **94**(1971), 447–493.

[6] Liao Shantao, Standard systems of differential equations, *Acta Math. Sinica*, **17**(1974), 100–109; 175–196; 270–295. (in Chinese)

[7] Liao Shantao, An existence theorem for periodic orbits, *Acta Sci. Natur. Univ. Pekinensis*, 1979, No.1, 1–20. (in Chinese)

[8] Liao Shantao, Obstruction sets (II), *Acta Sci. Natur. Univ. Pekinensis*, 1981, No.2, 1–36. (in Chinese)

[9] R. Mañé, An ergodic closing lemma, *Annals of Math.*, **116**(1982), 503–540.

[10] Ding Hongyu, Disturbance of the homoclinic trajectory and applications, *Acta Sci. Natur. Univ. Pekinensis*, 1986, No.1, 53–63. (in Chinese)

[11] Liao Shantao, Standard systems of differential equations and obstruction sets with applications to structural stability problems,

Proceedings of the 1983 Beijing Symposium on Differential Geometry and Differential Equations, Science Press, Beijing, 1986, 65–97.

[12] S. Smale, The Ω-stability theorem, *Proc. Symp. in Pure Math.*, Vol. XIV. AMS, 1970, 289–297.

[13] C. Pugh and M. Shub, The Ω-stability theorem for flows, *Inventiones Math.*, **11**(1970), 150–158.

[14] C.Robinson, Structural stability of C^1 flows, *Warwich 1974, Lecture Notes 468*, Springer-Verlag.

[15] R. Mañé, A proof of the C^1 stability conjecture, *Publ. Math. IHES*, 1988, 161–210.

[16] J. Palis, On the C^1 Ω-stability conjecture, *Publ. Math. IHES*, 1988, 211–215.

[17] M. Peixoto, Structural stability on two-dimensional manifolds, *Topology*, **1**(1962), 101–120.

[18] Liao Shantao, On the stability conjecture, *Chinese Annals of Math.*, **1**(1980), 9–29.

[19] Liao Shantao, On characterizations of structural stability, *Applied Math. and Mech.*, **5**(1984), 1745–1750.

Added in proof. After this paper was submitted, I received a preprint of Sen Hu "A proof of C^1 stability conjecture for 3 dimensional flows, 8 Dec 1989", in which a proof of the structural stability conjecture for C^1 vector fields of dimension 3 is provided by paving a way to apply Mañé's perturbation techniques, and a detailed discussion concerning this would appear in Hu's Princeton doctoral dissertation. His method is different from ours described above, that is based on obstruction sets.

A Survey of Developments of Mathematics Mechanization in China[*]

Wu Wen-tsun
Institute of Systems Science, Academia Sinica

Abstract

The present survey is devoted to activities in China on theorem-proving and equations-solving, the two main topics of MM (Mathematics Mechanization), restricted to works mainly due to the MM group in the Institute of Systems Science, Academia Sinica. The contributions on equations-solving in ancient China are also briefly described.

1 Mathematics Mechanization in Ancient Times

There are two main kinds of activities in mathematics, viz. equations-solving and theorem-proving. The latter one was typically of Greek origin and Euclid's Elements served as its representative work. In this classic a deductive system was established and the exposition bore the style of the form Definition-Axiom-Theorem-Proof. On the other hand the equations-solving was originated and well-developed in ancient China. In fact, the ancient Chinese were mainly devoted to solve concrete problems arising in practice which led naturally to problems of equations-solving. The representative works were the classic "Nine Chapters in Arithmetic" and its "Annotations" by Liu-Hwei, which, in accordance with the main spirit , adopted an exposition bearing the style of the form Example-Answer-Method & Result-Explanation. The item Method & Result was usually expressed in a form which now-a-days we would like to call it an

---

*The present paper is partially supported by NSFC Grant JI85312

ALGORITHM. The item Explanation usually gave the underlying reasons of the algorithm somewhat like a now-a-days correctness-proof. The sources of such equations were two-fold. One of such sources was rudimentary commerce or goods-exchange which led to the Excess-Deficiency Algorithm in Chapter 7 of "Nine Chapters" corresponding to, in modern language, the method of solving a system of two linear equations in two unknowns. It was developed into a general method of solving simultaneous linear equations which was the main concern of Chapter 8 of "Nine Chapters". This method was the same as the one bearing the name of Gaussian Elimination Method in modern writings, which, what seems to be legitimate, would be more accurately to be called Nine-Chapters-Elimination-Method. A second source of equations-solving was from problems of measurements. The simplest one of such problems consists in extracting square roots for the determination of the hypothenus (e.g. the sun-observer distance) from the known values of the two sides of a right-angled triangle (e.g. the sun-height and level-distance of sun-observer). It requires the extraction of square roots which is equivalent to the solving of a quadratic equation of the form $x^2 = A$. It is perhaps owing to this origin that throughout the incessant development of equations-solving up to the method of numerical solution of any higher degree algebraic equations during the Song dynasty the Chinese ancestors called equations-solving simply roots-extraction (with extra terms). The discovery of the notion of Tian-Yuan or Heavens-Element equivalent to the present-day notion of Unknown during the Song and Yuan dynasties furnished a systematic method of establishing equations for given problems. The two lines of developments of equations-solving were also rendered possible to merge into one which was closer to algebra in modern sense. It cultimated in the classic of Zhu Szejze bearing the title "The Jade Mirror of Four Elements" which gave an elimination method of solving systems of algebraic higher degree equations up to four unknowns. The following, expressed in modern languages, was one of the representative examples given in Zhu's classic:

Problem. For a right-angled triangle the hypothenus z and the two sides x and y are connected by the relations

$$x \cdot y \cdot z - x \cdot y^2 - z - x - y = 0, \qquad (1.1)$$

$$x \cdot z - x^2 - z - y + x = 0, \qquad (1.2)$$

$$x^2 + y^2 = z^2. \tag{1.3}$$

Find x, y, and z.

Method of Zhu (a little simplified).

Eliminate y from (1.1) by means of (1.2), (1.3) and remove the factor x one gets

$$-2 - z - 2 \cdot z^2 + x + x \cdot z + x \cdot z^2 + x^2 - x^2 \cdot z = 0. \tag{1.4}$$

Eliminate y from (1.2) by means of (1.3) and remove the factor x one gets

$$-2 \cdot z - 2 \cdot z^2 + 2 \cdot x + 4 \cdot x \cdot z + x \cdot z^2 - 2 \cdot x^2 - x^2 \cdot z + x^3 = 0. \tag{1.5}$$

Write (1.4) and (1.5) as equations in x and eliminate x^3, x^2 from (1.5) by means of (1.4), one gets successively equations in x^2 and x:

$$\begin{aligned}
(-3 - z + z^2) \cdot x^2 + (4 + 3 \cdot z - z^2 - z^3) \cdot x - 2 \cdot z + 2 \cdot z^3 &= 0, \\
(7 + 3 \cdot z - z^2) \cdot x - (6 + 7 \cdot z + 3 \cdot z^2 - z^3) &= 0.
\end{aligned} \tag{1.6}$$

Eliminate finally x from (1.6) and (1.4) one gets then an equation in z alone:

$$z^4 - 6 \cdot z^3 + 4 \cdot z^2 + 6 \cdot z - 5 = 0. \tag{1.7}$$

Solving, one gets one root $z = 5$. Substituting in (1.6) one finds $x = 3$. Substituting the values of x and z in (1.2), one finds $y = 4$. A solution of the problem is thus given by $(x, y, z) = (3, 4, 5)$.

The limitation of Zhu's method to utmost 4 unknowns was due to the fact that the Chinese ancients made all calculations on counting boards. Otherwise the method is entirely general and may be considered as a direct generalization of Nine-Chapters-Elimination-Method. Moreover, all such methods are highly mechanical in character and may be turned readily into computerized programs if one likes. This is indeed in high contrast with the theorem-proving dated from the ancient Greece which is exceedingly non-mechanical in character.

A turning point occurred in 17th century when Descartes, and also Fermat, created the method of analytic geometry. It seems that Descartes, as shown by his monumental classic "La Géomètrie", paid great emphasis on solving of concrete problems, which led naturally to equations-solving, while theorem-proving was somewhat neglected. In this respect the attitude of Descartes seems to be much in the same spirit of our Chinese ancestors. In any way the creation of analytic-geometrical method opened the way of mechanization, not only of equations-solving, but also theorem-proving, no matter how late such mechanization does begin to occur.

2   Discovery of Mechanization Methods for Theorem-Proving

The mathematics mechanization in a strict sense may be considered to begin since the appearance in 1899 of Hilbert's classic "Grunlagen der Geometrie". Hilbert was the doubtless founder of the modern axiomatization of mathematics. However, it was first time pointed out by Wu ([Wu3,5,6]) that Hilbert may also be considered as the pioneer of the actual mechanization of mathematics, especially in the realm of theorem-proving. In fact, in the end of the main body of the above classic, i.e. in the end of Chap 6, there was an obscure passage which was turned into the form of a theorem in the later editions, viz.

Theorem 62. Each pure intersection-point theorem in a Pascalian geometry, if true, can always be proved by the aid of suitably constructing auxiliary points and lines, as consequence of a combination of a finite number of Pascal configurations.

It seems that this theorem has never been taken into notice, in spite of the fact that Hilbert's classic was widely and seriously studied for all the time. It was Wu who first recognized the significance of this theorem and reformulated it in the following form:

Hilbert Mechanization Theorem. In a planar Pascalian geometry there is a mechanical procedure which permits to prove or disprove in a finite number of steps any pure intersection-point theorem of constructive type under certain subsidiary non-degeneracy conditions also generated mechanically during the procedure.

It would be better to elucidate the actual meaning of this theorem by a concrete example. Let us consider thus the following

Theorem of Gauss (Werke, Bd. 4, S. 385–392). For a complete quadrilateral consisting of 4 lines $L_1, \cdots, L_4$ intersecting in 6 points L_{ij}, the mid-points M_1, M_2, M_3 of the 3 diagonals $L_{12}L_{34}$, $L_{13}L_{24}$, and $L_{14}L_{23}$ are co-line.

The occurrence of the concept mid-point would surpass the range of applicability of the above mechanization theorem which does not matter, however. The mechanical procedure of Hilbert still works and runs as follows.

Take an arbitrary point L_{12} as the origin O. Take arbitrary lines L_1 and L_2 through L_{12} as the x- and the y-axis. Take on the x-axis L_1 arbitrary points L_{13} and L_{14} to be given the coordinates $(U_1, 0)$ and $(U_2, 0)$ with U_1 and U_2 as parameters. Take on the y-axis L_2 arbitrary points L_{23} and L_{24} to be given the coordinates $(0, U_3)$ and $(0, U_4)$. Draw now the line joining L_{13}, L_{23} and also the line L_4 joining L_{14} and L_{24}. Form the intersection point L_{34} of lines L_3 and L_4 and give it the coordinates (Y_1, Y_2) for which Y_1, Y_2 are no more arbitrary parameters but bounded with coordinates U of points already taken. The equations determining Y_1, Y_2 may be readily put in the following form:

$$Y_1 = U_1 \cdot U_2 \cdot (U_4 - U_3)/(U_1 \cdot U_4 - U_2 \cdot U_3), \qquad (2.1)$$
$$Y_2 = U_3 \cdot (U_1 - Y_1)/U_1. \qquad (2.2)$$

Draw now the line $L_{12}L_{34}$ and take the mid-point M_1 to be given the coordinates (Y_3, Y_4) so that

$$Y_3 = Y_1/2, \qquad (2.3)$$
$$Y_4 = Y_2/2. \qquad (2.4)$$

Draw next the line $L_{13}L_{24}$, take the mid-point M_2, and give it the co-ordinates (Y_5, Y_6) with

$$Y_5 = U_1/2, \qquad (2.5)$$
$$Y_6 = U_4/2. \qquad (2.6)$$

Finally draw the line $L_{14}L_{23}$ with mid-point $M_3(Y_7, Y_8)$ such that

$$Y_7 = U_2/2, \qquad (2.7)$$
$$Y_8 = U_3/2. \qquad (2.8)$$

The conclusion of the theorem is equivalent to $G = 0$, with the conclusion polynomial (abbr. henceforth as pol) G given by

$$G = Y_3 \cdot Y_6 + Y_5 \cdot Y_8 + Y_4 \cdot Y_7 - Y_3 \cdot Y_8 - Y_4 \cdot Y_5.$$

The proof of theorem now runs as follows.

Eliminate Y_8 in $G_8 = G$ by means of (2.8) to get a pol G_7 in U and $Y_1, \cdots, Y_7$ alone. Eliminate next Y_7 in G_7 by means of (2.7) to get a pol G_6 in U and $Y_1, \cdots, Y_6$ alone, etc. Finally, after eliminating Y_1 from G_1 by means of (2.1) one would get a pol G_0 in U alone which turns out to be identically 0. This proves the theorem. Moreover, it shows that the theorem is proved to be true only under the conditions $U_1 \neq 0$ and $U_1 \cdot U_4 - U_2 \cdot U_3 \neq 0$ which are the non-degeneracy ones as stated in the Mechanization Theorem and are found automatically with U_1, $U_1 \cdot U_4 - U_2 \cdot U_3$ as the denominators occurring in the equations (2.1)–(2.8).

The method of proof is a general one applicable to the whole class of theorems possessing the following properties:

Coordinates can be so chosen such that parameters $U_1, U_2, \cdots$ and bounded variables $Y_1, Y_2, \cdots$ will be introduced one by one with hypothesis of the theorem expressible in the form

$$Y_i = P_i/Q_i, \quad i = 1, 2, \cdots,$$

the P_i and Q_i being pols in U and Y_j alone for $j < i$. The conclusion is also expressible in the form $G = 0$ with the conclusion pol G one in U and Y. Eliminate now the Y's in G successively in the reverse order by means of these hypothesis equations. The theorem in question will then be seen to be true or un-true according to whether the final result is identically 0 or not. In the affirmative case the theorem is true GENERICALLY under the NON-DEGENERACY CONDITIONS $Q_i \neq 0$.

The method is seemingly a very trivial one which permits, however, to prove already quite non-trivial theorems. In year 1977 Wu discovered, being entirely ignorant of the above Hilbert's result at that time, a more general method of much wider applicability for mechanical geometry theorem-proving as follows (of. [Wu1,2,3])

Let K be a field of characteristic 0. For any pol in variables $U_1, \cdots, U_d$ and $Y_1, \cdots, Y_e$ on coefficients in K let c be the greatest subscript of Y such that Y_c actually occurs in G. One can then write G in the form

$$G = I \cdot Y_c^D + \text{lower degree terms in } Y_c,$$

in which all coefficients of Y_c-powers, if not 0, will be pols in U and Y_i with $i < c$ alone. We call then I the INITIAL of G and Y_c the

LEADING VARIABLE of G. Suppose now a theorem T be such that with coordinates suitably chosen, the hypothesis are expressible as $H_i = 0$, $i = 1, \cdots, e$ with H_i pols having Y_i as leading variables:

$$H_i = I_i \cdot Y_i \cdot D_i + \text{lower degree terms in } Y_i. \tag{2.9}$$

If in particular H_1 is irreducible as a pol in Y_1 on the field $K_0 = K(U_1, \cdots, U_d)$, H_2 is irreducible as a pol in Y_2 on the field $K_1 = K_0(Y_1)$ extended by adjoining Y_1 to K_0 by means of $H_1 = 0$, etc. Then the hypothesis set forming of $H_1, \cdots, H_e$ will be said to be IRREDUCIBLE. Suppose that $G = 0$ with G a pol in U and Y on K be the conclusion of the theorem T. To see whether T is true of not let us just divide G successively by $H_l, \cdots, H_2, H_1$ to remove the Y's as much as possible. We get then, for some non-negative integers S_i, and also pols A_i, a formula of the form

$$I_1^{S_1} \cdot \cdots \cdot I_e^{S_e} \cdot G = A_1 \cdot H_1 + \cdots + A_e \cdot H_e + R, \tag{2.10}$$

in which R is a pol in U and Y with degree in each Y_i less than the corresponding degree D_i in H_i. We call R the REMAINDER of the pol G w.r.t. the hypothesis-set $HYP = \{H_1, \cdots, H_e\}$ and write it as

$$R = \text{Remdr}(G/HYP\}.$$

From the remainder formula (2.10) we have the following

Mechanical Theorem-Proving Principle (abbr. MTP). If the remainder R off G w.r.t. the hypothesis-set HYP is identically 0, then the theorem T in question is GENERICALLY true under the non degeneracy conditions $I_i \neq 0$, $i = 1, \cdots, e$. Moreover, the condition $R = 0$ is not only sufficient, but also necessary for the theorem T to be generically true under the above non-degeneracy conditions in case the hypothesis-set is irreducible.

The above principle serves as the basis for proving, or even disproving, theorems in elementary geometries. We remark that theorems in elementary geometries can be usually, after some simple manipulations, put in the form of (2.10). Even trigonometrical identities can be transformed into the form of (2.10) by treating the various trigonometric functions as separate variables connected by algebraic relations. We refer to the examples given in the earliest paper [Wu1] of Wu on this subject which already show clearly how efficient the method could be. The method may even be served to discover new non-trivial ones since once one guesses some possible geometrical truth, one may just make verifications by the above principle to see whether it is really true or not. Among the earliest new theorems discovered in this way we may mention the Pascal Conic Theorem about chord intersections of 6 points on a conic,cf. e.g. [Wu2,3]. The following example is a further illustration of this principle ([Wu9]).

Example. Consider a plane moving over a fixed one and occupy 4 different positions. To move from i-th position to the j-th position there is associated a center of rotation of POLE P_{ij} and an angle of rotation R_{ij} mod $2 \cdot p_i$. Suppose that all such poles P_{ij} are situated in the finite part of the fixed plane so that we have a 6-pole-configuration. In some particular cases the configuration may degenerate into a complete quadrilateral with 3 pairs of diagonals $P_{12}P_{34}$, $P_{13}P_{24}$, and $P_{14}P_{23}$. In view of the theorem of Gauss we may ask ourselves whether this theorem remains true for the general 6-pole-configuration.

To see this let us first remark that the configuration will be completely determined by the positions of P_{12}, P_{13}, P_{14} and the 3 rotation angles R_{12}, R_{13}, and R_{14}. Let us take thus coordinates such that

$$P_{12} = (0,0), \ P_{13} = (U_1,0), \ P_{14} = (U_2,U_3),$$

and set

$$\mathrm{Tan}(R_{12}/2) = U_5 : U_4, \mathrm{Tan}(T_{13}/2) = U_7 : U_6, \mathrm{Tan}(R_{14}/2) = U_9 : U_8.$$

From fundamental theorems in planar kinematics we know that

$$\angle(P_{12}P_{13}, P_{12}P_{23}) = \angle(P_{12}P_{14}, P_{12}P_{24}) = R_{12}/2 \ \mathrm{mod} \ p_i, \ \mathrm{etc.}$$

Let the coordinates of P_{23}, P_{24}, P_{34} and the mid-points M_1, M_2, M_3 of the diagonals $P_{12}P_{34}$, $P_{13}P_{24}$, and $P_{14}P_{23}$ be given by

$$P_{23} = (Y_1,Y_2), \ P_{24} = (Y_3,Y_4), \ P_{34} = (Y_5,Y_6),$$

$$M_1 = (Y_7,Y_8), \ M_2 = (Y_9,Y_{10}), \ M_3 = (Y_{11},Y_{12}),$$

The geometrical conditions will then be turned into a hypothesis set consisting of 12 equations which, after simple manipulations, will be of the form $H_i = 0$, $i = 1, \cdots, 12$, with

$$
\begin{aligned}
H_1 &= +Y_1 \cdot U_7 \cdot U_4 - Y_1 \cdot U_6 \cdot U_5 - U_7 \cdot U_4 \cdot U_1, \\
H_2 &= -Y_2 \cdot U_7 \cdot U_4 + Y_2 \cdot U_6 \cdot U_5 + U_7 \cdot U_5 \cdot U_1, \\
H_3 &= -Y_3 \cdot U_9 \cdot U_4 + Y_3 \cdot U_8 \cdot U_5 - U_9 \cdot U_5 \cdot U_3 + U_9 \cdot U_4 \cdot U_2, \\
H_4 &= -Y_4 \cdot U_9 \cdot U_4 + Y_4 \cdot U_8 \cdot U_5 + U_9 \cdot U_5 \cdot U_2 + U_9 \cdot U_4 \cdot U_3, \\
H_5 &= -Y_5 \cdot U_9 \cdot U_6 + Y_5 \cdot U_8 \cdot U_7 - U_9 \cdot U_7 \cdot U_3 + U_9 \cdot U_6 \cdot U_2 \cdots \\
&\quad -U_8 \cdot U_7 \cdot U_2 + U_8 \cdot U_7 \cdot U_2 - U_8 \cdot U_7 \cdot U_1, \\
H_6 &= -Y_6 \cdot U_9 \cdot U_6 + Y_6 \cdot U_8 \cdot U_7 \cdots \\
&\quad +U_9 \cdot U_7 \cdot U_2 - U_9 \cdot U_7 \cdot U_1 + U_9 \cdot U_6 \cdot U_3, \\
H_7 &= 2 \cdot Y_7 - Y_1, \\
\cdots \quad & \cdots \quad \cdots \cdots \\
H_{12} &= 2 \cdot Y_{12} - Y_2 + U_3.
\end{aligned}
$$

The conclusion to be verified is $G = 0$ with the conclusion pol given by

$$
G = Y_7 \cdot Y_{10} + Y_9 \cdot Y_{12} + Y_8 \cdot Y_{11} - Y_{10} \cdot Y_{11} - Y_7 \cdot Y_{12} - Y_8 \cdot Y_9.
$$

It is readily verified, HYP being the polset $\{H_1\}$, that

$$
\mathrm{Remdr}(G/HYP) = 0.
$$

We have thus discovered a new theorem generalizing the theorem of Gauss. The non-degeneracy conditions are given by $I_i \neq 0$ with I_i the initial of H_i. It is readily seen that $I_i = 0$ corresponds to the degeneracy of the 6-pole-configuration into a complete quadrilateral and the theorem is already known to be true in this case owing to above theorem of Gauss.

In general, if we want to know whether a theorem remains to be true in some degenerate case $I = 0$, we may just add I to the original hyp-set to form a new one HYP' and try to see whether the conc-pol is still 0 at that time. Now it may happen, as is usually the case, that the hyp-set HYP' will be no more in the well-ordered form like (2.9). Some preliminary manipulations are then necessary to bring HYP' again to a well-ordered form. This and other considerations induce us to seek a more general setting of treatments as we shall see in the next section.

3   A General Method of Mechanization for Equations-Solving

Now there are theorems in elementary geometries for which it is rather complicate to bring the hypothesis set in some coordinate system to the form of (2.9). There are also occasions to consider problems dealing with sets of pols entirely arbitrary. It is also important to extend the whole theory to the case of differential geometries. Thus, after the completion of the work [Wu1], Wu saw the importance of the works of Ritt, at that time somewhat lying in the obscurity in the mathematics world, and began to study seriously the second book of Ritt, [R2]. The study was a fruitful one and brought about fertile results in diverse directions which will be explained below.

Let us consider a set of variables $X_1, X_2, \cdots, X_m$ and a field K of characteristic 0 fixed henceforth. Any polynomial in the ring

$$K[X_1, \cdots, X_m]$$

will be simply called a POL. A finite set of non-zero pols is then called a POLSET. Any point $A = (A_1, \cdots, A_m)$ with A_i in some extension field K' of K is called a K'-ZERO or simple a ZERO of a pool P if P becomes 0 when X_i in P are replaced by A_i. The set of K'-zeros of all pols in a polset PS will be denoted by K'-Zero(PS) or simply Zero(PS) if K' is taken to be arbitrary or unnecessary to be specified. If G is any other pol, then we shall also set

$$\mathrm{Zero}(PS/G) = \mathrm{Zero}(PS)\backslash\mathrm{Zero}(G).$$

We are now interested in solving the system of equations $P_i = 0$ with P_i running over the pols in a given polset PS, and in the study of the structure of the set Zero(PS). For this purpose we shall introduce the notions originated in Ritt's works as follows.

For any pol P let us set

$$
\begin{aligned}
c &= \text{greatest subscript of } X \text{ for which} \\
&\quad X_c \text{ actually appears in } P, \\
d &= \text{highest degree of } X_c \text{ in } P.
\end{aligned}
$$

For a non-zero constant pol we set simply $c = 0$ and $d = 0$. We call c the CLASS and d the DEGREE of P to be denoted by Cls(P) and Deg(P)

resp. A non-zero pol Q will be said to be REDUCED w.r.t P if P is not a constant one and, for $c = \text{Cls}(P) > 0$, the degree of Q in X_c is less than $\text{Deg}(P)$. We introduce now among the non-zero pols a partial ordering $<$ such that $P < Q$ if either $\text{Cls}(P) < \text{Cls}(Q)$ or $\text{Cls}(P) = \text{Cls}(Q)$ and $\text{Deg}(P) < \text{Deg}(Q)$. If neither is $P < Q$ nor is $Q < P$ in the above ordering then we say that P, Q are INCOMPARABLE in ORDER and we write then $P <=> Q$.

Consider now a finite sequence of non-zero pols

$$(P) \qquad P_1, P_2, \cdots, P_r.$$

We call (P) an ASCENDING SET if one of the following cases holds true:

(1) $r = 1$ and $\text{Cls}(P_1) = 0$ or P_1 is a non-zero constant in K. The asc-set in this case is then said to be TRIVIAL of CONTRADIC-TORY.

(2) $0 < \text{Cls}(P_1) < \cdots < \text{Cls}(P_r)$

and each P_i is reduced w.r.t. any preceding P_j with $j < i$.

We say that a pol G is REDUCED w.r.t. an asc-set (P) if G is reduced w.r.t. any pol in (P).

Let (P) be now a non-trivial asc-set as above and G any pol, then by successively dividing by pols of G in the reverse order, we get a formula of the form

$$J \cdot G = \sum_i A_i \cdot P_i + R,$$

in which J is a power-product of initials I_i of P_i, and A_i, R are pols with R reduced w.r.t. (P). We can choose exponents of I_i in J to be as small as possible. The pol R is then uniquely determined and will be called the REMAINDER of G w.r.t. (P), to be denoted by

$$R = \text{Remdr}(G/(P)).$$

We now introduce a partial ordering $<$ among all the asc-sets in the following way. A trivial asc-set will be always $<$ a non-trivial one. Let us consider, besides the non-trivial asc-set (P) as given above, a second non-trivial asc-set

$$(Q) \qquad Q_1, Q_2, \cdots, Q_s.$$

Then we say P is $< Q$ if one of the following conditions holds true:

$(1)'$ There is some k such that $P_i <=> Q_i$ for $i < k$, while $P_k < Q_k$.

$(2)'$ $r > s$ and $P_i <=> Q_i$ for $i = 1, \cdots, s$.

If for two asc-sets (P) and (Q) neither is $(P) < (Q)$ nor is $(Q) < (P)$, then we say that (P) and (Q) are INCOMPARABLE in ORDER and we write then $(P) <=> (Q)$.

Consider now any polset PS. Any asc-set of lowest order contained in PS will be called a BASIC-SET (abbr. bas-set) of PS. All such bas-set of PS are clearly incomparable in order each other. It is therefore legitimate to define for any two polsets PS_1 and PS_2 whether PS_1 is $< PS_2$ or PS_2 is $< PS_1$ according to their bas-sets.

For any polset PS we form now a certain particular asc-set CS according to the following scheme:

$$\begin{aligned}
PS = \quad & PS_0 \subset PS_1 \subset \cdots \subset PS_s \\
& BS_0 > BS_1 > \cdots > BS_s \ = CS \qquad\qquad \text{(S)} \\
& RS_0 \quad RS_1 \quad \cdots \quad RS_s \ = \text{Empty}
\end{aligned}$$

The above scheme (S) is formed in the following way:

Each BS_i is a bas-set of PS_i.

Each RS_i is the polset consisting of all non-zero remainders of pols in $PS_i \backslash BS_i$ w.r.t. BS_i.

Each polset PS_i is the union of PS_j and RS_j where $j = i - 1$:

$$PS_i = PS_j + RS_j, \quad j = i - 1.$$

We have then

$$BS_0 > BS_1 > \cdots$$

a sequence steadily decreasing in order and the construction should end in a finite number of steps so that in certain stage s we should have

$$RS_s = \text{Empty}.$$

DEF. The corresponding bas-set $BS_s = CS$ in the above scheme (S) is called a CHARACTERISTIC SET (abbr. char-set) of the given polset PS.

Well-Ordering Principle. Let I_i be the initials of pols in CS and J be the product of all these I_i. Then:

$$\text{Zero}(CS/J) \subset \text{Zero}(PS) \subset \text{Zero}(CS),$$

or more precisely,

$$\text{Zero}(PS) = \text{Zero}(CS/J) + \sum_i \text{Zero}(PS_i). \qquad \text{(I)}$$

In the formula (I) each PS_i is the enlarged polset obtained from PS by adjoining to it the initial I_i of the i-th pol of CS.

Now each polset PS_i in (I) is of lower order than PS. Hence if we apply (I) to each of PS_i and proceed in the same way again and again we should stop in a certain stage and arrive at the following

Zero Decomposition Theorem (Weak Form). There is an algorithmic procedure which permits to give for any polset PS a decomposition of the following form:

$$\text{Zero}(PS) = \sum_k \text{Zero}(CS_k/J_k). \qquad \text{(II)}$$

In the formula each CS_k is an asc-set and J_k is the product of all initials of pols in CS_k.

Consider now a non-trivial asc-set (P) as before. we shall say that (P) is REDUCIBLE at a certain stage i if we have a relation of the form

$$H_i \cdot P_i = P_i' \cdot P_i'' + \sum_j A_j \cdot P_j,$$

in which P_i', P_i'' are pols having same class as P_i while H_i is a pol with smaller class and is reduced w.r.t. the partial asc-set ($k = i - 1$)

$$(PP_k) \qquad P_1, \cdots, P_k.$$

The $\sum_j$ means summation over j from 1 to k with A_j some pols.

If (P) is not reducible at any stage, then (P) is said to be IRREDUCIBLE. In that case we can extend the field K to a field IK by successively adjoining $Z_1, Z_2, \cdots, Z_m$ to K with each Z_c either a transcendental element or an algebraic element defined by the equation $P_i = 0$, when c is the class of some pol P_i in (P). The point $Z = (Z_1, Z_2, \cdots, Z_m)$ in the space IK^m is then called a GENERIC POINT of the irreducible asc-set (P). We have now the following

Theorem. W.r.t. an irreducible asc-set (P) with a generic zero Z we have for any pol G:

$$\text{Remdr}(G/(P)) = 0 <===> Z \in \text{Zero}(G).$$

A point Z' in K'^m where K' is any extension field of K is called a SPECIALIZATION of Z if for any pol P,

$$Z \in \mathrm{Zero}(P) ===> Z' \in \mathrm{Zero}(P).$$

The set of all specializaitons of Z forms then an algebraic variety with Z as generic point in the usual sense. This variety is completely determined by Z, which, if it is a generic point of an irreducible asc-set (P), is in turn completely determined by (P). We shall call accordingly this variety as one ASSOCIATED TO (P), to be denoted by $\mathrm{Var}((P))$.

Consider now an asc-set (P) which is reducible at certain stage i as before. Then we have a decomposition of the form

$$\mathrm{Zero}((P)) = \mathrm{Zero}((P)'/H_i) + \mathrm{Zero}((P)''/H_i) + \mathrm{Zero}((P+)),$$

in which $(P)'$, $(P)''$ are the sequences obtained from (P) in replacing P_i of (P) by P' and P'' resp., while $(P+)$ is the polset enlarging (P) by adjoining to it the pol H_i. Applying now this procedure to the decomposition (II) whenever some asc-set CS_k is reducible and proceed in the same manner further and further as long as possible, we shall arrive at the following theorems:

Zero Decomposition Theorem (Strong Form). There is an algorithmic procedure which permits to give for any d-polset PS a decomposition of the form

$$\mathrm{Zero}(PS) = \sum_k \mathrm{Zero}(IRR_k/J_k), \qquad (III)$$

in which each IRR_k is an irreducible asc-set and each J_k is the product of all initials of pols in IRR_k.

Variety Decomposition Theorem. For any polset PS we have a decomposition of the form

$$\mathrm{Zero}(PS) = \sum_k \mathrm{Var}(IRR_k), \qquad (IV)$$

in which $\mathrm{Var}(IRR_k)$ is the algebraic variety associated to certain irreducible asc-set IRR_k. Moreover, this decomposition can be rendered uncontractible, and hence also unique, by mere computations.

The above decomposition theorems and the formulas (I)–(IV) give a complete overview of the structure of the zero-set of any polset. They form also the basis of a constructive theory of algebraic varieties. Moreover, since each asc-set possesses a form which permits to solve the corresponding equations one by one, they furnish us also a general method of solving arbitrary systems of algebraic equations which may be considered as a direct and more systematic generalization of the elimination method created by Zhu Szejze and his predecessors. Cf. [Wu9-11,16,20]. We remark however that there are various modified improved forms of the scheme (S) for the formation of char-set, cf.e.g.[Wu16,21]. We remark also that Gao in a paper [Gao4] has introduced the notion of Minimal Char-Set which is uniquely determined by the given polset.

The above theory has also been extended by Wu to the case of differential polsets, along the lines of development by M. Ch. Riquier, M. M. Janet , E. Cartan, J. M. Thomas, and particularly J. F. Ritt. The results are analogous to those in the above polynomial case but, of course, much more complicated in that integrability conditions or passivity in the sense of Riquier should be taken into consideration. We shall not enter into it here but refer to the relevant papers of Wu, particularly the paper [Wu20] in MM-Preprints No.3.

4 Basic Principles of Mechanization of Theorem Proving

The various decomposition theorems given in Sect 3 will naturally give rise to mechanization methods for the proving of theorems which are more general than the one in Sect 2 and are complete in a certain sense. Let us restrict ourselves first to the case of elementary geometries, i.e. geometries for which no notions of differentiations are involved. On the other hand it is immaterial whether the geometry concerned is the ordinary, the affine, the projective, or any one else. To make precise let us define a THEOREM in such geometries as a couple consisting of a polset called HYPOTHESIS SET (abbr. hyp-set) and a pol called CONCLUSION POL (abbr. conc-pol). We lay down now the following

DEF. Let the hyp-set and the conc-pol of theorem T be resp. HYP and $CONC$. Then we say:

(1) T is TRUE if

$$\text{Zero}(HYP) \subset \text{Zero}(CONC).$$

(2) T is GENERICALLY TRUE under NON-DEGENERACY CONDITIONs

$$DEG_i \neq 0$$

for DEGENERACY POLs (abbr. deg-pol) DEG_i if

$$\text{Zero}(HYP/\textstyle\prod DEG_i) \subset \text{Zero}(CONC).$$

(3) T is TRUE on a part PV of the algebraic variety $\text{Zero}(HYP)$ if

$$PV \subset \text{Zero}(CONC).$$

We remark that a zero in $\text{Zero}(HYP)$ is nothing but a geometrical configuration verifying the hypothesis of the theorem T and $\text{Zero}(HYP)$ is just the algebraic variety of all such geometrical configurations.

For the hyp-set HYP of a theorem T let us form now a char-set CS according to scheme (S) in Sect 3. Let the initials of pols in CS be I_i and J their product. Let HYP_i be the enlarged polsets obtained from HYP by adjoining to it I_i. Then by the well-ordering Principle we have:

$$\text{Zero}(HYP) = \text{Zero}(CS/J) + \sum_i \text{Zero}(HYP_i).$$

From now the remainder R of the conc-pol $CONC$ of theorem T w.r.t. CS:

$$R = \text{Remdr}(CONC/CS).$$

Suppose that $R = 0$ as a pol. Then we see from the remainder formula in Sect 3 that $HYP = 0$ would imply that $CONC = 0$ so far $J \neq 0$ or no I_i is 0, i.e.

$$\text{Zero}(HYP/J) \subset \text{Zero}(CONC).$$

The theorem T is thus seen to be generically true under the nondegeneracy conditions

$$I_i \neq 0.$$

If the asc-set CS is furthermore irreducible, then the generic zero Z of CS is a zero of CS for which no I_i is 0 and so it is also a zero of HYP.

If the theorem T is generically true under the nondegeneracy conditions $I_i \neq 0$ so that in particular Z is a zero of $CONC$ by the formula above, then it follows from the theorem about generic zero in Sect 3 that we have necessarily $R = 0$ as a pol. Hence we have:

Principle of Mechanical Theorem Proving or MTP-Principle I (Weak Form). If the remainder R of $CONC$ w.r.t. a char-set CS of HYP is 0, then the theorem T with hyp-set HYP and conc-pol $CONC$ is generically true under the non-degeneracy conditions $I_i \neq 0$ for which I_i are the initials of pols in CS. Moreover, if the char-set CS is irreducible, then the above condition $R = 0$ is also a necessary one for the theorem T to be generically true under the above non-degeneracy conditions.

Let us now apply the zero-decomposition theorem (weak form) (II) to the hyp-set HYP so that

$$\text{Zero}(HYP) = \sum_k \text{Zero}(CS_k/J_k),$$

in which each CS_k is an asc-set and J_k is the product of initials of pols in CS_k. Let R_k be the remainder of $CONC$ w.r.t. CS_k. Then we have the following

MTP-PRINCIPLE I' (General Weak Form). if $R_k = 0$, then the theorem T is true on the part $\text{Zero}(CS_k/J_k)$ of the whole variety $\text{Zero}(HYP)$ of geometrical configurations verifying the hyp-set HYP. The condition $R_k = 0$ is also necessary if CS_k is irreducible.

More generally, let us apply the variety-decomposition theorem to HYP so that we have

$$\text{Zero}(HYP) = \sum_k Var(IRR_k),$$

in which IRR_k are all irreducible asc-sets. Then we have the following

MTP-PRINCIPLE II (Strong Form). For a theorem T with hyp-set HYP and conc-pol $CONC$ to be true on the irreducible component $\text{Var}(IRR_k)$ of the variety $\text{Zero}(HYP)$, it is necessary and sufficient that the remainder of $CONC$ w.r.t. IRR_k is 0:

$$\text{Remdr}(CONC/IRR_k) = 0.$$

The last principle MTP-II is a complete one in that it can be completely decided whether a theorem is true or not true on this or that irreducible component of the whole variety of Zero(HYP). No more degeneracy conditions will be intervened. As an example we may cite the Feuerbach Theorem stating that the 9-point circle of a triangle is tangent to the 4 inscribed and escribed circles of that triangle. With lines and circles oriented we can establish the hyp-set HYP with Zero(HYP) decomposed into 8 irreducible components. The conclusion is seen to be true just on 4 of such components which reflects the geometrical fact that each of the two oriented 9-point circles is tangent in the same sense just to two of the four oriented in- or es-cribed circles. For this or other examples we refer to the paper [Wu15]. we should remark that, in practice the less general MTP-Principle I seems to be preferred which is already so efficient that a lot of non-trivial and difficult theorems have already been proved and even discovered.

The mechanization methods as given above for elementary geometries have also been extended to the differential case so that differential-geometrial theorems can be proved also in a mechanical way as in the case of elementary geometries. We refer again to the relevant papers of Wu, particularly [Wu20].

5   Further Developments and Applications

In the Institute of Systems Science, Academia Sinica, there was organized a small MM-Group (Mathematics-Mechanization Group) devoting mainly to the developments and applications of the mechanization method of mathematics as explained in previous sections and other intimately related subjects. The group edited also an irregular publication "MM-Res. Preprints" for which 2 issues have since already appeared and the third one is also in the press. We list below some of such studies in these years.

A1. Mechanical Theorem Proving and Discovering of Elementary Geometries

This was the beginning and also the most well-developed direction of the whole MM-method (mathematics mechanization method). It was proved by Wu in starting from axiomatic systems that theorems of various kinds of elementary geometries, whether ordinary, affine, projective, non-euclidean, line or circle ones, etc., can all be proved in a mechanical manner by the general method, so far no order relations are involved, cf. [Wu3]. Such results are formulated in the form of Mechanization theorem of this or that geometry, just like the Hilbert Mechanical Theorem in Sect 2. Gao has also made a detailed comparative study of various geometries connected with the parallelism axiom, cf. [Gao3]. For particular theorems we may mention the list of theorems proved or discovered by Wang and Gao as outlined in their joint paper [W-G]. The most extensive study has been undertaken by Chou and a book devoting to it has already been published, cf. [Ch1,2]. Furthermore, Wu has introduced the notion of oriented lines and circles to avoid or alleviate the difficulties of factorization in the case of ordinary geometry which has further increased the efficiency of the method and promise to discover much more difficult new theorems, cf. [Wu15]. Besides, some theorems in algebraic geometry and planar kinematics, have also been proved or discovered by Wu, cf. [Wu19]. In the case of ordinary geometry for which the hypothesis set is already in the well-ordered form, Hong and Tau-Zhang-Yang have proposed very interesting methods of mechanical theorem proving by means of a single example or a set of examples. cf. e.g. [H1,2] and papers of Tau et al to be published.

A2. Mechanical Theorem Proving and Discovering of Differential Geometries

In principle for differential geometries the same can be done as in the case of elementary geometries but much less have been actually carried out. For concrete results we may cite a detailed study of theorems about pairs of curves of Bertrand type in either ordinary or affine space, cf. [Wu14]. The same has also been partially carried out including a theorem generalizing the classical Baecklund theorem to the affine space which has been considered in recent years by S. Chern et al. (to be published).

A3. Automatic Determination of Unknown Relations

Wu showed how by well-arranging the variables involved our method will give rise automatically to the desired unknown relations. One of the simplest of such examples is furnished e.g. by the automatic discovery

of Chin-Heron formula of the area of a triangle in terms of its 3 sides, cf.
[Wu12,13]. It seems that the same method can furnish us with a formula
for the volume of a hyperbolic tetrahedron in terms of its 6 edges, a
problem which has greatly attracted the attention of mathematicians in
recent years. Note that the occurrence of highly transcendental functions
is not an insurmountable barrier for applying out method, see A6 below.

A4. Automatic Loci Determination

Again by well-arranging the variables involved our method will permit
to give automatically the equations of loci to be studied. A concrete
example is furnished by the determination of coupler-curve equation of
a 4-bar-linkage by Wu (unpublished).

A5. Problems about Polynomials

Hu and Wang have discovered a new method of factoring multivariate polynomials with coefficients in an arbitrary finite extension field
of rational number field which is based on our general MM-method, cf.
[H-W]. Gao has introduced the notion of general discriminant system
of a unary polynomial and gave also an algorithm of its determination
which has been improved by W.D.Wu, cf. [Gao1] and [WDWu]. Liu has
studied the complex root isolation and degree- depressing of polynomials
by decomposition instead of factorization, cf. [Liu1,2]. Li has discovered an interesting general phenomenon of occurrence of unavoidable
factors which generalizes the known theorem of Collins about successive
pseudo-divisions of two polynomials, cf. [Li1-3]. In connection with such
phenomenon Wu has proposed some alternative procedures in elimination for the formation of char-sets which would avoid the occurrence of
such un-desirable factors, cf. [Wu10]. Concrete examples have shown
that efficiency of computations will be raised to a much higher level and
our new implementation of programming has taken into account of this
advantage.

A6. Problems involving Transcendental Functions

At first sight it seems that the occurrence of transcendental functions
would be an insurmountable barrier to the application of our general

MM-method which is essentially algebraic in character. However, as already pointed out by Wu since his earliest writings, quite often it is rather the algebraic relations among the transcendental functions, but not the functions themselves, which will play an actual role in the problems to be solved. The transcendence may then be rendered disappear if one replaces the transcendental functions by new variables connected by algebraic relations or by applying rationalization process. It is for this reason that a lot of theorems involving trigonometric or hyperbolic functions, including trigonometric identities, have been proved or discovered by our MM-method. Recently Wu et al have shown how transcendental equations like the Kepler equation in astronomy can be turned into an algebraic (more exactly an algebrico-differential) ones so that one can solve it in an algebraic manner, in contrast to the generally used graphical or numerical methods of solution, cf. [W-W]. The method consists in replacing the transcendental function in question by a new function and a new differential equation determining that function. The method is again a general one and will be useful in many occassions.

A7. Studies on Limit Cycles

Some papers of Wang, e.g. [Wang1] have shown how our MM-method can be applied to study the Lyapounov functions at a singular point of a polynomial differential equation system in the plane. The Soviet and Chinese mathematicians have shown how such knowledge could be applied to the construction of limit cycles, whose study is seemingly the only one of the 23 problems of Hilbert which has little advance up to the present time.

A8. Automatic Derivation of Newton's Gravitational Laws from Kepler's Laws

In [Wu17] Wu has shown how Newton's gravitational laws can be automatically derived from the Kepler's Laws. This may serve as a simplest example of automatically formulating underlying mathematical principles from experimental facts in physical sciences.

A9. Inequalities Proving

A lot of inequalities occurring in elementary mathematics can be put into the form of equations-solving by means of method of Lagrange's multipliers. The latter one can then be dealt with by our general mechanization method of equations-solving. In [Wu18] Wu has shown how this can actually be carried out for various kinds of inequalities. There is also a paper of Wang [Wang2] dealing with the same subject.

A10. Non-linear Programming

The linear programming deals with the optimization of a linear function involving variables under linear constraints for which the variables and constraints are both large in number. On the contrary, the non-linear programming deals with optimization of a non-linear function involving variables under non-linear constraints for which the variables and constraints are both relatively few in numbers. The fewness of the number of constraints may be taken into great advantage by our method of inequalities-proving in A9 while the non-linearity of the function to be optimized and the constraints may be treated in the usual manner by our method. Wu has thus been able to solve in a new way certain problems in chemical engineering, viz. chemical equilibrium problem and chemical reaction optimization problem (to be published). Note that transcendental functions occurring quite often in non-linear programming can be dealt with again by the tactics in A6.

A11. Robotics

The solving of inverse kinematical equations (I.K.E.) of robot motions is usually considered as one of the most difficult problems in robotics studies. However, Wu has applied his MM-method of equations-solving to give simple solutions of I.K.E. for certain simple robots usually used in practice and is able to furnish the equations of continuous variation of control parameters meeting the continuous motion of robot hands (to be published). A complete solution in the general case is quite difficult but seems to be still possible and is yet in the course of study.

A12. Mechanical Design and Control Theory

In some early papers in Chinese Wu has indicated how our mechanization method of equations-solving can be used to some simple problems of linkage design, pole-assignment of control theory, etc. Now in the subsequent years the programming has been greatly improved and the MM-Group has been equipped with much more powerful computers than the ones available at that time. It seems that much harder problems in these realms will be successfully attacked in years to come.

References

[BL]   Bledsoe, W.W. and Loveland,D.W. (Ed.), *Automated Theorem Proving, after 25 Years*, Amer. Math. Soc.,(1984).

[CH1]   Chou,S.S., Proving elementary geometry theorems using Wu's algorithm, in [BL], 243–286.

[CH2]   —, *Mechanical geometry theorem-proving*, Reidel, (1988).

[Gao1]   Gao Xiao-shan, The discriminant systems of unary equations and their computation, *MM-Res. Preprints*, No.1(1987) 13–32.

[Gao2]   —, Transcendental functions and mechanical theorem proving in elementary geometries, *MM-Res. Preprints*, No.2 (1987) 37–47.

[Gao3]   —, Mechanical theorem proving in Cayley-Klein geometries, *MM-Res. Preprints*, No.3 (1989).

[Gao4]   —, The minimal characteristic set of polynomial ideal, *J. Sys. Sci. and Math. Scis.*, 2 (1989).

[H1]   Hong Jiawei, Can we prove geometry theorem by computing an example? *Scientia Sinica*, A29 (1986) 824–834.

[H2]   —, How fast the algebraic approximation can be? *ibid*, A29 (1986) 813–823.

[H-W]   Hu Sen and Wang Dongming, Fast factorization of polynomials over rational number field of its extension fields, *Kuxue Tongbao*, 31 (1986) 150–156.

[J-W]   Jin Xiaofan and Wang Dongming, On Kukles' condition for the existence of a center, *MM-Res. Preprints*, No.3 (1989).

[Li1]   Li Zi-ming, A new proof of Collins's theorem, *MM-Res. Preprints*, No.1 (1987) 33–37.

[Li2]   —, On the trianguration for any finite polynomial set (I), *MM-Res. Preprints*, No.2 (1987) 48–54.

[Li3]   —, Determinant polynomial sequences, *MM-Res. Preprints*, No.3, (1989).

[Liu1]   Liu Zhuo-jun, A method of isolating complex roots of polynomials — An application of Wu-Ritt Principle, *MM-Res. Preprints*, No.2 (1989) 55–61.

[Liu2]   —, Algorithm of decomposing high degree polynomials, *MM-Res. Preprints*, No.2 (1989) 62-67. also *Chinese Q. J. Math.* 3 (1988) 101–108.

[Ritt1]   Ritt,J.F., *Differential equations from the algebraic standpoint*, Amer. Math. Soc.,(1932).

[Ritt2]   —, *Differential algebra*, Amer. Math. Soc., (1950).

[Wang1]   Wang Dong-ming, Mechanical manipulation of differential systems, *MM-Res. Preprints*, No.1(1987) 38–52.

[Wang2]   —, A decision method for definite polynomial, *MM-Res. Preprints*, No.2 (1987) 68–74.

[Wang3]   —, On Wu's method for proving constructive geometric theorems, *MM-Res. Preprints*, No.3 (1989).

[Wang4]   —, Proving-by-examples method and inclusion of varieties, *Kexue Tongbao*, 33 (1988) 2015–2018.

[W-G]   Wang Dong-ming and Gao Xiao-shan, Geometry theorems proved mechanically using Wu's method — Part on Euclidean geometry, in *MM-Res Preprints*, No.2, (1987) 75–106.

[WDWu]   Wu Wen-da, A note to the discriminant systems for the unary equations, *MM-Res. Preprints*, No.3 (1989).

[Wu1]   Wu Wen-tsun, On the decision problem and the mechanization of theorem-proving in elementary geometry, *Scientia Sinica* 21 (1978) 159–172, re-published in [BL], 213–234.

[Wu2]   —, Some recent advance in mechanical theorem-proving of geometries, in [BL], 235–242.

[Wu3]   —. *Basic Principles of Mechanical Theorem Proving in Geometries (Part on Elementary Geometries)*, (in Chinese), Science Press, (1984).

[Wu4]   —, Basic principles of mechanical theorem-proving in elementary geometries, *J. Sys. Sci. and Math. Scis.*, 4(1984) 207–235. Republished in *J. of Automated Reasoning*, 2(1986) 221–252.

[Wu5]   —, Toward mechanization of geometry — Some comments on Hilbert's "Grundlagen der Geometrie", *Acta Math. Scientia*, 2 (1982) 125–138.

[Wu6]   —, Some remarks on mechanical theorem-proving in elementary geometry, *Acta Math. Scientia* 3 (1983) 357–360.

[Wu7]   —, On the mechanization of theorem-proving in elementary differential geometry, *Scientia Sinica, Math. Supplement (I)*, 94–102 (1979). (in Chinese).

[Wu8]   —, Mechanical theorem proving in elementary geometry and differential geometry, in *Proc.1980 Beijing DD-Symposium*, Beijing, v.2, 1073–1092 (1982).

[Wu9]   —, A constructive theory of differential algebraic geometry, in Differential Geometry and Differential Equations, *Lect. Notes in Math.* No. 1255, Springer (1984) 173–189.

[Wu10]   —, On zeros of algebraic equations — an application of Ritt principle, *Kewue Tongbao* 31 (1986) 1–5.

[Wu11]   —, A mechanization method of geometry I. Elementary geometry, *Chinese Q. J. Math.*, 1(1986) 1–14, Errata and addenda, *ibid*, 1 (1987) 20.

[Wu12]   —, A mechanization method of geometry and its applications, I. Distances, areas, and volumes, *J. Sys. Sci. and Math. Scis.*, 6(1986) 204–216.

[Wu13]   —, A mechanization method of geometry and its applications, I. Distances, areas, and volumes in euclidean and non-euclidean geometries, *Kuxue Tongbao* 32 (1986) 436–440.

[Wu14]   —, A mechanization method of geometry and its applications, 2. Curve pairs of Bertrand type, *Kuxue Tongbao* 32(1987) 585–588.

[Wu15]   —, On reducibility problem in mechanical theorem proving of elementary geometries, *Chinese Quarterly J. of Math.*, 2 (1987) also in *MM-Res. Preprints*, No.2, (1987) 18–36.

[Wu16]   —, A zero structure theorem for polynomial-equations- solving and its applications, *MM-Res. Preprints*, No.1(1987) 2–12.

[Wu17]   —, Mechanical derivation of Newton's Gravitational Laws from Kepler's Laws, *MM-Res. Preprints*, No.1, (1987) 53–61.

[Wu18]   —, A mechanization method of geometry and its applications, 3. Mechanical proving of polynomial inequalities and equations-solving, *MM-Res. Preprints*, No.2, (1987) 1–17, also in *J. Sys. Sci. and Math. Scis.*, Inst. of Systems Science, 2 (1989).

[Wu19]   —, A mechanization method of geometry and its applications, 4. Some theorems in planar kinematics, *J. Sys. Sci. and Math. Scis.*, Inst. of SYstems Science, 2 (1989) 97–109.

[Wu20]   —, On the foundation of algebraic differential geometry, *MM-Res. Preprints*, No.3 (1989), also *J. Sys. Sci. and Math. Scis.*

[Wu21]   —, Some remarks on characteristic-set formation, *MM-res. Preprints*, No.3 (1989).

[W-W]   Wu Wen-tsun and Wu Tianjiao, A mechanization method of geometry and its applications, 5. Solving transcendental equations by algebraic methods, *MM-Res. Preprints*, No.3 (1989).

Critical Groups, Morse Theory and Application to Semilinear Elliptic BVPs

Chang Kung-ching
Institute of Math., Peking Univ., China

Abstract

An unified framework, which contains various critical point theorems, is presented. Applications are given to show the advantage for this treatment.

In this paper, we study the Morse Theory as a tool in dealing with multiple solutions of differential equations arising in the calculus of variations. It consists of two parts: local and global. In the local theory, a sequence of groups, which are called the critical groups, is studied. They are used in distinguishing isolated critical points. As to the global theory, the relative homology groups of two certain level sets according to the special structure of the functional are estimated. The mini-max principle for homology classes is applied to obtain critical points and to estimate the associated critical groups. In this account, many critical point theorems obtained by the mountain pass lemma, the saddle point theorem, the Ljustenik Schninelman Theory, the index theory and the pseudo-index theory etc. are treated in a unified manner.

The paper consists of three parts. §1 is a review of the basic properties of the critical groups obtained previously. §2 is devoted to the global theory. In particular, we prove the linking theorem and the equivariant Mountain Pass Theorem by our homology method. In §3, one may find several interesting applications of the unified theory to the semilinear elliptic boundary value problems.

1 Critical groups — the local theory

Let M be a Banach Finsler manifold (we always assume that M is complete), and let $f \in C^1(M, \mathbf{R}^1)$. The critical set of f is denoted by K, and $f^{-1}(c) \cap K$ is denoted by K_c, for each $c \in \mathbf{R}^1$. The set $f_c = \{x \in M \mid f(x) \leq c\}$ is called the level set. According to Brézis, Nirenberg, we call the function f satisfying the $(PS)_c$ condition, if

Any sequence $x_j \in M$, along which $f(x_j) \to c$ and $df(x_j) \to \theta$ possesses a convergent subsequence.

For an isolated critical point p, there exists an open neighbourhood U of p such that $U \cap K = \{p\}$. The critical groups for f at p are defined as follows:

$$C_q(f,p) = H_q(f_c \cap U, (f_c \backslash \{p\}) \cap U, G), \qquad (1.1)$$

$q = 0, 1, 2, \cdots$, $c = f(p)$, where G is an Abelian coefficient group, and $H_*(.,.,G)$ stands for the singular relative homology groups with coefficients in G. Due to the excision property, the critical groups are well defined. They are a series of topological invariants which characterize the local behaviour of the function f at the critical point p.

The following propositions for critical groups are of fundamental importance:

(1) If p is an isolated local minimum, then $C_q(f,p) = \delta_{q0}G$.

(2) If $f \in C^{2-0}(M, \mathbf{R}^1)$ satisfies $(PS)_c$, $c = f(p)$, and if p is an isolated critical point, which is not a local minimum, then $C_0(f,p) = 0$.

Combining (1) with (2), under suitable conditions of f, we conclude that an isolated critical point p is a local minimum if and only if

$$C_q(f,p) = \delta_{q0}G.$$

(3) If p is a nondegenerate critical point with Morese index m, where M is a Riemannian Hilbert manifold, then $C_q(f,p) = \delta_{qm}G$.

Moreover, the following splitting theorem which is a generalization of Morse lemma holds in a small neighbourhood of an isolated critical point.

Theorem 1.1 (Splitting). *Let U be a neighbourhood of p in a Riemannian Hilbert manifold M. Let $f \in C^2(U, \mathbf{R}^1)$, and let p be the unique critical point of f. Let N be the kernel of $A = d^2f(p)$ in $T_p(M)$. Suppose that 0 is either an isolated point of spectrum $\sigma(A)$ or not in $\sigma(A)$. Then there exist a neighborhood $V \subset U$ of p, a local homomorphism*

$\Phi : B(\theta,\rho) \to V$ *(where $B(\theta,\rho)$ is a ball centered at θ with radius $\rho > 0$ in $T_p(M)$), and a C^1 mapping $h : B(\theta,\rho) \cap N \to N^\perp$ such that*

$$f \circ \Phi(y+z) = \frac{1}{2}(Az,z) + f \circ \exp_p(h(y)+y), \quad \forall x \in B(\sigma,\rho),$$

where $y = Px$ and $z = P^\perp x$, P and $P^\perp$ are the orthogonal projections onto N and $N^\perp$ respectively.

The piece of submanifold $\phi(B(\theta,\rho) \cap N)$ is called the characteristic manifold of f at p. Let

$$f_0(y) = f \circ \exp_p(h(y) + y), \tag{1.2}$$

which is, loosely speaking, a function defined on the characteristic manifold.

One concludes

(4) Assume that $f \in C^2(M, \mathbf{R}^1)$ satisfies $(PS)_c$, where $c = f(p)$, and p is an isolated critical point with Morse index m. Then we have

$$C_q(f,p) = C_{q-m}(f_0,\theta).$$

A critical point p is called a Mountain pass point, if $C_1(f,p) \neq 0$.

(5) Let p be a mountain pass point. If $d^2 f(p)$ is Fredholm, and if $\dim \ker d^2 f(p) \leq 1$, then

$$C_q(f,p) = \delta_{q1} G.$$

(6) (*Homotopy invariance*) Suppose that $\{f_\sigma \in C^2(M, \mathbf{R}^1) \mid \sigma \in [0,1]\}$ is a family of functions satisfying the (PS) condition on a Riemannian Hilbert manifold M. Suppose that there exists an open set U such that f_σ has unique critical point p_σ in U, $\forall \sigma \in [0,1]$, and that $\sigma \to f_\sigma$ is continuous in $C^1(U)$ topology. Then $C_*(f_\sigma, p_\sigma)$ is independent of σ.

(7) (*Relation to the Leray Schander index*) Let H be a real Hilbert space, and let $f \in C^2(H, \mathbf{R}^1)$ satisfy the (PS) condition. If $df(x) = x - T(x)$ where T is a compact mapping, and if p is an isolated critical point of f then we have

$$\text{LS-ind}(df,p) = \sum_{q=0}^{\infty}(-1)^q \text{ rank } C_q(f,p) \tag{1.3}$$

where LS-ind denotes the Leray Schander index of the vector field $df(x)$ at its zero p.

The proposition (7) reveals that the critical groups provide more information than the Leray Schander index in describing isolated critical points.

The proofs of all the above propositions can be found in K.C. Chang [Ch 1]. In fact, propositions (3), (4) and (6) had already known under stronger conditions by Gromoll Meyer [GM 1] in 1969, while propositions (1) and (7) were also obtained by E. Rothe [Ro 1, 2] with some extra-assumptions. The above definition of critical groups was introduced by E. Rothe [Ro2], a different one was given by Gromoll Meyer. Indeed, under suitable conditions, these two definitions are equivalent, as shown in [Ch 1].

Remark. The context made above can be extended to isolated critical orbits if the function f is invariant under a compact Lie group action. cf K.C. Chang [Ch 2]. and Z.Q. Wang [Wan 1].

One can also define the critical groups by singular cohomology. Namely,

$$C^q(f, p_0) = H^q(f_c \cap U, (f_c \backslash \{p_0\}) \cap U; G).$$

where U, c and G have the same meaning as in (1.1). $C^q(f, p_0)$ enjoys all basic properties for $C_q(f, p_0)$, $\forall q$.

(8) If $n = \dim M < \infty$, and if f satisfies $(PS)_d$ $\forall d \in [c - \delta, c + \delta]$ for some $\delta > 0$, where $c = f(p_0)$; then

$$C^q(f, p_0) = C_{n-q}(-f, p_0), \quad \text{for } q = 0, 1, \cdots, n.$$

Proof. We find a closed neighbourhood U of p_0 and $\varepsilon > 0$ satisfying

(1) $K \cap U = \{p_0\}$, U has arbitrarily small neighbourhoods which are homotopy equivalent to U.

(2) The set $f^{-1}[c - \varepsilon, c + \varepsilon] \cap \tilde{U}$ is in a local chart containing p_0, where $\tilde{U} = \bigcup_{t \in \mathbf{R}^1} \eta(t, U)$ is the smallest gradient flow invariant set containing U.

Let $f^d = \{x \in M \mid f(x) \geq d\}$. We have $f^d = (-f)_{-d}$. Thus

$$C_{n-q}(-f, p_0) \cong H_{n-q}\left((-f)_{-c+\varepsilon} \cap \tilde{U}, (-f)_{-c-\varepsilon} \cap \tilde{U}, G\right)$$

$$\cong \overline{H}^q\left((-f)^{-c-\varepsilon}\cap \tilde{U},(-f)^{-c+\varepsilon}\cap \tilde{U};G\right)$$

$$\cong \overline{H}^q\left(f_{c+\varepsilon}\cap \tilde{U},f_{c-\varepsilon}\cap \tilde{U};G\right),$$

provided by the Alexander duality theorem and the excision property, where $\overline{H}$ is the Alexander cohomology. (We embed the local chart, which contains $f^{-1}[c-\varepsilon,c+\varepsilon]\cap \tilde{U}$, in S^n).

According to the condition (1),

$$\overline{H}^q\left(f_{c+\varepsilon}\cap \tilde{U},f_{c-\varepsilon}\cap \tilde{U};G\right) \cong H^q\left(f_{c+\varepsilon}\cap \tilde{U},f_{c-\varepsilon}\cap \tilde{U};G\right),$$

$$\cong C^q(f,p_0).$$

2 Morse theory — the global theory

It is a well-known fact, that $\forall a < b$, if $K\cap f^{-1}[a,b]=\emptyset$, and if f satisfies $(PS)_c$ $\forall c\in[a,b]$, then f_a is a strong deformation retract of f_b. The basic idea in the Morse theory is to prove the existence of critical points by verifying the nontriviality of the topological pair (f_b,f_a) and to estimate the multiplicity of critical points by counting the richness of the topology of (f_b,f_a). A natural candidate in measuring the topology of (f_b,f_a) is the singular relative homology group, because on one hand it is easier to compute, and on the other hand, it is consistent to the critical groups.

First, we set up the Morse relation, which is a connection between the local theory and the global theory.

Lemma 2.1. *Suppose that $f\in C^1(M,\mathbf{R}^1)$ satisfies $(PS)_d$ $\forall d\in[c-\varepsilon,c+\varepsilon]$ for an isolated critical value c, and $\varepsilon>0$ such that c is the unique critical value in $[c-\varepsilon,c+\varepsilon]$. Let $K_c=\{z_1,z_2,\cdots,z_\ell\}$. Then*

$$H_*\left(f_{c+\varepsilon},f_{c-\varepsilon};G\right)\cong \bigoplus_{j=1}^{\ell}C_*(f,z_j).$$

Assume that $\{c_i\}_1^\infty\subset[a,b]$ (b may be $+\infty$) are only critical values, with

$$K_{c_i}=\left\{z_j^i\right\}_{j=1}^{m_i},\quad i=1,2,\cdots.$$

Let us define

$$M_q(a,b)=\sum_i\sum_{j=1}^{m_i}\mathrm{rank}C_q(f,z_j^i)$$

to be the qth Morse type number. $q = 0, 1, 2, \cdots$.

As a direct consequence of Lemma 2.1 and the sub additivity of the functors for pairs (X, Y):

$$S_q(X, Y) = \sum_{j=0}^{q} (-1)^{q-j} \operatorname{rank} H_j(X, Y)$$

$q = 0, 1, 2, \cdots$. we obtain the Morse inequalities:

Theorem 2.2. *Suppose that $f \in C^1(M, \mathbf{R}^1)$ satisfies $(PS)_c \ \forall c \in [a, b]$, and that f has only isolated critical points. Then there is a formal series $Q(t)$ with nonnegative coefficients such that*

$$\sum_{q=0}^{\infty} M_q t^q = \sum_{q=0}^{\infty} \beta_q t^q + (1+t)Q(t)$$

where $M_q = M_q(a, b)$, and $\beta_q = \beta_q(a, b) = \operatorname{rank} H_q(f_b, f_a, G)$.

In the following, we shall give a unified way to study various known results in the critical point theory, namely, we prove these theorems by observing the relative homology groups for certain pairs of level sets. As we have seen in lemma 2.1, the advantage of this approach is to provide some information on critical groups for critical points so obtained.

Definition 2.3. *Let D be a k-topological ball in M, and let S be a subset. We say that $S^{k-1} = \partial D$ and S homologically link, if $\partial D \cap S = \emptyset$ and $|\tau| \cap S \neq \emptyset$ for each singular k chain τ, with $\partial \tau = \partial D$, where $|\tau|$ is the support of τ.*

Theorem 2.4. *Assume that ∂D and S homologically link. If*

$$f \in C(M, \mathbf{R}^1)$$

satisfies

$$f(x) > a, \qquad \forall x \in S,$$
$$f(x) \leq a, \qquad \forall x \in \partial D;$$

then $H_k(f_b, f_a; G) \neq 0$, for any coefficient group G, where $b > \operatorname{Max}\{f(x) \mid x \in \overline{D}\}$.

Proof. We want to show that the singular homology class $[\partial D]$, which is generated by the closed chain ∂D, is trivial in $H_{k-1}(f_b)$, but not trivial in $H_{k-1}(f_a)$. Indeed, since $D \subset f_b$, $[\partial D]$ is trivial in $H_{k-1}(f_b)$. However, $\forall$ singular k chain τ, with $\partial \tau = \partial D$, $\mathrm{Max}_{x \in |\tau|} f(x) > a$, because $|\tau| \cap S \neq \emptyset$. This means that ∂D cannot be the boundary of any singular k chain τ with support $|\tau| \subset f_a$, i.e., $[\partial D]$ is nontrivial in $H_{k-1}(f_a)$.

We observe the exact sequence

$$\cdots \to H_k(f_b, f_a) \xrightarrow{\partial_*} H_{k-1}(f_a) \xrightarrow{i_*} H_{k-1}(f_b) \to \cdots,$$

where $i : f_a \to f_b$ is the inclusion, and ∂ is the boundary operator. It follows

$$[\partial D] \in \ker i_* = \mathrm{Im}\partial_*.$$

Consequently, $H_k(f_b, f_a)$ is nontrivial.

The following result provides examples of homologically linking sets.

Theorem 2.5. *Let D be a k-dimensional topologically ball in M, and let S be a path connected submanifold satisfying $S \cap D = \{a\ single\ point\}$, and codimension of $S = k$. Assume that there exists a tubular neighbourhood N of S such that $N \cap D$ is homeomorphic to D. Then ∂D and S homologically link.*

Proof. Let $N = S \times B^k$ be a tubular neighbourhood of S, by excision, $H_*(M, M \backslash S) \cong H_*(N, N \backslash S)$. After a standard deformation,

$$H_*(N, N \backslash S) \cong H_*(N, \partial N) \cong H_*(S \times B^k, S \times S^{k-1}).$$

According to the Künneth formula,

$$H_*(S \times B^k, S \times S^{k-1}) \cong H_*(S) \otimes H_*(B^k, S^{k-1}).$$

Since S is path connected, $H_0(S) \neq 0$, therefore

$$H_k(M, M \backslash S) = G.$$

Let $[\tau]$ be the generator of $H_k(M, M \backslash S) \cong H_k(N, \partial N)$. We may choose a representation $\tau \in [\partial \tau]$ such that $\partial \tau = \partial N \cap D$ by homotopy. According to the assumptions that $D \cap S = \{a\ single\ point\}$ and that $N \cap D \cong D$, we find that ∂D and $\partial \tau$ are homologous in $H_k(M, M \backslash S)$.

Therefore $[D] = [\tau]$ is nontrivial in $H_k(M, M\backslash S)$. Obviously $\partial D \cap S = \emptyset$ and $|\tau| \cap S \neq \emptyset$ for each τ in $[D]$, i.e., ∂D and S homologically link.

Example 1. (Mountain pass) Let Ω be a ball with center z_0 in a Banach space X, and let $z_1 \notin \overline{\Omega}$. Set $D =$ the segment joining z_0 and z_1, and $S = \partial\Omega$. Then $\partial D = \{z_0, z_1\}$ and S homologically link.

Example 2. Let X be a Banach space. Let $X = X_1 \oplus X_2$ be a direct sum decomposition, where $k = \dim X_1 < +\infty$. Set $D = X_1 \cap B_1$ where B_1 is the unit ball centered at θ, and $S = X_2$. Then ∂D and S homologically link.

Example 3. Let $X = X_1 \oplus X_2$ be as in Example 2. Let $e \in X_2$ with $\|e\| = 1$ and let $R_1, R_2, \rho > 0$ satisfy $\rho < R_1$. Set $D = \{x + se \mid x \in X_1 \cap B_{R_2}, s \in [0, R_1]\}$ and $S = X_2 \cap \partial B_\rho$. Then ∂D and S homologically link.

For any nontrivial homology class $[\tau] \in H_k(f_b, f_a, G)$, one defines

$$c = \inf_{\tau \in [\tau]} \sup_{X \in |\tau|} f(x).$$

Theorem 2.6. *If f satisfies $(PS)_c$, then c is a critical value. Furthermore, we conclude that there exists $p \in K_c$ such that*

$$C_k(f, p) \neq 0,$$

if K_c is isolated.

This is a recent result due to J.Q. Liu [Li 1]. Actually Theorems 2.4 and 2.5 for the special case, Example 2, were treated by him And for the special case $k = 1$, (i.e. Example 1) it was a result due to G. Tian [Ti 1].

Remark. Theorem 2.5 is somewhat related to V. Benci [Be 1] and X.F. Yang [Ya 1].

Different homology classes may not create different critical points / values. In order to estimate the number of critical points, the following definition is needed.

Definition 2.7. *Let (X, Y) be a topological pair, and let $[\sigma_1], [\sigma_2] \in H_*(X, Y; G)$ be nontrivial. We say $[\sigma_1]$ is subordinate to $[\sigma_2]$, denoted by $[\sigma_1] < [\sigma_2]$, if there exists a cohomology class $\omega \in H^*(X, G)$ with $\dim \omega > 0$ such that $[\sigma_1] = [\sigma_2] \cap \omega$ where $\cap$ is the cap product.*

Set

$$L(X, Y; G) \;=\; \mathrm{Max}\{\ell \in \mathbf{N} \mid \exists \text{ nontrivial classes}$$
$$[\sigma_1], [\sigma_2], \cdots, [\sigma_\ell] \in H_*(X, Y; G)$$
$$\text{such that } [\sigma_1] < [\sigma_2] < \cdots < [\sigma_\ell]\}.$$

Theorem 2.8. *Suppose $f \in C^1(M, \mathbf{R}^1)$, and that $a < b$ are regular values. Assume that $[\sigma_1] < [\sigma_2]$ are in $H_*(f_a, f_b; G)$. Let*

$$c_i = \inf_{\tau \in [\sigma_i]} \sup_{X \in |\tau|} f(x), \quad i = 1, 2.$$

If $(PS)_{c_i}$, $i = 1, 2$, hold, and if $\sharp K_{c_2} < \infty$, then $c_1 < c_2$ are two distinct critical values.

Corollary 2.9. *If $f \in C^1(M, \mathbf{R}^1)$ satisfies $(PS)_c$ $\forall c \in [a, b]$, where a and b are regular values. Then f has at least $L(f_b, f_a)$ critical points with critical values in $[a, b]$.*

If we drop out the condition $\#K_{c_2} < \infty$ in Theorem 2.8, then the conclusion turns to the following.

Theorem 2.10. *Suppose that $f \in C^1(M, \mathbf{R}^1)$ has regular values $a < b$. Assume that $[\sigma_1], [\sigma_2], \cdots, [\sigma_\ell]$ are subordinate classes in $H_*(f_b, f_a; G)$. Let*

$$c_i = \inf_{\tau \in [\sigma_i]} \sup_{X \in |\tau|} f(x), \quad i = 1, 2, \cdots, \ell$$

If $c = c_1 = c_2 = \cdots = c_\ell$ and if f satisfies $(PS)_c$. Then

$$\mathrm{cat}(K_c) \geq \ell.$$

Remark. Theorems 2.8 to 2.10 are extensions of the Ljusternik Schnirelmann theory. The extensions focus on functions not necessarily bounded from below, so that the relative homology of pairs of level sets are considered. The proofs can be found in [Ch 1] and [Li 2].

We shall continue the homology method in studying critical point theorems for invariant functionals which were obtained by some kinds of index theory or pseudo index theory cf. Ambrosetti Rabinowitz [AR 1]. Rabinowitz [Ra 1], Fadell Rabinowitz [FR 1] and V. Benci [Be 1]. Not only this approach provides a clear picture about how these critical points are created, but also gives more information on the critical groups for these critical points.

Let X be a Banach space, and let $f \in C^1(M, \mathbf{R}^1)$ is even.

Theorem 2.11. *Assume*

(f_1) $\exists$ *regular values* $a < b$ *such that* $(PS)_c$ *holds for all* $c \in [a, b]$.

(f_2) $\exists$ *linear subspaces* X_+ *and* X_- *with* $j = \mathrm{codim}X_+ < m = \dim X_- < +\infty$ *such that*

    (1) $f(x) > a, \quad \forall x \in X_+,$

    (2) $f(x) \le b, \quad \forall x \in X_- \cap S_\rho$ *for some* $\rho > 0,$

    (3) $f(\theta) > b.$

Then $L(f_b/Z_2, f_a/Z_2; Z_2) > m - j$. *Moreover, there exist* $m - j$ *pairs of critical points* $\pm x_\ell$ *with*

$$C_\ell(f, \pm x_\ell) \ne 0 \quad \ell = j, j+1, \cdots, m-1,$$

if f *has only isolated critical points in* $f^{-1}[a, b]$.

(In this statement, since $\theta \notin f_b$, f_b/Z_2 is the quotient space of f_b under the group action $Z_2 = \{\mathrm{id}, \tau\}$, where $\tau x = -x \ \forall x \in X$. The same meaning for f_a/Z_2.

Proof. By the assumption (f_2), $f_a \subset X\backslash X_+$ and $X_- \cap S_\rho \subset f_b$. We have the injections i, j and k as follows:

$$(X_- \cap S_\rho, \emptyset)/Z_2 \xrightarrow{\ i\ } (f_b, f_a)/Z_2$$

$$k \searrow \qquad \downarrow j$$

$$(X\backslash\{\theta\}, X\backslash X_+)/Z_2$$

Let us show that

$$H_*\left((X\backslash\{\theta\})/Z_2, (X\backslash X_+)/Z_2; Z_2\right) \cong H_*\left(P^{n-1}, P^{j-1}, Z_2\right),$$

where $n = \dim X$ may be $+\infty$, and P^{n-1} and P^{i-1} are real projective spaces. Indeed, $\forall x \in X$, we have a direct sum decomposition $x = y + z$ where $z \in X_+$. Define a deformation

$$\eta(t, x) = y + tz \quad t \in [0, 1]$$

which is Z_2 equivalent, and deforms $X\backslash X_+$ into $X_+^\perp\backslash\{\theta\}$ where $X_+^\perp$ is the direct complement of X_+.

Since $H_*((X_- \cap S_\rho)/Z_2; Z_2) \cong H_*(P^{m-1}; Z_2)$, we obtain the commutative diagram:

$$H_*(P^{m-1}, \emptyset; Z_2) \xrightarrow{\;i_*\;} H_*(f_b/Z_2, f_a/Z_2, Z_2)$$

$$k_* \searrow \qquad\qquad \downarrow j_*$$

$$H_*(P^{n-1}, P^{j-1}, Z_2)$$

Similarly, for the cohomology rings, we have

$$H^*(P^{m-1}; Z_2) \xleftarrow{\;i^*\;} H^*(f_b/Z_2; Z_2)$$

$$k^* \nwarrow \qquad \nearrow j^*$$

$$H^*(P^{n-1}; Z_2)$$

Noticing that $H^*(P^{n-1}; Z_2) = \sum_{j=0}^{n-1} \omega^j$, where ω is the generator with $\dim \omega = 1$, and that $k^* \omega^2 = (k^* \omega)^2$, we find nontrivial classes $[z_\ell] \in H_\ell(P^{m-1}, \emptyset, Z_2)$, $\ell = 0, 1, 2, \cdots, m-1$, satisfying $[z_{\ell-1}] = [z_\ell] \cap k^* \omega$, for $\ell = 1, 2, \cdots, m-1$.

We shall prove that $k_*[z_\ell] \neq 0$ for $\ell = j, j+1, \cdots, m-1$. Indeed, it suffices to prove that $|z_\ell| \cap X_+ \neq \emptyset \ \forall z_\ell \in [z_\ell]$, for $\ell = j, j+1, \cdots, m-1$. If there exists $\ell_0 \in [j, m-1]$ such that $|z_{\ell_0}| \cap X_+ = \emptyset$, then $|z_{\ell_0}|$ is deformed to $(X_+^\pm \cap S_1)/Z_2$, which implies $\dim |z_{\ell_0}| \leq j-1$. It is impossible.

Since $k^* = i^* \circ j^*$, we have

$$i_*[z_{\ell-1}] = i_*([z_\ell] \cap k^* \omega) = i_*[z_\ell] \cap j^* \omega.$$

From $\dim j^* \omega = 1$, we obtain $m-j$ subordinate classes $i_*[z_j] < i_*[z_{j+1}] < \cdots < i_*[z_{m-1}]$, i.e., $L(f_b/Z_2, f_a/Z_2; Z_2) \geq m-j$. Moreover, $i_*[z_\ell] \in H_\ell(f_b/Z_2, f_a/Z_2; Z_2)$. Then there exists $\xi_\ell = $ the class $\{x_{\ell'} - x_\ell\} \in f^{-1}[a, b]/Z_2$ such that $C_\ell(\hat{f}, \xi_\ell) \neq 0$, $\ell = j, j+1, \cdots, m-1$, where $\hat{f}$ is the reduction of f on $f^{-1}[a, b]/Z_2$. However, by definition, $C_\ell(\hat{f}, \xi_\ell) = C_\ell(f, x_\ell) \oplus C_\ell(f, -x_\ell)$, and $C_\ell(f, -x_\ell) \cong C_\ell(f, x_\ell)$. We conclude $C_\ell(f, \pm x_\ell) \neq 0$.

A dual version of theorem 2.11 relies on a Galerkin approximation scheme. Assume that there is a sequence of finite dimensional linear subspace

$$X^1 \subset X^2 \subset \cdots \subset X^n \subset \cdots, \quad \dim X^n = n,$$

with $\bigcup_{n=1}^{\infty} X^n = X$.

In replacing the (PS) condition, we need

$[((PS)^n$ condition$)$ $\exists n_0 \in N$ such that $\forall n \geq n_0$, the restriction $f^n = f \mid_{X^n}$ satisfies the $(PS)_c$ condition $\forall c \in [a, b].]$

$[((PS)^*$ condition$)$ For any sequence $x_n \in X^n$, $n = 1, 2, \cdots$, along which $a \leq f(x_n) \leq b$ and $df^n(x_n) \to \theta$ imply a convergent subsequence.]

Theorem 2.12. *Assume*

(f'_1) *$f \in C^1(X, R^1)$ is even, and has regular values $a < b$. The $(PS)^n$ and $(PS)^*$ conditions hold with respect to a sequence of finite dimensional linear subspaces $\{X^n\}_1^\infty$.*

(f'_2) *$\exists$ linear subspaces X_+ and X_- with*

$$j = \operatorname{codim} X_+ < m = \dim X_- < +\infty$$

such that

(1) *$f(x) \geq a$, $\forall x \in X_+ \cap S_\rho$ for some $\rho > 0$,*
(2) *$f(x) < b$, $\forall x \in X_-$,*
(3) *$f(\theta) < a$.*

Then f has at least $m - j$ pairs of critical points.

Proof. First, we change f to $-f$, $-a$ to b and $-b$ to a. Then the assumption (f'_2) is changed to be the following:

(1) $f(x) \leq b$, $\forall x \in X_+ \cap S_\rho$,

(2) $f(x) > a$, $\forall x \in X_-$,

(3) $f(\theta) > b$.

According to (f'_1), we choose n large enough such that both X_+^+ and X_- are included in X^n. If we restrict ourselves on the functions f^n, then (1) to (3) turn out to be the following:

$$f^n(x) < b, \quad \forall x \in (X^n \cap X_+) \cap S_\rho,$$
$$f^n(x) > a, \quad \forall x \in X_-,$$
$$f^n(\theta) > b.$$

Let $X_-^n = X_+ \cap X^n$, and $X_+^n = X_-$, we have $\text{codim} X_+^n = n - m$ and $\dim X_-^n = n - j$. We apply Theorem 2.11 to f^n, and conclude that there are $(m - j)$ subordinate classes:

$$[z_j^n] < [z_{j+1}^n] < \cdots < [z_{m-1}^n].$$

which correspond to critical values

$$a \leq c_j^n \leq c_{j+1}^n \leq \cdots \leq c_{m-1}^n \leq b \quad \text{for } n \text{ large.}$$

Let

$$c_\ell = \lim_{n \to \infty} c_\ell^n, \ \ell = j, j+1, \cdots, m-1$$

(may omit a subsequence). Then

$$a \leq c_j \leq c_{j+1} \leq \cdots \leq c_{m-1} \leq b.$$

According to the $(PS)^*$ condition, c_ℓ are all critical values for $j \leq \ell \leq m - 1$. It remains to prove that if

$$c = c_k = c_{k+1} = \cdots = c_\ell, \quad j \leq k \leq \ell \leq m - 1,$$

then $s = \#K_c \geq \ell - k + 1$.

We choose closed neighborhoods U_i of p_i, $i = 1, 2, ..., s$, where $\{p_1, \cdots, p_s\} = K_c$, such that $U_i \cap U_j = \emptyset \ \forall i \neq j$.

Denote $M = f^{-1}[a, b]/Z_2$ and $M^n = (f^n)^{-1}[a, b]/Z_2$. In the following, we omit the subscripts. Let $U^n = M^n \cap U$, and let p^n be the nearest point to p in U^n. One may assume that the geodesic segment connecting p and p^n is in U for n large.

We choose suitable coordinates (Y, ϕ), $Y = Y_1 \oplus Y_2$, where $\dim Y_1 = n$, and $\phi : Y \cap B_1 \to U$ satisfying

$$\phi(Y_1 \cap B_1) \subset U^n,$$

$$\phi(\theta) = p^n, \text{ and}$$

$$\exists \overline{y}_2 \in Y_2 \text{ such that } p = \phi(\overline{y}_2) \text{ and } \phi(t\overline{y}_2) \in U \ \forall t \in [0, 1].$$

For any $\delta > 0$, we construct

$$\begin{aligned} V_\delta = \ &\{(y_1, y_2) \in Y_1 \oplus Y_2 \mid \|y_1\| < \delta, \exists t \in [0, 1], \\ &\text{such that } \|y_2 - t\overline{y}_2\| < \delta\}. \end{aligned}$$

Consequently, $\exists \bar{\delta} > 0$ such that $N = \phi(V_{\bar{\delta}}) \subset U$.

The neighborhood N possesses the following property:

$$\exists \text{ a deformation retract } \eta : N \to N^n = M^n \cap N.$$

Indeed, $\eta = \phi \circ \pi \circ \phi^{-1}$, where π is the projection onto Y_1. We prove

     $1°$ $\text{Cat}_{N^n}(N^n) = 1$.

In fact, $V_{\bar{\delta}}$ is contractible, so is N (in N). It follows that N^n is contractible in N, i.e. $\exists \xi \sim \text{id}_N$ with $\xi(N^n) = q^n \in N$. Let us define $\tau = \eta \circ \xi$, then $\tau \sim \text{id}_{N^n}$, $\tau(N^n) = \eta(q^n) \in N^n$ and the homotopy is in N^n, i.e. N^n is contractible in itself.

     $2°$ Let

$$\mathcal{N} = \bigcup_{i=1}^{s} N_i, \text{ and } \mathcal{N}^n = \bigcup_{i=1}^{s} N_i^n.$$

We shall prove that $\forall \delta > 0, \exists \bar{b} > 0, \bar{\varepsilon} > 0$ and a closed neighborhood $\mathcal{N}'$ of K_c such that

$$\text{dist}(\partial \mathcal{N}, \mathcal{N}') \geq \delta,$$
$$\|df^n(x)\| \geq \bar{b} \qquad \forall x \in (f^n)_{c+\bar{\varepsilon}} \backslash ((f^n)_{c-\bar{\varepsilon}} \cup \mathcal{N}'),$$
$$0 < \bar{\varepsilon} < \text{Min}\{2\delta\bar{b}^2, \delta\bar{b}\}.$$

Indeed, $\mathcal{N}'$ is easy to construct. Once it is fixed, we prove the next two conclusions. If not, one can find $x_n \in M^n \backslash \mathcal{N}'$ satisfying

$$f(x_n) \to c \text{ and } \|df^n(x_n)\| \to 0.$$

By the $(PS)^*$ condition, $\exists x^* \in K_c \backslash \mathcal{N}'$. This is impossible.

     $3°$ $\forall \varepsilon \in (0, \bar{\varepsilon}/3), \exists n_0 \in N$ such that $\forall n \geq n_0$,

$$c - \varepsilon < c_k^n \leq c_{k+1}^n \leq \cdots \leq c_\ell^n < c + \varepsilon.$$

We prove that

$$\text{Cat}_{M^n}(\mathcal{N}'^n) \geq k - \ell + 1, \text{ where } \mathcal{N}'^n = M^n \cap \mathcal{N}'.$$

In fact, we have known $L((f^n)_{c+\varepsilon}, (f^n)_{c-\varepsilon}) \geq k - l + 1$. If

$$r = \text{Cat}_{M^n}(\mathcal{N}'^n) < k - \ell + 1,$$

then there is a closed covering $\bigcup_{j=1}^{r} B_j \supset \mathcal{N}'^n$, such that each B_j is contractible in M^n. Now we choose

$$[\sigma_1], \cdots, [\sigma_r] \in H_*((f^n)_{c+\varepsilon}, (f^n)_{c-\varepsilon}; \mathbf{Z}_2), \text{ and}$$
$$[\omega_2], \cdots, [\omega_r] \in H^*((f^n)_{c+\varepsilon}, \mathbf{Z}_2) \text{ satisfying}$$
$$[\sigma_{j-1}] = [\sigma_j] \cap [\omega_j], \ \dim \omega_j > 0, \ j = 2, 3, \cdots, r.$$

Choosing $\sigma_j \in [\sigma_j]$ and $\omega_j \in [\omega_j]$ such that $|\omega_j| \cap B_j = \emptyset$. We decompose

$$\sigma_1 = \tau_1 + \tau_2 + \cdots + \tau_r$$

such that

$$|\tau_j| \subset B_j, \ j = 2, \cdots, r, \text{ and } |\tau_1| \subset (f^n)_{c+\varepsilon} \backslash \mathcal{N}'.$$

By the deformation lemma, $\exists \eta \sim$ id such that $\eta(|\tau_1|) \subset (f^n)_{c-\varepsilon}$. Therefore

$$c_k^n \leq \sup_{x \in \eta(|\tau_1|)} f(x) \leq c - \varepsilon.$$

This is a contradiction.

In summary, we have

$$s = \sum_{j=1}^{s} \mathrm{Cat}_{N_j^n}(N_j^n) \geq \mathrm{Cat}_{M^n}(\mathcal{N}'^n) \geq k - \ell + 1.$$

Remark. Theorems 2.11, 2.12 can be extended to functions f invariant under the S^1 group action, cf [Ch 2].

3 Applications to semilinear elliptic BVPs

We shall present here few examples showing how the above theories work. In the following, we consider the problem

$$\begin{cases} -\Delta u(x) = g(u(x)), & x \in \Omega \subset \mathbf{R}^n, \\ u \mid \partial\Omega = 0 \end{cases} \tag{3.1}$$

where Ω is a smooth bounded domain in $\mathbf{R}^n$. We assume

(g_1) $g \in C^1(\mathbf{R}_+^1)$ and $g(0) \geq 0$,

(g_2) $\exists\, 0 < a_1 < a_2 < \cdots < a_m$ such that $g(a_i) = 0$, $i = 1, 2, \cdots, m$,

(g_3) Let $G(t) = \int_0^t g(s)ds$,

$$\text{Max}\{g(t) \mid t \in [0, a_{i-1}]\} < G(a_i), \quad i = 2, 3, \cdots, m;$$

and $G(a_1) > 0$.

Theorem 3.1. *Under the assumptions* $(g_1), (g_2)$ *and* (g_3), *the nonlinear eigenvalue problem*

$$\begin{cases} \Delta u = \lambda g(u) & in \ \Omega, \\ u|\partial\Omega = 0 \end{cases} \tag{3.2}$$

possesses at least $(2m - 1)$ *nontrivial solutions for* λ *large enough. Furthermore, if* $g(0) = 0$ *and* $g'(0) \leq 0$, *then* (3.2) *possess at least* $2m$ *nontrivial solutions for large* λ.

Proof. By a cut off technique, we define

$$\tilde{g}_i(t) = \begin{cases} g_i(0) & , \quad t < 0, \\ g_i(t), & \quad t \in [0, a_i], \\ 0, & \quad t > a_i, \end{cases}$$

then $\tilde{g}_i \in C^{1-0}(\mathbf{R}^1)$. The solutions of

$$\begin{cases} -\Delta u = \lambda \tilde{g}(u) & in \ \Omega, \\ u|\partial\Omega = 0 \end{cases}$$

are also solutions of (3.2), provided by the Maximum Principle.

Define

$$J^\lambda(u) = \int_\Omega \left[\frac{1}{2}(\nabla u)^2 - \lambda G(u)\right] \quad \text{and}$$

$$J_i^\lambda(u) = \int_\Omega \left[\frac{1}{2}(\nabla u)^2 - \lambda \tilde{G}_i(u)\right]$$

on the Sobolev space $H_0^1(\Omega)$, where

$$\tilde{G}_i(t) = \int_0^t \tilde{g}_i(s)ds.$$

Let K^λ and K_i^λ denote the critical sets of J^λ and J_i^λ respectively. It is easily seen that

$$K_i^\lambda \subset K_{i+1}^\lambda \subset \cdots \subset K^\lambda, \quad \text{and that}$$
$$\forall u \in K_i^\lambda \quad 0 \leq u(x) \leq a_i, \quad \forall x \in \overline{\Omega}.$$

The functionals J^λ and J_i^λ are all bounded from below, and satisfy the (PS) condition. Consequently, there exists the global minimum $u_i^\lambda \in K_i^\lambda$ for each $i = 1, 2, \cdots, m$.

However, it follows from P. Hess [He1], that there exists $\lambda_i^* > 0$ and $\omega \in H_0^1(\Omega)$ with $0 \leq \omega(x) \leq a_i$ such that $J_i^\lambda(\omega) < J_{i-1}^\lambda(u_{i-1}^\lambda) \, \forall \lambda > \lambda_i^*$, $i = 2, 3, \cdots, m$. This means $u_i^\lambda \notin K_{i-1}^\lambda$.

First. We claim that $C_q(J^\lambda, u_i^\lambda) = \delta_{q0} G$. Indeed by the strong Maximum Principle, u_i^λ is in the interior of the ordered interval $\{\omega \in C_0^1(\Omega) | 0 \leq \omega(x) \leq a_i\}$, so u_i^λ is the local minimum of J^λ in the $C_0^1(\overline{\Omega})$ topology. But $C_0^1(\overline{\Omega})$ is dense in $H_0^1(\Omega)$, and J^λ is continuous on $H_0^1(\Omega)$, therefore u_i^λ is a local minimum of J^λ. By the Prop. 1 in I, we obtain the conclusion.

Next in case $g(0) = 0$, we have $\theta \in K_i^\lambda$, $\forall i$. But we want to show that $C_*(J_i^\lambda, \theta) = 0$ if $\lambda g'(0) > \lambda_1$, where λ_1 is the 1st eigenvalue of the Laplacian. Indeed. let us define

$$g_\sigma(t) = g(t) + \sigma \quad \text{with} \quad \sigma > 0.$$

We show that there is no solution in a small neighbourhood of 0 for the following equations with parameter $\sigma \in (0, 1]$:

$$\begin{cases} -\Delta u = \lambda g_\sigma(u), & \text{in } \Omega, \\ u|\partial\Omega = 0 \end{cases} \tag{3.3}$$

if $\lambda g'(0) > \lambda_1$.

Claim. By the assumption $\lambda g'(0) > \lambda_1$, we find $\delta > 0$ such that

$$\lambda g(t) > \lambda_1 t \quad \text{for } 0 \leq t \leq \delta.$$

For fixed λ, if u is a solution of (3.3) with $\max u \leq \delta$, then we have

$$\lambda_1 \int u \cdot \varphi_1 = \int -\Delta u \cdot \varphi_1 = \int \lambda g_\sigma(u) \cdot \varphi_1 > \lambda_1 \int u \cdot \varphi_1$$

where $\varphi_1 > 0$ is the first eigenfunction. This is impossible. However, by the L^p estimates and the embedding theorem

$$\begin{aligned}
\|u\|_{C(\overline{\Omega})} &\leq C_1\|u\|_{W_p^2(\Omega)} \leq C_{2,p}\|\lambda g_\sigma(u)\|_{L^p(\Omega)} \\
&\leq \lambda C_3\|u\|_{H_0^1(\Omega)}
\end{aligned}$$

where we choose $p > \frac{n}{2}$. We see that there is no solution of (3.3) in a small H_0^1 neighbourhood U of θ. Let

$$J^{\lambda,\sigma}(U) = \int \left[\frac{1}{2}(\nabla u)^2 - \lambda G_\sigma(u)\right], \quad \sigma \in (0,1].$$

We choose a Gromall Meyer pair (W, W_-) of J^λ at θ (cf. [Ch 1]) such that $W \subset U$, then by the homotopy invariance (Prop. 6, I), (W, W_-) is also a Gromoll Meyer pair of $J^{\lambda,\sigma}$ at θ for $\delta > 0$ small. Therefore $H_*(W, W_-) = 0$, which implies that $C_*(J^\lambda, \theta) = 0$ and then $C_*(J_i^\lambda, \theta) = 0$, $\forall i$.

Third, the Morse relation for J_i^λ reads as

$$\sum_{j=0}^\infty (-1)^j M_j^{\lambda,i} = 1$$

where $M_j^{\lambda,i}$ is the jth Morese type number for J_i^λ, $j = 0, 1, 2, \cdots$. From $u_i^\lambda \in K_i^\lambda \backslash K_{i-1}^\lambda$, it follow $\#(K_i^\lambda \backslash K_{i-1}^\lambda) \geq 2$. For otherwise the balance will be lost.

Therefore

$$\#(K_m^\lambda \backslash \{\theta\}) = \# \bigcup_{i=2}^m (K_i^\lambda \backslash K_{i-1}^\lambda) \bigcup (K_i^\lambda \backslash \{\theta\}) \geq 2m - 1.$$

Furthermore, if $g(0) = 0$ and $g'(0) \leq 0$, then θ is a local minimum of J_1^λ, but not the global minimum u_1^λ, for λ large. We have $\#(K_1^\lambda \backslash \{\theta\}) \geq 2$. The conclusion follows.

Remark. The assumption (g_3) is essential. In fact, we have the following example: the equation

$$\begin{cases} -\Delta u = \lambda \cos u & \text{on the unit ball } B \text{ in } \mathbf{R}^3 \ \lambda > 0, \\ u|_{\partial B} = 0 \end{cases} \tag{3.4}$$

possesses a unique positive solution u^λ, which satisfies $0 \leq u^\lambda(x) \leq \dfrac{\pi}{2}$.

Claim. If we presume $0 \leq u^\lambda(x) \leq \frac{\pi}{2}$, then the solution $u^\lambda(x)$ is unique. Indeed, let u_i $i = 1, 2$ be two such solutions. Let $v(x) = u_1(x) - u_2(x)$, then $\exists \xi(x) \in (0, 1)$ for $x \in B$, such that

$$-\Delta v(x) = -\lambda \sin(\xi(x)u_1(x) + (1 - \xi(x))u_2(x))v(x).$$

It follows

$$\int_B (\nabla v)^2 = -\lambda \int_B \sin(\xi(x)u_1(x) + (1 - \xi(x))u_2(x))v^2(x) \leq 0,$$

thus $v \equiv 0$.

Next, we shall prove that if u is a positive solution of (3.4), then $\text{Max} u(x) \leq \frac{\pi}{2}$. In fact, if $\text{Max} u(x) > s > \frac{\pi}{2}$, then the set $\Omega_s = \{x \in B | u(x) > s\}$ is an nonempty ball, provided by a theorem due to Gidas, Ni, Nireberg [GNN 1]. Let

$$F(t) = \int_s^t \cos \tau d\tau = \sin t - \sin s.$$

By the Pohozaev identity, we have

$$3 \int_{\Omega_s} F(u(x))dx = \frac{1}{2} \int_{\Omega_s} |\nabla u|^2 dx + \frac{1}{2} \int_{\partial \Omega_s} \left(\frac{\partial u}{\partial v}\right)^2 d\sigma > 0.$$

It implies that there exists a point $x_0 \in \Omega_s$ such that $\sin u(x_0) > \sin s$. However, by the strong Maximum Principle, if $\text{Max} u \leq \frac{\pi}{2} + 2k\pi$, for some positive integer k, then $\text{Max} u < \frac{\pi}{2} + 2k\pi$. Let k_0 be an natural number such that

$$\frac{\pi}{2} + 2k_0\pi < \text{Max} u < \frac{\pi}{2} + 2(k_0 + 1)\pi.$$

We choose $s \in (\frac{\pi}{2} + 2k_0\pi, \text{Max} u)$ such that $\sin s > \sin(\text{Max} u)$. By the above conclusion, there exists $x_0 \in B$ satisfying $u(x_0) > s$ and $\sin u(x_0) > \sin s > \sin(\text{Max} u)$. This is impossible.

Remark. The special case where $g(0) > 0$ was already studied by K.J. Brown, H.Budin [BB 1] and P. Hess [He 1].

Remark. The C^1 condition for the function g is only used in the case where $g(0) = 0$ and $g'(0) > 0$.

Now we turn to (3.1), where

$$g(u) = \lambda_\infty u + \phi(u).$$

For simplicity, we assume

$$(\phi_1) \qquad \phi \in C^1(\mathbf{R}^1) \text{ is bounded, and } \phi(0) = 0.$$

In case $\lambda_\infty \notin \sigma(-\Delta)$ the spectrum of $-\Delta$, the existence of a nontrivial solution for (3.1) was studied by Amann Zehnder [AZ 1]. Let $\lambda_0 = \lambda_\infty + \phi'(0)$, a sufficient condition for the existence is that there exists an eigenvalue of $-\Delta$ lying between λ_0 and λ_∞. If $\lambda_\infty \in \sigma(-\Delta)$, then the problem is called a resonance problem. Under the Landesman Lazer condition. The author extended the above result in [Ch 3]. According to (Ward [War 1] and Mawhin [Ma 1], a more sophisticate case is the so-called strong resonance problem in which we assume

$$(\text{SR}) \qquad \int_\Omega \phi\left(u(x) + \sum \xi_i e_i(x)\right) v(x)dx \to 0 \quad \text{and}$$

$$\int_\Omega \Phi\left(u(x) + \sum \xi_i e_i(x)\right) v(x)dx \to 0 \quad \text{as}$$

$$|\xi| \to \infty, \quad \forall u, v \in H_0^1(\Omega),$$

where $\xi = (\xi_1, \xi_2, \cdots, \xi_{m_0})$, $\text{span}\{e_1, e_2, \cdots, e_{m_0}\} = \ker(-\Delta - \lambda_\infty I)$, and $\Phi(t)$ is a primitive of ϕ (not necessarily assume that $\Phi(t) = \int_0^t \Phi(x)dx!$).

In the variational approach, the functional

$$J(u) = \int_\Omega \frac{1}{2}[(\nabla u)^2 - \lambda_\infty u^2] - \Phi(u).$$

loses the $(PS)_0$ condition. Apart from the works of Ward [War 1], Solimini [So 1], in Chang, Liu [CL 1], we compactify the underlying space by adding some "∞" points. Namely, we decompose

$$H_0^1(\Omega) = H_+ \oplus H_0 \oplus H_-$$

according to the spectrum of $A = -\Delta - \lambda_\infty I$, where H_* $(* = +, 0, -)$ is the (positive, zero, negative) invariant space. Adding a "∞" point on H_0, it turns out to be a sphere S^{m_0}. We extend the functional J to $\hat{J}$ on the enlarged space by

$$\hat{J}(w, \xi) = \begin{cases} J(w + \xi), & (w, \xi) \in H^\perp \times H_0, \\ \dfrac{1}{2}\displaystyle\int_\Omega (\nabla w)^2 - \lambda_\infty w^2, & (w, \xi) \in H^\perp \times \{\infty\}. \end{cases}$$

where $H^\perp = H_+ \oplus H_-$.

The functional J is continuous, but not differentiable. We call $\hat{K} = K \cup \{\hat{\infty}\}$ the fake critical set of J, where $\hat{\infty} = (\theta, \infty)$ in $H^{\perp} \times S^{m_0}$, and K is the critical set of J. The point $\{\hat{\infty}\}$ is called a fake critical point. Although the (PS) condition does not hold, the deformation theorem, and then the critical point theorems discussed above all remain true Cf [CL 1]. Other points in $\hat{K}$ are called genuine.

According to a theorem due to the author [Ch 3],

$$H_q(H^{\perp} \times S^{m_0}, \hat{J}a) \cong H_{q-m_-}(S^{m_0})$$

for $-a$ large enough, where $m_- = \dim H_-$. Moreover, one has a pair of subordinate classes $[\sigma_1] < [\sigma_2]$ with $[\sigma_1] \in H_{m_-}(H^{\perp} \times S^{m_0}, \hat{J}a)$, and $[\sigma_2] \in H_{m_-+m_0}(H^{\perp} \times S^{m_0}, \hat{J}a)$. Hence we have critical values $c_1 < c_2$ and $u_1, u_2 \in \hat{K}$ satisfying

$$C_{m_-}(\hat{J}, u_1) \neq 0, \quad C_{m_-+m_0}(\hat{J}, u_2) \neq 0,$$
$$\text{and} \qquad \hat{J}(u_i) = c_i \qquad i = 1, 2.$$

In order to obtain nontrivial "genuine" critical points, our main task is to single out u_1 and u_2 from θ and $\hat{\infty}$. The following theorem was already known by Capozzi, Lupo, Solimini [CLS 1] and Chang, Liu [CL 1].

Theorem 3.2. *Under the assumptions (ϕ_1) and (SR), if one of the following three cases occurs, then there exists a nontrivial solution of* (3.1).

(1) $\Phi(0) = 0$,

(2) $\Phi(0) > 0$ *and* $\lambda_0 \notin [\underline{\lambda}, \lambda_{\infty}]$,

(3) $\Phi(0) < 0$ *and* $\lambda_0 \notin [\lambda_{\infty}, \overline{\lambda}]$

where $\underline{\lambda} < \lambda_{\infty} < \overline{\lambda}$ *are consecutive eigenvalues of* $-\Delta$.

Proof. Since $\hat{J}(\theta) = \hat{J}(\hat{\infty}) = 0$. if $\Phi(0) = 0$. So $\exists c_i \neq 0$, it corresponds to a genuine critical point.

If $\Phi(0) > 0$ and $\lambda_0 \notin [\underline{\lambda}, \lambda_{\infty}]$, then $C_{m_-}(\hat{J}, \theta) = 0$, So, $u_1 \neq \theta$, and $\hat{J}(\theta) < 0$. If $\hat{J}(u_1) \geq 0$ then $\hat{J}(u_2) > 0$ so $u_2 \neq \theta$ and $\hat{\infty}$. If $\hat{J}(u_1) < 0$; then $u_1 \neq \theta$ and $\hat{\infty}$.

Similarly, we prove the case (3).

However, if we replace the condition (SR) by a stronger one:

(RD) $\qquad \Phi(t) \to 0$ and $|t|^{\alpha} \phi'(t) \to 0$ as $|t| \to \infty$ for $\alpha > 2$.

Then we may improve the above result to the following.

Theorem 3.3. *Under the assumptions* (ϕ_1) *and* (RD). *We assume that* $\Phi(t)$ *changes sign at* ∞. *The equation* (3.1) *possesses at least two non-trivial solutions if one of the following cases occurs*

(1) $\lambda_0 \notin [\underline{\lambda}, \overline{\lambda}]$ *or*

(2) $\Phi(0) = 0$, $\lambda_0 = \lambda_\infty$ *and* $\Phi(t)$ *changes sign at* 0.

Proof. Firstly, we claim that (RD) implies (SR). In fact, from $\phi(t) = \int_0^t \phi'(s)ds + \phi(0)$, and $|t|^\alpha \phi'(t) \to 0$ as $|t| \to \infty$, $\alpha > 1$, it follows $\phi(t)$ has limits as $t \to 0$ as $t \to \pm\infty$. But $\Phi(t) \to 0$, as $t \to \pm\infty$, which implies that $\phi(t) \to 0$ as $t \to \infty$. Applying the Lebesque dominance theorem, (SR) follows.

Next. We claim that J is C^2 at $\hat{\infty}$, with

$$d^2\hat{J}(\hat{\infty})((w,\eta),(w,\eta)) = \frac{1}{2}\int (\nabla w)^2, \forall (w,\eta) \in H^\perp \times T_{\hat{\infty}} S^{m_0}$$

i.e. We want to show that the functional

$$I(w,\xi) = \begin{cases} \displaystyle\int_\Omega \Phi\left(w(x) + \sum_{i=1}^{m_0} \xi_i e_i(x)\right)dx, & (w,\xi) \in H^\perp \times H_0 \\ 0, & (w,\xi) \in H^\perp \times \{\infty\} \end{cases}$$

is C^2 at $\hat{\infty}$, with 0 Hessian. For this purpose, we introduce a local coordinate at $\hat{\infty}$. Define

$$\eta = \frac{\xi}{|\xi|\ln|\xi|}, \ |\xi| \text{ large}, \quad \text{and}$$
$$H(w,\eta) = I(w,\xi).$$

H is well defined in a small neighbourhood of (θ,θ) in $H^\perp \times H_0$. A direct computation verifies $d^2H(\theta,\theta) = 0$.

Finally. From the assumption that $\Phi(t)$ changes sign at ∞, it follows that H changes sign in a small neighbourhood of (θ,θ). According to propositions 3, 4 and 8 in I. we conclude that

$$C_{m_-}(\hat{J},\hat{\infty}) = C_{m_0+m_-}(\hat{J},\hat{\infty}) = 0,$$

provided $\hat{\infty}$ is neither a local minimum nor a local maximum of the function $\hat{J}$ restricted on its characteristic manifold.

Both cases (1) and (2) imply that

$$C_{m_-}(\hat{J},\hat{\theta}) = C_{m_0+m_-}(\hat{J},\hat{\theta}) = 0,$$

provided by the same reason. Thus $u_i \neq \hat{\theta}$ and $\hat{\infty}$, $i = 1, 2$. We have two nontrivial solutions.

Remark. We say $\Phi(t)$ changes sign at ∞, it means that $\Phi(t)$ has different signs as $t \to +\infty$ and $t \to -\infty$. The same meaning for $\Phi(t)$ changing sign at 0.

As the last example, we consider the equation

$$\begin{cases} -\Delta u = \lambda_\infty u + te^{-|u|}\sin^3 u & \text{in } \Omega \subset \mathbf{R}'', \\ u|_{\partial\Omega} = 0. \end{cases} \tag{3.5}$$

Theorem 3.4. *For $\lambda_\infty > \lambda_2$ and $t \geq \frac{3\pi}{2}\lambda_\infty e^{\frac{3\pi}{2}}$, the equation possesses at least* $\dim\ker(-\Delta - \lambda_\infty I) + 2$ *pairs of nontrivial solutions.*

Proof. We consider the functional

$$J(u) = \frac{1}{2}\int_\Omega (\nabla u)^2 - \lambda_\infty u^2 + tI(u)$$

where $I(u) = -\int_\Omega \int_0^{u(x)} e^{-|s|}\sin^3 s\, ds dx$. Since $J(u)$ is even, and satisfies the condition (RD). The $(PS)_c$ condition holds for $c \neq 0$.

On one hand, let $E_\lambda = \oplus\{M_{\lambda_j} | \lambda_j < \lambda, \lambda_j \in \sigma(-\Delta)\}$ where M_{λ_j} is the eigenspace of $(-\Delta)$ corresponding to the eigenvalue λ_j. We have

$$J|_{E_{\lambda_\infty} \cap S_\rho} < 0 \quad \text{for some } \rho > 0,$$

and

$$J|_{E_\lambda^\perp} \geq M \quad \text{(a constant)}.$$

According to Theorem 2.11. There exists at least $\dim M_{\lambda_\infty}$ pairs of nontrivial solutions $\pm z_\ell$ with $C_\ell(J, \pm z_\ell) \neq 0$, $\ell = m_-, m_- + 1, \cdots, m_- + m_0 - 1$, where $m_- = \dim E_\lambda > 1$ and $m_0 = \dim M_{\lambda_\infty}$.

On the other hand, the function

$$g_t(s) = \lambda_\infty s + te^{-|s|}\sin^3 s \leq 0 \quad \text{at } s = \frac{3\pi}{2} \text{ for } t \geq \frac{3\pi\lambda_\infty}{2}e^{\frac{3\pi}{2}}.$$

We may easily construct a pair of positive sub and super-solutions: $\varepsilon\varphi_1$ and $(-\Delta)^{-1}(\frac{3\pi}{2}\lambda_\infty)$ for $\varepsilon > 0$ small enough, where φ_1 is the first eigenfunction .

Consequently, there is a positive local minimum z_0 of J. Since $C_q(J, z_0) = \delta_{q0}G$, $z_0 \neq z_\ell$, $\ell \in [m_-, \cdots, m_- + m_0 - 1]$.

Moreover. J is unbounded from below, we find (Cf [Ch 4]) a mountain pass point $z_1 \neq z_0, z_\ell, \forall \ell \in [m_-, \cdots, m_- + m_0 - 1]$, and $z_1 \neq \hat{\infty}, \theta$. Therefore, we obtain $m_0 + 2$ pairs of nontrivial solutions.

Remark. Another interesting applications to superlinear BVP is referred to Z.Q. Wang [Wan 2].

References

[AR 1]   A. Ambrosetti, P.H. Rabinowitz; *J.Funct. Analysis*, 14,(1973), 349–381.

[AZ 1]   H. Amann, E. Zehnder; *Ann. Scu norm. Sup. Pisa.* 7,(1980), 539–603.

[BB 1]   K.J. Brown, H. Budin; *SIAM J. Math Analy.* 10, (1979), 875–883.

[Be 1]   V. Benci;*Conf. Semin. Mat Univ. Bari*, 218(1987), 1–32.

[Be 2]   V. Benci; *Comm. Pure Appl. Math* 34 (1981), 393–432.

[Ch 1]   K.C. Chang; *Infinite Dimensional Morse Theory and applications*, Univ. de Montreal, 97, (1985).

[Ch 2]   K.C. Chang; *Infinite Dimensional Morse Theory and Multiple Solutions*, (in preparation) Birkhauser Boston.

[Ch 3]   K.c. Chang, *Proc. Symp. Pure Math* AMS (1986), 253–262.

[Ch 4]   K.C. Chang; *Scientia Sinica, Ser A.*, 26 (1983), 1256–1265.

[CL 1]   K.C. Chang, J.Q. Liu; *Chinese Annales of Math.*, 11 B.2(1990), 191–210.

[CLS 1]   Capozzi, D. Lupo, S. Solimini; *Nonlinear Analysis,* TMA 13, (1989), 151–163.

[FR 1]   E. Fadell, P.H. Rabinowitz; *Invent. Math.* 45(1978), 139–174.

[GM 1]   D. Gromoll, W. Meyer; *Topology* 8 (1969), 361–369.

[GNN 1] B. Gidas, W. M. Ni, L. Nirenberg; *Comm. Math Phys.* 68 (1979), 209–243.

[He 1]   P. Hess; *Comm. PDE.* 6, (1981), 951–961.

[Li 1]   J.Q. Liu; A Morse index of a saddle point, (*preprint*).

[Li2]   J.Q. Liu; *J. Diff. Equation*,(1989).

[Ma 1]   J. Mawhim; *Problems de Dirichlet, variationnels nonlineaires*, Univ. de Montreal, 104,(1987).

[Ra 1]   P.H. Rabinowitz; Minimax methods in critical point theory with applications to differential equations. *AMS. Reg. Conf. Ser. in Math.* 65 AMS (1986).

[Ro 1]   E. Rothe; *Rocky Mojuntain Jomath.* 3, (1973), 251–274.

[Ro 2]   E. Rothe; *J. Math. Analy. Appl.* 88, (1952), 265–269.

[So 1]   S. Solimini; *J. Math. Analy. Appl.* 117, (1986), 138–152.

[Ti 1]   G. Tian; *Kexue Tongbao* 29 (1984), 1150–1154.

[Wan 1]   Z.Q. Wang; *Lecture Notes in Math 1306*, Springer (1988), 202–221.

[Wan 2]   Z.Q. Wang; On a superlinear elliptic equations, *Preprint* (1989).

[War 1]   J.R. Ward; *Nonlinear Analysis*, TMS 10, (1986), 207–213.

[Ya 1]   X.F. Yang, The Morse Critical groups of minimax theorem, *Preprint*.

Some Notes on Kähler-Einstein Metrics with Positive Scalar Curvature

Gang Tian

School of Mathematics, Institute for Advanced Study
Princeton, NJ 08540
and
Institute of Mathematics, Peking University
Beijing, China

Abstract

In this note, we study the generalized Calabi's problem, namely, the existence of Kähler-Einstein orbifold metric on Kähler orbifolds with the first Chern class positive. We give some analytic obstructions to the existence. We construct a 2-dimensional Kähler orbifold with positive first Chern class and without holomorphic vector field and Kähler-Einstein metric, either. Following the author's solution of Calabi's problem for complex surfaces, we also outline a program to solve this generalized Calabi's problem.

1 Introduction

The main theorem in [T2] states that a compact Kähler surface with positive first Chern class admits a Kähler-Einstein metric iff its Lie algebra of holomorphic vector fields is reductive. This solves a long-standing conjecture of Calabi in case of complex surfaces with positive first Chern class. The proof of the main theorem in [T2] is based on a new partial C^0-estimate the author developed in [T2], the previous work in [T1],[TY],[Y]. Next it is interesting to compactify the moduli space of Kähler-Einstein metrics with positive scalar curvature on complex surfaces. Then we are bound to study singular Kähler-Einstein metrics of special type. In this note, we give some discussions on Kähler orbifolds with positive first Chern class. The original goal of this note is to generalize the method in [T2] to prove the existence of Kähler-Einstein orbifold metrics on Del-Pezzo orbifolds (cf. section 1). My belief is that these Kähler-Einstein metrics on Del-Pezzo orbifolds should be sufficient to compactify the moduli space of Kähler-Einstein metrics on complex surfaces with positive first Chern class. However, because there are some technical difficulties we can not overcome at this moment, we decide to write this note first to indicate the difficulties of the problem and some very hopeful methods to solve the problem.

2 Kähler orbifolds with positive first Chern Class

A n-dimensional complex orbifold M is a topological space satisfying: (1) each point x in M admits an open neighborhood U_x which is a quotient space of an open subset $\tilde{U}_x \subset C^n$ by a finite group G_x and G_x is trivial for generic x in M; (2) for any two points x, y in M with $U_x \bigcap U_y \neq \emptyset$, there is an equivariant biholomorphism $\phi_{x,y}$ from $\pi_x^{-1}(U_x \bigcap U_y)$ onto $\pi_y^{-1}(U_x \bigcap U_y)$ with respect to the groups G_x and G_y where π_x, π_y are natural projections from $\tilde{U}_x, \tilde{U}_y$ onto U_x, U_y; (3) this $\phi_{x,y}$ descends to the identity map on U_x i.e., the diagram

$$
\begin{array}{ccc}
\tilde{U}_x & \overset{\phi_{x,y}}{\longrightarrow} & \tilde{U}_y \\
\pi_x \downarrow & & \downarrow \pi_y \\
U_x & \overset{\text{id}}{\longrightarrow} & U_y
\end{array}
\tag{2.1}
$$

is commutative; (4) for x, y, z in M with $U_x \bigcap U_y \bigcap U_z \neq \emptyset$ we have $\phi_{y,z} \circ \phi_{x,y} = \phi_{x,z}$ on $\pi_x^{-1}(U_x \bigcap U_y \bigcap U_z)$. Sometimes, by abusing the notation, we write $\phi_{x,y}$ as $\pi_y^{-1} \circ \pi_x$. We define Reg(M) to the set of those points x with G_x being trivial and Sing$(M)= M\backslash$Reg(M). Then Reg(M) is a complex manifold and Sing(M) is a subvariety of M. We call the point in Sing(M) the singular point of M. In particular, if Sing$(M) = \emptyset$, then M is just a complex manifold. In general, the Hausdorff dimension of Sing(M) is not greater than $2n - 2$. In case the Hausdorff dimension of Sing(M) is less than or equal to $2n - 4$, the complex orbifold M is normal. This normality of M is characterized by Hartog's property: any holomorphic function on $U_x \bigcap$ Reg(M) can be lifted to a holomorphic function on $\tilde{U}_x$.

In this note, we always assume that the complex orbifolds considered are normal.

It is obvious that we can define differential operators $\partial, \overline{\partial}$ and hermitian metrics, etc. on a complex orbifold M as we did for a complex manifold. In particular, a Kähler metric g on M is a Kähler metric on $\text{Reg}(M)$ in the usual sense such that for each point x in M, the pull-back $\pi_x^* g$ on $\tilde{U}_x \bigcap \pi_x^{-1}(\text{Reg}(M))$ can be extended to be a Kähler metric on $\tilde{U}_x$.

Definition 2.1. A Kähler orbifold (M, g) is a complex orbifold M with a Kähler metric g.

From now on let (M, G) be a Kähler orbifold. Then there is an associated Kähler form ω_g on M, i.e., for any local uniformization $\tilde{U}_x$ and coordinates $(z_1, \cdots, z_n)$ in $\tilde{U}_x \subset C^n$, if $\pi_x^* g$ is represented by hermitian matrices $(g_{i\overline{j}})_{1 \le i, \overline{j} \le n}$ then in $\tilde{U}_x$,

$$\pi_x^* \omega_g = \frac{\sqrt{-1}}{2\pi} \sum_{i,j=1}^{n} g_{i\overline{j}} dz_i \bigwedge d\overline{z}_j \tag{2.2}$$

It is clear that ω_g defines a cohomology class in $H^2(M, R)$. As usual, we define the Ricci curvature form $\text{Ric}(g)$ as

$$\pi_x^*(\text{Ric}(g)) = -\frac{\sqrt{-1}}{2\pi} \partial\overline{\partial} \log \det(g_{i\overline{j}})_{1 \le i, j \le n} \quad \text{in} \quad \tilde{U}_x \tag{2.3}$$

Then this Ricci form $\text{Ric}(g)$ is also a cohomology class in $H^2(M, R) \bigcap H^{1,1}(M, C)$. In fact, this cohomology class is independent of the choice of Kähler metric g and is called the first cohomology class of M, denoted by $C_1(M)$. Now we say that M has positive first Chern class if there is a Kähler metric g such that the form $\text{Ric}(g)$ is pointwise positive definite. The algebraically geometric correspondence of this definition is that the anticanonical line bundle K_M^{-1} on M is ample, in other words, there is an integer $m > 0$ such that any basis of $H^0(M, K_M^{-1})$ embeds M into some projective space CP^{N_m} (cf. [Bai]), where $H^0(M, K_M^{-1})$ consists of all holomorphic global sections of K_M^{-m} on M.

The simplest Kähler orbifolds with positive first Chern class are those quotient spaces of Kähler manifolds with positive first Chern class by a finite automorphic group having no divisor as fixed points. Now let us describe a class of 2-dimensional Kähler orbifolds with positive first Chern class. Let $\tilde{M}$ be a complex surface obtained by blowing up CP^2 successively. Precisely, there is a sequence of complex surfaces

$$\tilde{M} = M_l \xrightarrow{\pi_l} \longrightarrow \cdots \xrightarrow{\pi_1} M_0 = CP^2 \qquad (2.4)$$

where each $M_j (1 \le j \le l)$ is obtained by blowing up M_{j-1} at one point and π_j is the natural projection. Then by blowing down some exceptional divisors in $\tilde{M}$, one can obtain a Kähler orbifold with positive first Chern class. In particular, if $l \le 8$ and there are l points $\{x_1, \cdots, x_l\}$ in CP^2 such that no three of $\{x_1, \cdots, x_l\}$ are colinear, no six of them are on a common quadratic curve and the restriction of π to $\tilde{M} \setminus \pi^{-1}(\{x_j\}_{1 \le j \le l})$ is a biholomorphism where $\pi = \pi_1 \circ \cdots \circ \pi_l$, then we have a complex surface with positive first Chern class. More generally, if neither four of $\{x_1, \cdots, x_l\}(l \le 8)$ are colinear nor seven of them are on a quadratic curve , we can have a complex surface $\tilde{M}$ with its first Chern class nonnegative in the sense that there is a nonnegative (1,1)-form representing $C_1(\tilde{M})$. Such a surface $\tilde{M}$ is usually called almost Del-Pezzo surface. By blowing down the (-2)-curves in an almost Del-Pezzo surface $\tilde{M}$, we obtain a Kähler orbifold with only rational double points as singular points. A rational double point is the singular point of the quotient space C^2/Γ with $\Gamma \subset SU(2)$ a finite group (cf. [BPV]).

Generalized Calabi's problem. Is there a Kähler-Einstein metric on a compact Kähler orbifold with positive first Chern class?

Here we call an orbifold metric Kähler-Einstein if its Ricci curvature form is the constant multiple of its Kähler form. The importance of Kähler-Einstein metrics lies in the compactification of the moduli space of Kähler-Einstein metrics on compact Kähler manifolds (cf. [An], [Na], [T2] in case of complex surface).

3   Some obstructions to the existence of Kähler-Einstein metrics

Let M be a normal compact Kähler orbifold with positive first Chern class $C_1(M)$ and g be a Kähler metric on M with its Kähler form representing $C_1(M)$ in the second cohomology class. Then $\mathrm{Sing}(M)$ is a subvariety of M with codimension ≥ 2. A holomorphic automorphism of M is that of $\mathrm{Reg}(M)$ which can be lifted to be a local automorphism in the local uniformization of each singular point in $\mathrm{Sing}(M)$. Let $\mathrm{Aut}(M)$ be the automorphism group of M consisting of all automorphisms on M, and $\eta(M)$ be its Lie algebra. Recall the definition of Futaki invariant $F : \eta(M) \longrightarrow C$,

$$F(X) = \int_M X(f)\omega_g^n, \quad x \in \eta(M) \tag{3.1}$$

where ω_g is the Kähler form of g, and f is given by the equation $\mathrm{Ric}(g) - \omega_g = \frac{\sqrt{-1}}{2\pi}\partial\bar{\partial}f$. Note that X is a holomorphic vector field on M.

Theorem 3.1. (Matsushiam[Ma], [Futaki] [Fu1]). Suppose that (M, g) admits a Kähler- Einstein metric. Then
(1) The Lie algebra $\eta(M)$ of all holomorphic vector fields is reductive.
(2) The Futaki invariant F is identically zero on $\eta(M)$.

Remark. The original version of this proposition in [Ma] and [Fu1] is stated on smooth Kähler-Einstein manifolds with positive first Chern class. But there is no additional difficult if the manifolds are replaced by orbifolds.

One can compute $F(X)$ by residue type arguments if X satisfying some nondegeneracy conditions. On smooth Kähler-Einstein manifolds, a formula for computing $F(X)$ was given by Futaki himself in [Fu2] in terms of X along its zero submanifolds and the first Chern forms of the normal bundles of these submanifolds. The proof resembles Chern's proof for generalized Gauss-Bonnet formula or that of Bott's residue formula. It is local in nature. Therefore this formula for computing $F(X)$ can be easily generalized to Kähler orbifolds. For simplicity, we here only generalize this to 2-dimensional Kähler Einstein orbifold (M, g). The readers may easily work out the formula by themselves in the general case. We will state the formula in 2-dimensional case more explicitly than that in [Fu2], so that it can be easily applied in our following examples.

Let X be a holomorphic vector field on M, i.e. $X \in \eta(M)$. We say that X is nondegenerate if the zero set Z of X satisfies the followings:

1. Z is a union of distinct smooth complex submanifolds in M, write $Z = \bigcup_{\lambda=1}^{\nu} Z_\lambda$ each Z_λ is an irreducible component, such that $\dim_C Z_\lambda = 0$ for $1 \leq \lambda \leq \nu_0$; $\dim_C Z_\lambda = 1$ for $\nu_0 \leq \lambda \leq \nu$.

2. For any λ between 1 and ν, any $z \in Z_\lambda$, there is either a local coordinate system (z_1, z_2) or a local uniformization $\pi_z : \tilde{U}_z \to U_z \subset M$ with coordinates (z_1, z_2) on $\tilde{U}_z$ according to whether z is a regular point of M or not, such that if $1 \leq \lambda \leq \nu_0$, then either X or $\pi_z^* X$ is of form $a_\lambda z_1 \frac{\partial}{\partial z_1} + b_\lambda z_2 \frac{\partial}{\partial z_2} +$(high order terms) near z or $\pi_z^{-1}(z)$ where $a_\lambda b_\lambda \neq 0$, if $\lambda > \nu_0$, then z_2 is tangent to Z_λ or $\pi_z^{-1}(Z_\lambda)$ and either X or $\pi_z^* X$ is of form $a_\lambda z_1 \frac{\partial}{\partial z_1} +$(high order terms) near z, where $a_\lambda \neq 0$. For such a nondegenerate holomorphic vector field X on M_∞, we define divergence $\mathrm{div}_{Z_\lambda}(X)$ of X along Z_λ to be $a_\lambda + b_\lambda$ if $\dim_C Z_\lambda = 0$, or a_λ if $\dim_C Z_\lambda = 1$. It is easy to see that for any λ, the divergence $\mathrm{div}_{Z_\lambda}(X)$ is a constant along Z_λ. In case $\lambda \leq \nu_0$, we also define $\det_{Z_\lambda}(X) = a_\lambda b_\lambda$.

Remark. In general, in the definition of nondegeneracy, we may allow Z_λ to be orbifolds, too.

Lemma 3.1. Let X be a nondegenerate holomorphic vector field defined as above on a 2-dimensional Kähler-Einstein orbifold (M, g). We further assume that each 1-dimensional component Z_λ does not pass any singular point of M. Then by adopting the notations in the above definition of nondegeneracy, we have

$$F(X) = \sum_{\lambda=1}^{\nu_0} \frac{1}{\deg(\pi_{Z_\lambda})} \cdot \frac{(\mathrm{div}_{Z_\lambda}(X))^3}{\det_{Z_\lambda}(X)}$$

$$+ \sum_{\lambda=\nu_0+1}^{\nu} (\mathrm{div}_{Z_\lambda}(X))(2\deg(Z_\lambda) + 2 - 2g(Z_\lambda))$$

$$= 0 \tag{3.2}$$

where π_{Z_λ} is the local uniformization of point Z_λ, the degree $\deg Z_\lambda$ of the curve Z_λ is taken with respect to ω_g and $g(Z_\lambda)$ is genus of Z_λ for each λ.

Proof. It follows from the generalized version of Theorem 5.2.8 in [Fu2] for 2 dimensional Kähler orbifolds with the help of the Gauss equations of curvature forms for subbundles (cf. [GH], p78). We remark that the proof in [Fu2] for Theorem 5.2.8 can be modified without additional difficulty to the case of orbifolds. We refer readers to [Fu2] for details.

Example 3.1. Let $\tilde{M}$ be the surface obtained by blowing up CP^2

at $[1,0,0]$, $[0,1,0]$, $[1,1,0]$, $[0,0,1]$. Then $\tilde{M}$ has a unique rational curve with self intersection -2, i.e., the quadratic transformation of the line $\{[x_0, x_1, x_2] \in CP^2 \,|\, x_2 = 0\}$ in $\tilde{M}$. By blowing down this (-2)-curve, we obtain a 2-dimensional Kähler orbifold M with a unique singular point p of type $C^2 \setminus \{\sigma_2\}$, where $\sigma_2 : (z_1, z_2) \to (-z_1, -z_2)$ in the coordinates (z_1, z_2) of C^2. Then one can easily check that the identity component $\mathrm{Aut}_0(M)$ in $\mathrm{Aut}(M)$ is just that in $\mathrm{Aut}(\tilde{M})$ and so is equal to $\{\tau \in \mathrm{Aut}(CP^2) \,|\, \tau$ fixes the line $\{x_2 = 0\}$ and $[0,0,1]\}$. Therefore, $\eta(M)$ is reductive. Let X be the holomorphic vector field induced by the one-parameter family of automorphisms $[x_0, x_1, x_2] \to [x_0, x_1, e^t x_2]$ on CP^2. Then X is nondegenerate in M and has the point p and an exceptional curve E over $[0,0,1]$ as singular points. An easy computation shows

$$F(X) = \frac{1}{\deg(\pi_p)} \frac{(\mathrm{div}(X))^3}{\det_p(X)} + 4\mathrm{div}_E(X) \neq 0 \qquad (3.3)$$

Therefore M does not admits any Käahler-Einstein orbifold metric.

Both obstructions of Matsushima and Futaki are obviously trivial in case $\mathrm{Aut}(M)$ is discrete. E.Calabi conjectured that M would admit a Kähler-Einstein metric if M is smooth and $\mathrm{Aut}(M)$ is discrete. While this conjecture for complex surfaces has been recently affirmed by the author [T2], the general case is wildly open. However, if we drop the smoothness of M, we can find some other obstructions even if $\mathrm{Aut}(M)$ is discrete. Let us show a simple one of them.

Theorem 3.2. Let (M, g) be a n-dimensional Kähler-Einstein orbifold with positive first Chern class. Then for any singular point x in M,

$$\#(G_x) \leq \frac{(2n-1)^n \mathrm{Vol}(S^{2n})}{C_1(M)^n} \qquad (3.4)$$

where G_x is the local uniformization group of x in M, $C_1(M)^n$ is the volume of M with respect to any Kähler metric on M with the associated Kähler form in $C_1(M)$, and $\mathrm{Vol}(S^{2n})$ is the volume of S^{2n} with respect to the standard metric.

Proof. First we remark that Bonnet-Myers theorem [CE] and Bishop's volume comparison theorem [Bi] still hold on compact Kähler-Einstein orbifolds. The proofs for them are just the modifications of those in case of smooth Riemmanian manifolds in [CE] and [Bi]. Since g is Kähler-Einstein with $\mathrm{Ric}(g)=\omega_g$, by Bonnet-Myers theorem, we have

$$\mathrm{diam}(M,g) \leq \sqrt{2n-1}\pi \tag{3.5}$$

where $n = \dim_C M$.

Then by Volume Comparison Theorem, we have

$$\frac{\mathrm{Vol}_g(B_r(x))}{\sqrt{2n-1}\mathrm{Vol}(B_{\frac{r}{\sqrt{2n-1}}}(o,S^{2n}))} > \frac{\mathrm{Vol}_g(M)}{\sqrt{2n-1}\mathrm{Vol}(S^{2n})} \quad \text{for any } r \text{ small} \tag{3.6}$$

where x is any point in M, $B_r(x)$ is the geodesic ball in M with center at x and radius r and $B_{\frac{r}{\sqrt{2n-1}}}(o,S^{2n})$ is the geodesic ball of S^{2n} with respect to the standard metric. Now take x to be a singular point, then the estimate (3.4) follows from (3.6) by letting r go to zero. The theorem is proved.

Corollary 3.1. Let (M,g) be a 2-dimensional Kähler-Einstein orbifold with $C_1(M) > 0$. Then for any singular point x at M, we have.

$$\#(G_x) < \frac{48}{C_1(M)^2} \tag{3.7}$$

The following example shows that Theorem 3.2 indeed gives a nontrivial obstruction to the existence of Kähler-Einstein metrics besides those in Theorem 3.1.

Example 3.2. We define a rational surface $\tilde{M}$ by successive blowing-ups as described in last section, i.e.,

$$\tilde{M} = M_4 \xrightarrow{\pi_4} M_3 \xrightarrow{\pi_3} M_2 \xrightarrow{\pi_2} M_1 \xrightarrow{\pi_1} M_0 = CP^2 \tag{3.8}$$

The blowing-ups $\pi_i(1 \leq i \leq 4)$ are defined as follows. Let $p_1 = [0,0,1], p_2 = [0,1,0], p_3 = [0,0,1], F_{0j}$ be the line in CP^2 connection p_j, p_{j+1} (say $p_4 = p_1$), where $j = 1,2,3$. Then M_1 is the surface obtained by blowing up CP^2 at p_1, p_2, p_3. Let E_{1j} be the exceptional curve over p_j and F_{1j} be the quadratic image of F_{0j} in $M_1(j = 1,2,3)$. In particular, F_{11} intersects E_{11} at a unique point q_{11}. The surface M_2 is the blowing up of M_1 at q_{11}. Let F_{21}, E_{21} be the quadratic images of F_{11}, E_{11} under the projection π_2 at q_{11} and E_{22} be the exceptional curve over q_{11}. The $F_{21}^2 = E_{21}^2 = -2$ and E_{22} intersects E_{21} at a unique point q_{21}. The surface M_3 is the blow-up of M_2 at q_{21}. Let F_{31}, E_{31}, E_{32} be the quadratic images of F_{21}, E_{21}, E_{22} respectively. Then $F_{31}^2 = -2, E_{31}^2 = -3, E_{32}^2 = -2, F_{31} \cdot E_{32} = 1$ and E_{31} does not intersect with F_{31}, E_{32}. Moreover, the new exceptional curve E_{33} of π_3 over q_{21} intersects E_{31}, E_{32} at one point, respectively. Now $\tilde{M} = M_4$ is the surface obtained by blowing up M_3 at a point in $F_{31} \setminus E_{32}$ and a point in $E_{33} \setminus E_{31} \bigcup E_{32}$. Let $F_{41}, E_{4j}(j = 1, 2, 3)$ be the quadratic images of F_{31}, E_{3j}, respectively. Then $D = F_{41} + \sum_{j=1}^{3} E_{4j}$ is a Hirzebruch-Jung string[BPV], precisely, $F_{41}^2 = -3 = E_{41}^2, E_{42}^2 = E_{43}^2 = -2, F_{41} \cdot E_{4j} = 1$ if $j = 2;$. Otherwise, $E_{41} \cdot E_{42} = E_{42} \cdot E_{43} = 1, E_{41} \cdot E_{43} = 0$.

Our 2-dimensional Kähler orbifold M is obtained by collapsing the Hirzebruch-jung string D to a point. One can show that M has positive first Chern class. It follows from the construction of $\tilde{M}$ that the automorphism group $\text{Aut}(\tilde{M})$ is finite. In fact, the identity component $\text{Aut}_0(M_3)$ of $\text{Aut}(M_3)$ is an one-parameter group, i.e., the subgroup of $\text{Aut}(CP^2)$ consisting of those elements fixing p_3 and the line connecting p_1, p_2. Now $\text{Aut}_0(\tilde{M})$ is the subgroup of $\text{Aut}(M_3)$ fixing the rational curve E_{33}. It implies that $\text{Aut}_0(\tilde{M})$ is trivial, i.e., $\text{Aut}(M)$ is finite.

We claim that M dose not admits any Kähler-Einstein metric. Let $\pi : \tilde{M} \to M$ be the natural projection, then $\pi^* C_1(M) = C_1(\tilde{M}) - \frac{1}{2}D$, therefore,

$$C_1(M)^2 = \pi^* C_1(M)^2 = C_1(\tilde{M})^2 - C_1(\tilde{M}) \cdot D + \frac{1}{4}D^2$$
$$= 1 - C_1(\tilde{M}) \cdot D = 3$$

Let x be the unique singular point of M and G_x be its local uniformization group. Then by Corollary 3.1, if M admits a Kähler-Einstein metric, we have

$$\#G_x < \frac{48}{C_1(M)^2} = 16 \tag{3.9}$$

On the other hand, the singular point x is of type C^2/σ, where $\sigma : (z_1, z_2) \to (e(\frac{1}{m})z, e_2(\frac{l}{m})z_2)$ for the euclidean coordinates z_1, z_2 on C^2, $e(\frac{l}{m}) = \exp(\frac{2\pi l\sqrt{-1}}{m})$ and m, l are determined by the self-intersection numbers of F_{41} and $E_{4j}(1 \leq j \leq 4)$ as follows (cf. [BPV])

$$\frac{m}{l} = 3 - \frac{1}{2 - \frac{1}{2 - \frac{1}{3}}} = \frac{16}{7}$$

Therefore, the order of G_x is at least 16. It contradicts to (3.9). So M does not admit any Kähler-Einstein metric.

Remark. The above example indicates that holomorphic vector fields may not be only difficulty in solving Calabi's conjecture. This may justify the belief through [T1],[T2] that the existence of Kähler-Einstein metrics with positive scalar curvature should be closely related to the geometry of pluri-anti-canonical divisors.

4 Some remarks on the existence of Kähler-Einstein metrics

Let (M, g) be a compact Kähler orbifold with a Kähler metric g satisfying : $\omega_g \in C_1(M) > 0$. In this section, we will discuss the existence of Kähler-Einstein metrics on M. By our assumption on g, there is a smooth function f on M such that

$$\mathrm{Ric}(g) - \omega_g = \frac{\sqrt{-1}}{2\pi}\partial\bar{\partial}f \text{ on } M \quad \text{and} \quad \int_M e^f \omega_g^n = \mathrm{Vol}_g(M) \quad (4.1)$$

Note that a function on M is smooth iff it is smooth in every local uniformization. Let $C^\infty(M, R^1)$ be the collection of all real valued C^∞ smooth functions on M. Then the existence of a Kähler-Einstein metric on M is equivalent to the solvability of the following complex Monge-Ampére equations

$$\begin{cases} (\omega_g + \dfrac{\sqrt{-1}}{2\pi}\partial\bar{\partial}\varphi)^n = e^{f-t\varphi}\omega_g^n \\[2mm] (\omega_g + \dfrac{\sqrt{-1}}{2\pi}\partial\bar{\partial}\varphi) > 0 \end{cases} \quad \text{on} \quad M \qquad (4.2)_t$$

where $\varphi \in C^\infty(M, R^1)$, $0 \leq t \leq 1$. As in the case of compact Kähler manifolds with positive first Chern class, the continuity method is applied to solve $(4.2)_t$ and the main difficulty is how to obtain the a priori C^0-estimate for the solution of $(4.2)_t$. Therefore, we will concentrate on C^0-estimate for $(4.2)_t$ in the following discussions. Let G be a maximal compact subgroup in $\mathrm{Aut}(M)$. Without losing generality, we may assume that the Kähler metric g and the function φ in $(4.2)_t$ are G-invariant. Define

$$P_G(M, g) = \{\varphi \in C^\infty(M, g) \mid \varphi \text{ is } G - \text{invariant},$$

$$\omega_g + \frac{\sqrt{-1}}{2\pi}\partial\overline{\partial}\varphi \geq 0 \text{ on } M, \ \sup_M\varphi = 0\} \quad (4.3)$$

Then as in [T1], we can introduce a holomorphic invariant

$$\alpha(M) = \sup\{\alpha \,|\, \exists C > 0, \text{ s.t. } \int_M e^{-\alpha\varphi}dV_g \leq C \text{ for}$$
$$\text{all } \varphi \text{ in } P_G(M,g)\} \quad (4.4)$$

One can easily see that $\alpha(M)$ is independent of the choices of g and G. By the identical arguments as those in [T1], one can prove that

Theorem 4.1. Let (M,g) be a n-dimensional compact Kähler orbifold with positive Chern class. Then M admits a Kähler-Einstein metric whenever $\alpha(M) > \frac{n}{n+1}$.

This theorem can be used to produce Kähler-Einstein manifold with positive first Chern class (cf. [T1], [T2], [TY]). But one can easily find a Kähler orbifold with positive first Chern class and $\alpha(M) < \frac{n}{n+1}$.

Example 4.1. Let $\tilde{M}$ be the surface obtained by blowing up CP^2 at six points $[1,0,0]$, $[0,1,0,]$, $[0,0,1]$, $[1,1,0]$, $[1,0,1]$, and another generic point. Then $\tilde{M}$ has exactly two rational curves with self-intersection -2. Our orbifold M is obtained by blowing down those two (-2)-curves in $\tilde{M}$. Then M has two singular points of type C^2/Z_2 and positive first Chern class. The automorphism group $G = \text{Aut}(M)$ of M is trivial. Let S be the anticanonical section on M induced by the union of three lines intersecting at $[1,0,0]$ and passing through all other blowing-up points once. Then we have smooth functions $\varphi_\varepsilon = \log(\|\,S\,\|_g^2 + \varepsilon)$ in $P_G(M,g)$ for $\varepsilon > 0$. One can easily compute

$$\lim_{\varepsilon \to 0} \int_M e^{-\alpha(\varphi_\varepsilon - \sup_M\varphi_\varepsilon)}dV_g = +\infty \quad \text{for any } \alpha \geq \frac{1}{2} \quad (4.5)$$

It implies that $\alpha(M) \leq \frac{1}{2}$. However, we do expect the existence of Kähler-Einstein metric on M(cf. Example 4.2 below).

Therefore, Theorem 4.1 is not the sharpest result. We need to improve it. In [T2], in order to solve Calabi's conjecture for complex surfaces, the author proposed and proved the partial C^0-estimate for the solution of $(4.2)_1$. Then the partial C^0-estimate for $(4.2)_1$ is applied to improve Theorem 4.1 in case of complex surfaces (cf. Theorem 2.1 in

[T2]). Here we will outline a refinement of Theorem 4.1 by using the idea in [T2]. For simplicity, we will only consider 2-dimensional Kähler orbifolds or complex surfaces. The general case will be discussed in the forthcoming paper.

The proposed partial C^0-estimate is as follows. There are $m > 0$ and $C > 0$, depending only on the Chern characteristics of M, such that for solution φ of $(4.2)_t$, there are an orthonormal $\{S_\nu^m\}_{0 \le \nu N_m}$ basis of $H^0(M, K_M^{-m})$ and positive numbers $\lambda_0 \le \lambda_1 \le \cdots \le \lambda_{N_m} = 1, N_m + 1 = \dim_C H^0(M, K_M^{-m})$ satisfying

$$\left| \varphi - \sup_M \varphi - \frac{1}{m} \log \left(\sum_{\nu=0}^{N_m} \lambda_\nu^2 \parallel S_\nu^m \parallel_g^2 \right) \right| \le C \qquad (4.6)$$

where $\parallel \cdot \parallel_g$ is the induced hermitian metric on K_M^{-m} by the metric g.

In [T2], the author proved (4.6) for the solutions of $(4.2)_1$ on complex surfaces with positive first Chern class. Furthermore, he can choose m as small as six. The methods in [T2] can be easily generalized to 2-dimensional Kähler orbifolds to prove the partial C^0-estimate (4.6) for $(4.2)_1$ We omit the details. Now we come to see how to use (4.6) to improve Theorem 4.1. As we remarked before, it suffices to C^0-estimate for the solutions of $(4.2)_t$. Take a sequence of solutions $\{\varphi\}_{i>1}$ of $(4.2)_{t_i}$ with $\lim_{i\to\infty} t_i = \tilde{t} < 1$. We assume that the partial C^0-estimate holds for $(4.2)_t$ for $0 \le t \le 1$, then we have orthonormal bases $\{S_\nu^m(i)\}_{0 \le \nu \le N_m(i \ge 1)}$ of $H^0(M, K_M^{-m})$ with respect to g such that

$$\sup_M \left| \varphi_i - \sup_M \varphi_i - \frac{1}{m} \log \left(\sum_{\nu=0}^{N_m} \lambda_\nu^2(i) \parallel S_\nu^m(i) \parallel_g^2 \right) \right| \le C \qquad (4.7)_i$$

where $0 < \lambda_0(i) \le \cdots \le \lambda_{N_m}(i) = 1$. By taking the subsequence, we may assume that $\lim_{i\to\infty} \lambda_\nu(i) = \lambda_\nu(\infty) \ge 0$, $\lim_{i\to\infty} S_\nu^m(i) = S_\nu^m(\infty)$ and $\{S_\nu^m(\infty)\}$ is also an orthonormal basis of $H^0(M, K_M^{-m})$ with respect to g. Furthermore, we may assume that $\lim_{i\to\infty} \lambda_\nu(i)/\lambda_{\nu-1}(i) = +\infty$ or $c_\nu > 0$.

Theorem 4.2. Let (M, g) be a 2-deminsional Kähler orbifold given as above. Suppose

1. Define

$$\alpha_l = \sup \left\{ \alpha \,\Big|\, \exists C > 0, \text{s.t.} \int_M \frac{dV_g}{\left(\displaystyle\sum_{\nu=\nu_l}^{N_m} \| S_\nu^m(i) \|_g^2 \right)^{\frac{2\alpha}{m}}} \leq C \text{ for all } i \right\} \quad (4.8)$$

where $l = 1,2$, $\nu_1 = N_m - 1$, $\nu_2 = N_m$. Then $\alpha_1 > \frac{2}{3}$.

2. Let D be the 1-dimensional component in the base locus of $S_{N_m}^m(\infty)$. We denote by d the degree of D with respect to the anticanonical class ω_g. Define $k = \max\{2 - md\frac{3\alpha_1 - 2}{\alpha_1}, 0\}$. Then either $\lim_{i\to\infty} \lambda_{N_m-1}(i)/\lambda_{N_m}(i) > 0$ or $\alpha_2 > \frac{k}{k+1}$.

Then the solutions φ_i are in fact uniformly bounded.

Proof. For simplicity, put $N = N_m$.

Claim 1. For any $\varepsilon > 0$, there is a constant C_ε independent of i such that

$$\sup_{M_i} \varphi_i \leq -\left(\frac{\alpha_1}{m(3\alpha_1 - 2)} - \varepsilon\right)\log(\lambda_{N-1}-(i))^2 + C_\varepsilon \quad \text{for all } i \quad (4.9)$$

Let ε' be a small positive number, determined later. Then by our assumption and $(4.7)_i$, there is a constant $\tilde{C}_{\varepsilon'}$ such that

$$\int_M e^{-(\alpha_1-\varepsilon')(\varphi_i - \sup_M \varphi_i)} dV_g$$

$$\leq \lambda_{N-1}(i)^{-\frac{2(\alpha_1-\varepsilon')}{m}} \int_M \frac{dV_g}{(\|S_{N-1}^m(i)\|^2 + \|S_N^m(i)\|^2)^{\frac{2(\alpha_1-\varepsilon')}{m}}}$$

$$\leq \tilde{C}_{\varepsilon'} \cdot (\lambda_{N-1}(i))^{-\frac{2(\alpha_1-\varepsilon')}{m}}$$

By the concavity of logarithmic functions, it follows

$$(\alpha_1 - \varepsilon')\sup_M \varphi_i + (t_i - \alpha_1 + \varepsilon')\int_M \varphi_i \left(\omega_g + \frac{\sqrt{-1}}{2\pi}\partial\bar{\partial}\varphi_i\right)^2$$

$$\leq \log\left(\int_M e^{(\alpha_1-\varepsilon')\sup_M \varphi_i + (t_i-\alpha_1+\varepsilon')\varphi_i - f} e^{f - t_i\varphi_i} dV_g\right)$$

$$\leq \sup_M f + \log\tilde{C}_{\varepsilon'} - \frac{(\alpha_1 - \varepsilon')}{m}\log(\lambda_{N-1}(i))^2 \quad (4.10)$$

On the other hand, by a result in [T1], we have

$$-\mathrm{inf}_M\varphi_i \leq 2\mathrm{sup}_M\varphi_i + C' \qquad (4.11)$$

Then (4.9) follows from (4.10) and (4.11) by taking ε' sufficiently small. We introduce two functionals first considered by Aubin in [Au] (see also [BM],[T1],

$$I(u) \;=\; \int_M u\left(\omega_g^2 - \left(\omega_g + \frac{\sqrt{-1}}{2\pi}\partial\bar\partial u\right)^2\right)$$

$$J(u) \;=\; \int_0^1 \frac{I(su)}{s}\,ds$$

Then it is proved in [T1] that

$$-\mathrm{inf}_M\varphi_i - C'' \leq \int_M (-\varphi_i)\left(\omega_g + \frac{\sqrt{-1}}{2\pi}\partial\bar\partial\varphi_i\right)^2 \leq I(\varphi_i) - J(\varphi_i)$$

$$= -\frac{2}{3}\int_M \varphi_i\left(\omega_i + \frac{\sqrt{-1}}{2\pi}\partial\bar\partial\varphi_i\right)^2 + \frac{2}{3}\int_M \varphi_i\omega_g^2$$

$$-\frac{1}{3}\int_M \frac{\sqrt{-1}}{2\pi}\partial\varphi_i\wedge\bar\partial\varphi_i\wedge\omega_g \qquad (4.12)$$

where C'' is a uniform constant. Therefore,

$$-\mathrm{inf}_M\varphi_i \leq 2\mathrm{sup}_M\varphi_i - \frac{\sqrt{-1}}{2\pi}\int_M \partial\varphi_i\wedge\bar\partial\varphi_i\wedge\omega_g \qquad (4.13)$$

Claim 2. For any $\varepsilon > 0$, there is a constant C'_ε independent of i, such that

$$\frac{\sqrt{-1}}{2\pi}\int_M \partial\varphi_i\wedge\bar\partial\varphi_i\wedge\omega_g \geq -(d_2 - \varepsilon)\log(\lambda_{N-1}(i))^2 - C'_\varepsilon \qquad (4.14)$$

Proof. Write $D_2 = \sum_{j=1}^J \mu E_j$, where each E_j is irreducible and μ is the multiplicity of E_j in D_2. Let η be small. Choose a finite covering $\{U_\beta\}$ of $\mathrm{D}_{\mathrm{red}} \setminus \mathrm{Sing}(\mathrm{D}_{\mathrm{red}})$, where $\mathrm{D}_{\mathrm{red}} = \sum_{j=1}^J E_j$, such that (1) $U_\beta \cap U_{\beta'} = \emptyset$ for $\beta \neq \beta'$; (2) for each β, there is a coordinate system (z_β, ω_β) on an open neighborhood of $\overline{U}_\beta$ in M such that $U_\beta = \{|z_\beta| < \eta, |\omega_\beta| < \eta\}$, $\mathrm{D}_{\mathrm{red}} \cap U_\beta = \{\omega_\beta = 0\}$ and z_β is tangent to $\mathrm{D}_{\mathrm{red}} \cap U_\beta$; (3) for a sufficiently small $\delta = \delta(\varepsilon)$, we have

$$\sum_{j=1}^J\sum_\beta \mu_j \int_{E_j\cap U_\beta} \omega_g \geq (d_2 - \delta) \qquad (4.15)$$

In the following, we always denote by $O(1)$ a bounded quantity independent of i. Let f_{β_i} be local holomorphic defining function of $S_N^m(i)$ in U_β with leading coefficient 1. Then for i sufficiently large, by partial C^0-estimate $(4.7)_i$ one can show

$$\frac{\sqrt{-1}}{2\pi}\int_M \partial\varphi_i \wedge \bar\partial\varphi_i \wedge \omega_g$$

$$= O(1) + \frac{\sqrt{-1}}{2\pi}\int_M \partial\log\left(\sum_{\nu=0}^N \|\lambda_\nu(i)S_\nu^m(i)\|_g^2\right)$$

$$\wedge\bar\partial\log\left(\sum_{\nu=0}^N \|\lambda_\nu(i)S_\nu^m(i)\|_g^2\right)\wedge\omega_g$$

$$\geq O(1) + (1-\delta)\frac{\sqrt{-1}}{2\pi}\sum_\beta\int_{\{(\lambda_{N-1}(i))^{1-\delta}\leq|f_{\beta_i}|\leq\eta^2\}\bigcap U_\beta}\frac{\partial f_{\beta_i}\wedge\bar\partial f_{\beta_i}\wedge\omega_g}{|f_{\beta_i}|^2}$$

$$\geq O(1) + (1-\delta)^2(-\log(\lambda_{N-1}(i)))^2 - O(1))\cdot\sum_\beta\int_{\{S_N^m(i)=0\}\bigcap U_\beta}\omega_g$$

$$\tag{4.16}$$

By (4.15) and the fact that $\lim_{i\to\infty} S_N^m(i) = S_N^m(\infty)$, for i sufficiently large, we have

$$\sum_\beta\int_{\{S_N^m(i)=0\}\bigcap U_\beta}\omega_g \geq (d_2 - 2\delta) \tag{4.17}$$

It follows from (4.17) and (4.16),

$$\frac{\sqrt{-1}}{2\pi}\int_M \partial\varphi_i\wedge\bar\partial\varphi_i\wedge\omega_g \geq (1-4\delta)^2(d_2-\delta)(-\log(\lambda_{N-1}(i))^2) + O(1)$$

Suppose that δ is so small such that $(1-4\delta)^2(d_2-\delta) \geq d_2 - \varepsilon$, then our claim is proved. Now by (4.13) and Claim 2, the assumption (4) and taking ε sufficiently small, we have

$$-\inf_M\varphi_i \leq 2\sup_M\varphi_i + (d_2-\varepsilon)\log(\lambda_{N-1}(i))^2 + O(1)$$

$$\leq \left(2 - (d_2-\varepsilon)\left(\frac{\alpha_1}{m(3\alpha_1-2)} - \varepsilon\right)^{-1}\right)\sup_M\varphi_i + O(1)$$

$$< \left(\frac{\min\{1,\alpha_2\}}{1-\min\{1,\alpha_2\}} - \varepsilon\right)\sup_M\varphi_i + O(1) \tag{4.18}$$

On the other hand, by (4.8),

$$\int_M e^{-\alpha(\varphi_i-\sup_M\varphi_i)}dV_g \leq C''\int_M\frac{dV_g}{\|S_N^m(i)\|^{\frac{2\alpha}{m}}} = O(1),$$

$$\text{for any } \alpha < \alpha_2 \tag{4.19}$$

It follows from (4.19) and the concavity of logarithmic functions that

$$\sup_M \varphi_i \leq \frac{1-\alpha}{\alpha}(-\inf \varphi_i) + O(1) \qquad (4.20)$$

Choose $\alpha < \alpha_2$ such that $\frac{\alpha}{1-\alpha} > \frac{\min\{1,\alpha_2\}}{1-\min\{1,\alpha_2\}} - \varepsilon$, then (4.18) and (4.20) implies that the solutions φ_i are uniformly bounded. The theorem is proved.

Example 4.2. Let (M,g) be the 2-dimensional Kähler orbifold defined in Example 4.1. We have already known that $\alpha(M) \leq \frac{1}{2}$. In fact, one can prove that $\alpha(M) = \frac{1}{2}$ Furthermore, one can compute that for any section S in $H^0(M, K_M^{-1})$, either

$$\int_M \frac{dV_g}{\| S \|^{2\alpha}} < +\infty, \quad \text{for some } \alpha > \frac{2}{3} \qquad (4.21)$$

or for any other section S' in $H^0(M, K_M^{-1})$ linearly independent of S, we have

$$\int_M \frac{dV_g}{(\| S \|^2 + \| S' \|^2)^\alpha} < +\infty, \quad \text{for any } \alpha < \frac{5}{6} \qquad (4.22)$$

Suppose that we have the partial C^0-estimate $(4.7)_i$ for a sequence of solutions $\{\varphi_i\}$ of $(4.2)_{t_i}$ with $\lim_{i\to\infty} t_i = \tilde{t} \leq 1$. For simplicity, we further suppose that $m = 1$ in $(4.7)_i$. Then by adopting the notations in Theorem 3.2, it follows from (4.21) and (4.22) that either $L = 1$, $\alpha_1 > \frac{2}{3}$ or $L = 2$, $\alpha_1 \geq \frac{5}{6}$, $\alpha_2 = \frac{1}{2}$. On the other hand, in case $L = 2$, $\alpha_2 = \frac{1}{2}$, $d_2 = 3$ so $2 - 3\frac{5/6}{3(5/6)-2} < 0$. Then Theorem 3.2 implies that φ_i are uniformly bounded. Therefore, the existence of Kähler-Einstein metric on M will follows from the partial C^0-estimate $(4.7)_i$ for $m = 1$. Note that the restriction on m is not crucial.

References

[An]   M. Anderson, Ricci curvature bounds and Einstein metrics on compact manifolds, preprint.

[Au]   T.Aubin, Réduction du cas positif de l'equation de Monge-Ampére sur les varietés Kählerinnes compactes á la démonstration d'un intégalité, *J.Funct. Anal.* 57(1984), 143–153.

[Bai] W.Baily, On the embedding of V-manifolds in projective space, *Amer.J.Math.*,79 (1957), 403–430.

[Bi] R.L.Bishop and R.J.Crittended, Geometry of Manifolds, *Pure and Applied Math.*, Vol. XV, Academic Press, New York-London, 1964.

[BM] S.Bando and T.Mabuchi, Uniqueness of Einstein Kähler metrics modulo connected group actions, *Tohoku Math.J.*, (1987).

[BPV] W.Barth, C.Peters,A.Van de Ven, Compact complex surfaces, Spinger-Verlag, Berlin- Heideberg 1984.

[CE] J. Cheeger and D.G.Ebin, Comparison theorems in Riemannian geometry, North- Holland Mathematical Library, 9(1975).

[Fu1] S.Futaki, An obstruction to the existence of Einstein-Kähler metrics,*Inv. Math.*, 73 (1983), 437–443.

[Fu2] S.Futaki, Lecture Notes in Math. Vol.1304(1988), Springer.

[GH] P.Griffith and j. Harris, Principles of Algebraic Geometry, Wiley, New York, 1978.

[Ma] Y. Matsushima, Sur la structure du group d'homeomorphismes analytiques d'une. certaine varietie Kaehlerinne, *Nagoya Math. J.*,11(1957), 145–150.

[Na] H.Nakajima, Hausdorff convergence of Einstein 4-manifolds, *J. Fac. Sci. Univ. Tokyo*, 35(1988), 411–424.

[T1] G. Tian, On Kähler-Einstein metrics on certain Kähler manifolds with $C_1(M) > 0$, *Inv. Math.*, 89(1987),225–246.

[T2] G.Tian, On Calabi's Conjecture for Complex Surfaces with Positive First Chern Class, To appear.

[TY] G.Tian and S.T.Yau, Kähler-Einstein metrics on complex surfaces with $C_1(M)$ positive, *Comm. Math. Phy.*, 112(1987).

[Y] S.T.Yau, On the Ricci curvature of a compact Kähler manifold and the complex Monge-Ampére equation, I^*, *Comm. Pure Appl. Math.*, 31(1978), 339–411.

A Brief Introduction to Kolmogorov Complexity and Its Applications *

Ming Li
Institute of Computing Technology,
Academy of Sciences,Beijing China
and
Department of Computer Science,
York University,Ontario,Canada

Paul M.B. Vitányi
Centrum voor Wiskunde en Informatica
and
Faculteit Wiskunde en Informatica,
Universiteit van Amsterdam
Amsterdam, The Netherlands

Abstract

Kolmogorov Complexity is a new rising branch of applied mathematics and computer science. Since its introduction in the sixties, Kolmogorov complexity has gained a rich mathematical theory and abundant applications in many fields of science.

This exposition, based on our forthcoming book [41], the first textbook on Kolmogorov complexity, gives a brief introduction to the main ideas of this field and demonstrates, through examples, how these ideas can actually be applied in mathematics and computer science.

---

*The work of the first author was partially performed at Aiken Computation Lab, Harvard University and supported by NSF Grant DCR-8606366, Office of Naval Rearch Grand N00014-85-k-0045, Army Research Office Grand DAAL03-86-K-0171, and NSERC Operating Grand OGP0036747. Current Address: Department of Computer Science, York University, Ontario M3J 1P3, Canada. Address for the second author: CWI, Kruislaan 413,1098 SJ Amsterdam, The Netherlands.

1   Introduction

Intuitively, the amount of information in a finite string is the size (i.e., number of binary digits or *bits*) of the smallest program that, without additional data, computes the string and then terminates. A similar definition can be given for infinite strings, but in this case the program produces element after element forever. Thus, 1^n (a string of n ones) contains little information because a program of size about $\log n$ outputs it. Likewise, the transcendental number $\pi = 3.1415\cdots$ an infinite sequence of seemingly "random" decimal digits, contains $O(1)$ information. (There is a short program that produces the consecutive digits of π forever.) Such a definition would appear to make the amount of information in a string depend on the particular programming language used. Fortunately, it can be shown that all choices of programming languages(that make sense) lead to quantification of the amount of information that is invariant up to an additive constant.

This elegant notion has turned out to be applicable to many different areas. It often allows us to give rigor to our straightforward intuition why a certain result should hold, where previously our myopic point of view in terms of mathematically justified techniques necessitated a tortuous meandering proof, seemingly unrelated to the original intuition. In other cases we are able to obtain completely new results which we were unable to obtain by other methods. Finally, of course, when such a powerful new tool becomes available to use in mathematical and scientific analysis we also find new problems and insights which could not be formulated before. It turns out that many different aspects of this versatile notion can be used in applications. For instance, using the fact that some strings are extremely compressible; using the compressibility of strings as a selection criterion; using the fact that many strings are not compressible at all; and using the fact some strings may be compressed, but that it takes a lot of effort do so.

The theory dealing with the quantity of information in individual objects goes by names such as "Algorithmic Information theory", "Kolmogorov complexity", "Kcomplexity", "Kolmogorov-Chaitin randomness", "Algorithmic complexity theory", "Stochastic complexity", "Minimum description length principle", "Descriptive complexity", "Program-Size complexity", and others. Each such name may represent a variation of the basic underlying idea or a different point of departure. The mathematical formulation in each case tends to reflect the particular traditions of the field from that which gave birth to it, be it probability theory, information theory, statistics or artificial intelligence.

There is a central theme that is perhaps basic to even the pursuit of science in the larger sense, and that goes back to the principle known as 'Occam's razor': if presented with a choice between indifferent alternatives one ought to select the simplest one. Unconsciously or explicitly, informal applications of this principle in science and mathematics abound. The unique feature of the current developments is that we can formalize and quantify this principle in an objective form using invariant notions of computability theory. Thus, we seem to be in the extremely unusual situation that the choice of mathematical axioms of an applicable theory is not more or less arbitrary but is of an unavoidable nature. In our particular case they are justified by what is known in computability theory as 'Church's Thesis'.

The conglomerate of different research threads drawing on this approach seems to us to be part of a single emerging discipline, that will become a major applied science like information theory or probability theory for a long time to come. We shall try to present in a unified treatment an introduction to the central ideas and its applications.

This raises the question about the proper name for the area. Although there is a good case to be made for each of the alternatives listed above, and a name like "Solomonoff-Kolmogorov-Chaitin complexity" would give proper credit to the spiritual founders, R. J. Solomonoff[57], A. N. Kolmogorov[30] and G.J. Chaitin [5] (in chronological order), we regard "Kolmogorov complexity" as well entrenched and commonly understood,and use it hereafter.

The mathematical theory of Kolmogorov complexity contains deep and sophisticated mathematics. Yet the amount of this mathematics one needs to know to apply the notions fruitfully in widely divergent areas, from recursive function theory to chip technology, is very little. However, formal knowledge does not necessarily imply the wherewithal to apply it, perhaps especially so in the case of Kolmogorov complexity. It is our purpose to develop the minimum amount of theory needed, and outline a scale of illustrative applications. In fact, while the pure theory of the subject will have its appeal to the selected few, the surprisingly large field of its applications will, we hope, delight the multitude.

Intuitively, the amount of information in a finite string is the size (literally, number of bits) of the shortest program that, starting with a blank input, computes the string and then terminates. A similar definition can be given for infinite strings, but in this case the program produces element after element forever. Thus, 1^n(a string of n ones) contains little information because the following program of size about $O(\log n)$ outputs it:

$$for\ i := 1\ to\ n\ do\ print(\text{'}1\text{'}).$$

Likewise, the transcendental number $\pi = 3.1415\cdots$, an infinite sequence of seemingly "random" decimal digits, contains $O(1)$ information. (There is a short program that produces the consecutive digits of π forever.) Such a definition would appear to make the amount of information in a string depending on the particular programming language used. Fortunately, it can be shown that all choices of programming languages(that make sense) lead to quantification of the amount of information that is invariant up to an additive constant.

In the past two decades, Kolmogorov complexity has been successfully applied to many research areas. Although Kolmogorov complexity contains rich and deep mathematics, the amount of this mathematics one needs to know to apply the notion fruitfully is little. However, formal knowledge does not necessarily imply the wherewithal to apply it, perhaps especially so in the case of Kolmogorov complexity. It is the purpose of this exposition to develop the minimum amount of theory needed, and give several illustrative applications in computer science.

2   A Simplified Mathematical Theory

In order to give a simplest but sufficient treatment of Kolmogorov complexity for the applications, we will deal with only finite strings. We will also stick with the most basic type of Kolmogorov complexity and discuss no other variants. Although the mathematical theory of Kolmogorov complexity extends far beyond our treatment given here, we will treat only the essential ideas and useful facts that are used in the applications to come. we refer those readers who are interested in a more complete treatment of Kolmogorov complexity or a detailed survey of applications of Kolmogorov complexity to [35] or [37], and to our forthcoming book[41] where one can also find all the proofs that are missing here. The classical early survey in the area is Levin and Zvonkin's treatment[67].

We are interested in defining the complexity of a concrete individual finite string of zeros and ones. Unless otherwise specified, all strings will be binary and of finite length. All logarithms in this paper are based 2, unless it is explicitly noted they are not. If x is a string, then $|x|$ denotes the *length* (number of zeros and ones) of x. We identify throughout the xth finite binary string with the natural number x, according to the correspondence:

$$(\varepsilon, 0), (0, 1), (1, 2), (00, 3), (01, 4), (10, 5), (11, 6), (000, 7), \cdots.$$

If A is a set, then $d(A)$ is the *cardinality* (the number of elements) of A. Intuitively, we want to call a string simple if it can be described in a few words, like "the string of a million ones"; A string is considered complex if it cannot be so easily described, like a "random" string which does not follow any rule and hence we do not know how to describe apart from giving it literally. A description of a string may depend on two things, the decoding method (the machine which interprets the description) and outside information available (input to the machine). We are interested in descriptions which are effective, and restrict the decoders to Turing machines. Without loss of generality, our Turing machines use binary input strings which we call *programs*. More formally, fixing a Turing machine T, we would like to say that p is a description of x if, on input p, T outputs x. It is also convenient to allow T to have some extra information y to help to generate x. We write $T(p, y) = x$ to mean that Turing machine T with input p and y terminates with output x.

Definition 1.   The descriptional complexity K_T of x, relative to Turing machine T and binary string y, is defined by

$$K_T(x \mid y) = \min\{|p| : p \in \{0, 1\}^* \& T(p, y) = x\}.$$

The complexity measure defined above is useful and makes sense only if the complexity of a string does not depend on the choice of T. Therefore the following simple theorem is vital.

Invariance Theorem (Solomonoff [57], Kolmogorov [30], Chaitin [5]). *There exists a universal Turing machine U, Such that, for any other Turing machine T, there is a constant c_T such that for all strings x, y, $K_U(x \mid y) \leq K_T(x \mid y) + c_T$.*

proof. Fix some standard enumeration of Turing machines $T_1, T_2, \cdots$. Let U be the Universal Turing machine such that when starting on input $0^n 1p, p \in \{0, 1\}^*$, U simulates the nth Turing machine T_n on input p. For convenience in the proof, we choose U such that if T_n halts, then U first erases everything apart from the halting contents of T_n's tape, and also halts. By construction, for each $p \in \{0, 1\}^*$, T_n started on p eventually halts iff U started on $0^n 1p$ eventually halts. Choosing $c_T = n + 1$ for T_n finishes the proof.   □

Clearly, the Universal Turing machine U that satisfies the Invariance Theorem is *optimal* in the sense that K_U minorizes each K_T up to a fixed additive constant (depending on U and T). Moreover, for each pair of Universal Turing machines U and U', satisfying the Invariance Theorem, the complexities coincide up to an additive constant (depending only on U and U'), for all strings x, y:

$$|K_U(x \mid y) - K_{U'}(x \mid y)| \leq c_{U,U'}.$$

Therefore, we set the canonical conditional Kolmogorov complexity $K(x \mid y)$ of x under condition of y equal to $K_U(x \mid y)$, for some fixed optimal U. We call U the *reference* Turing machine. Hence the Kolmogorov complexity of a string does not depend on the choice of encoding method and is well-defined. Define the *unconditional Kolmogorov* complexity of x as $K(x) = K(x \mid \varepsilon)$, where ε denotes the empty string ($|\varepsilon| = 0$).

Definition 2. A Turing machine such that the set of programs for which it halts is a prefix-code is called a *prefix machine*. (Recall that if p and q are two code words of a *prefix-code*, then p is not a proper prefix of q.) We can give an effective enumeration of all prefix machines. Then the *self-delimiting* descriptional complexity of $x \in \{0, 1\}^*$, with respect to prefix machine M, and binary string y, is similarly defined as

$$K_M(x \mid y) = \min\{|p| : p \in \{0, 1\}^* \& M(p, y) = x\}.$$

One can prove an Invariance Theorem for self-delimiting complexity, and define the conditional and unconditional *self-delimiting Kolmogorov complexity*, by fixing some reference optimal prefix machine.

Remark. The self-delimiting Kolmogorov complexity of string x, is the length of the shortest self-delimiting program that outputs x. Mathematically, self-delimiting Kolmogorov complexity has nicer properties. In this exposition, we will use $K(x)$ to denote both regular and self-delimiting Kolmogorov complexity of x. They differ by an $O(\log|x|)$ additive factor. In some applications this does not make any difference. But in some other applications, for example inductive inference, this is vital. In the latter case, we will state clearly that we are using the self-delimiting version.

Definition 3. A binary string x is *incompressible* if $K(x) \geq |x|$.

Remark. Since Martin-Lof has shown that incompressible strings pass all effective statistical tests for randomness, we will also call incompressible strings *random* strings.

We now start to prove several useful and simple properties. We name those properties that will be needed for later applications as Facts, and some other properties, that are just "nice" to know, are listed as Examples.

Example 1. For each finite binary string x we have $K(xx) \leq K(x) + O(1)$. Namely, let T compute x from program p. Now fix a universal machine V which, on input $0^{n(T)}1p$, simulates T just like the reference machine U in the proof of the Invariance Theorem, but additionally V doubles Ts output before halting. Now V started on $0^{n(T)}1p$, computes xx, and therefore U started on $0^{n(V)}10^{n(T)}1p$, computes xx. Hence, for all $x, K(xx) \leq K(x) + n(V) + 1$.

Example 2. Let us define $K(x, y) = K(\langle x, y \rangle)$ with $\langle \cdot, \cdot \rangle$ is a standard one-one mapping of pairs of natural numbers to natural numbers. I.e., $K(x, y)$ is the length of a shortest program that outputs x and y and a way to tell them apart. It is seductive to conjecture $K(x, y) \leq K(x) + K(y) + O(1)$, the obvious (but false) argument running as follows. Suppose we have a shortest program p to produce x, and a shortest program q to produce y. Then with $O(1)$ extra bits to account for some Turing machine T that schedules the two programs, we have a program to produce x followed by y. However, any such T will have to know where to divide its input to identify p and

q. We can separate p and q by prefixing pq by a clearly distinguishable encoding r of the length $|p|$ in $O(\log|p|)$ bits (see next Section on self-delimiting strings). Consequently, we have at best established $K(x,y) \leq K(x) + K(y) + O(\log(\min(K(x), K(y))))$. In general this cannot be improved.

Fact 1. For each n, there exists a random string of length n.

Proof. Since there are 2^n binary strings of length n, but only $2^n - 1$ possible shorter descriptions d, it follows that, for all n, there is a binary string x of length n such that $K(x) \geq n$. $\square$

Fact 2. If $x = uvw$, then $K(v) \geq K(x) - |uw| - O(\log|x|)$. Hence if x is random, i.e. $K(x) \geq |x|$, then $K(v) \geq |v| - O(\log|x|)$.

Proof. A string $x = uvw$ can be specified by a description of v, literal descriptions of the binary representation of $|u|$ and the concatenation uw. Additionally, we need information of how to tell these three items apart. Such information can be provided in $O(\log|x|)$ bits. Thus,

$$K(x) \leq K(v) + O(\log|x|) + |uw|.$$

Rearranging above proves the fact. Further, for random string x with $K(x) \geq |x|$ we obtain

$$K(v) \geq |v| - O(\log|x|).$$

      $\square$

This fact shows that all long substrings of a random string are also almost incompressible. It can be shown that this is optimal - a substring of an incompressible string may be compressible by $O(\log|x|)$ bits. This conforms to a fact we know from probability theory: every sufficiently long random string must contain long runs of zeros.

Fact 3. If $K(x) \geq |x|$, then no substring of length longer than $3\log|x|$ occurs in x more than once.

Proof. Let $x = uvw, |x| = n$, and v of length $3\lceil \log n \rceil$ occurs twice in uv. To describe x, we only need to concatenate the following information, all items encoded so that we can tell them apart (cf. also Example 6):

- this discussion;

- the location of the beginning of the first instance of v in uv, using at most $\lceil \log n \rceil + 2 \log \log n$ bits;

- the location of the beginning of the second instance of v in uv, using at most $\lceil \log n \rceil + 2 \log \log n$ bits;

- the length of v in at most $2 \log \log n$ bits;

- the length of u in at most $2 \log \log n$ bits;

- the shortened word uw, using exactly $|uw|$ bits.

Altogether this description requires at most $n - \log n + 8 \log \log n + O(1)$ bits. For n large enough we have $K(x) < n$, a contradiction.　　□

Example 3. Apart from showing that complexity is an attribute of the finite object alone, the Invariance Theorem has also another most important consequence: it gives an upper bound on the complexity. Namely, there is a fixed constant c such that for all x of length n we have

$$K(x) \leq n + c.$$

This is easy to see. If T is a machine that just copies its input to its output, then $p = 0^{n(T)}1x$ is a program for the reference machine U to output x. This says that $K(x)$ is bounded above by the length of x modulo an additive constant. (For the self-delimiting Kolmogorov complexity one finds $K(x) \leq |x| + 2 \log |x|$.)

Example 4. Define $p(x)$ as a shortest program for x. We show that $p(x)$ is incompressible. There is a constant $c > 0$, such that for all strings x, we have $K(p(x)) \geq c \cdot |p(x)|$. For suppose the contrary, and there is a program $p(p(x))$ that generates $p(x)$ with $|p(p(x))| \leq c \cdot |p(x)|$. Define a universal machine V that works just like the reference machine U, except that V first simulates U on its input to obtain an output, and then uses this output as input on which to simulate U once more. But then, U with input $0^{n(V)}1p(p(x))$ computes x, and therefore $K(x) \leq c \cdot |p(x)| + n(V) + 1$. However, this yields $(1-c)K(x) \leq n(V) + 1$, for all x, which is impossible by a trivial counting argument. Similarly we can show that there is a $c > 0$ such that for all strings x, we have $K(p(x)) \geq |p(x)| - c$.

Example 5. It is easy to see that $K(x \mid x) \leq |T| + 1$, where T is a machine that just copies the input to the output. However, it is more interesting that $K(p(x) \mid x) \leq \log K(x) + O(1)$, which cannot be improved in general. Hint: later we show that K is a noncomputable function. This rules out that we can compute $p(x)$ from x. However, we can dovetail the computation of all programs shorter than $|x| + 1$: run the 1st program 1 step, run the 1st program 1 step and 2nd program 1 step, and so on. This way we will eventually enumerate all programs that output x. However, since some computations may not halt, and the halting problem is undecidable, we need to know the length of a shortest program $p(x)$ to recognize such a program when it is found.

A natural question to ask is: how many strings are incompressible? It turns out that virtually strings of given length n are incompressible. Namely, there is at least one x of length n that cannot be compressed to length $< n$ since there are 2^n strings of length n and but $2^n - 1$ programs of length less than n; at least $1/2$ of all strings of length n cannot be compressed to length $< n - 1$ since there are but $2^{n-1} - 1$ programs of length less than $n - 1$; at least $3/4$ of all strings of length n cannot be compressed to length $< n - 2$, and so on.

Generally, let $g(n)$ be an integer function. Call a string x of length n, *gincompressible* if $K(x) \geq n - g(n)$. There are 2^n binary strings of length n, and only $2^{n-g(n)} - 1$ possible descriptions shorter than $n - g(n)$. Thus, the ratio between the number of strings x of length n with $K(x) < n - g(n)$ and the total number of strings of length n is at most $2^{-g(n)+1}$, a *vanishing fraction* when $g(n)$ increases unboundedly with n. In general we loosely call a finite string x of length n *random* if $K(x) \geq n - O(\log n)$.

Intuitively, incompressibility implies the absence of regularities, since regularities can be used to compress descriptions. Accordingly, we like to identify incompressibility with absence of regularities or *randomness*. In the context of finite strings randomness like incompressibility is a matter of degree: it is obviously absurd to call a given string random and call nonrandom the string resulting from changing a bit in the string to its opposite value. Thus, we identify c-incompressible strings with c-*random* strings.

Fact 4. Fix any consistent and sound formal system F. For all but finitely many random strings x, the sentence "x is random" is not provable in F.

Proof. Assume that the fact is not true. Then given F, we can start to exhaustively search for proof that some string of length $n \gg |F|$ is random, and print is once we find such a string x. This procedure to print x of length n is only $O(\log n) + |F| \ll n$. But x is random by the proof. Hence F is not consistent. $\qquad\qquad\square$

The above fact shows that though most strings are random, it is impossible to effectively prove them random. This in a way explains why Kolmogorov complexity is so successful in many applications when effective construction of a required string is difficult by other methods. The fact that almost all finite strings are random but cannot be proved to be random amounts to an information-theoretic version of Gödel's theorem and follows from the noncomputability of $K(x)$. Strings that are not incompressible are *compressible* or *nonrandom*.

Fact 5 (Kolmogorov). (a) *The function $K(x)$ is not partial recursive. Moreover, no partial recursive function $\phi(x)$, defined on an infinite set of points, can coincide with $K(x)$ over the whole of its domain of definition.*

(b) *There is a (total) recursive function $H(t,x)$, monotonically decreasing in t, such that $\lim_{t\to\infty} H(t,x) = K(x)$. I.e., we can obtain arbitrary good estimates for $K(x)$ (but not uniformly).*

Proof. (a) Every infinite r.e. set contains an infinite recursive subset, Theorem 5-IV in [54]. Select an infinite recursive set A in the domain of definition of $\phi(x)$. The function $f(m) = \min\{x : K(x) \geq m, x \in A\}$ is (total) recursive (since $K(x) = \phi(x)$ on A), and takes arbitrary large values. Also, by construction $K(f(m)) \geq m$. On the other hand, $K(f(m)) \leq K_f(f(m)) + c_f$ by definition of K, and obviously $K_f(f(m)) \leq |m|$. Hence, $m \leq \log m$ up to a constant independent of m, which is false.

(b) Let c be a constant such that $K(x) \leq |x| + c$ for all x. Define $H(t,x)$ as the length of the smallest program p, with $|p| \leq |x| + c$, such that the reference machine U with input p halts with output x within t steps. $\qquad\qquad\square$

Example 6 (Self-delimiting Descriptions). Let the variables x, y, x_i, y_i, $\cdots$, denote strings in $\{0,1\}^*$. A *description* of x, $|x| = n$, can be given as follows.

(1) A piece of text containing several formal parameters $p_1,\cdots,p_m$. Think of this piece of text as a formal parametrized procedure in an algorithmic language like PASCAL. It is followed by

(2) an ordered list of the actual values of the parameters.

The piece of text of (1) can be thought of as being encoded over a given finite alphabet, each symbol of which is coded in bits. Therefore, the encoding of (1) as prefix of the binary description of x requires $O(1)$ bits. This prefix is followed by the ordered list (2) of the actual values of $p_1, \cdots, p_m$ in binary. To distinguish one from the other, we encode (1) and the different items in (2) as self-delimiting strings, an idea used already by C.E.Shannon.

For each string $x \in \{0, 1\}^*$, the string $\bar{x}$ is obtained by inserting a "0" in between each pair of adjacent letters in x , and adding a "1" at the end. I.e.,

$$\overline{01011} = 0010001011.$$

Let $x' = \overline{|x|}x$ (the length of x in binary followed by x in binary). The string x' is called the *self-delimiting* version of x. So '100101011' is the self-delimiting version of '01011'. (Note that according to our convention "10" is the 5th binary string.) The self-delimiting binary version of a positive integer n requires $\log n + 2 \log \log n$ bits, and the self-delimiting version of a binary string w requires $|w| + 2 \log |w|$ bits. For convenience, we denote the length $|n|$ of a natural number n by "$\log n$".

Example 7 (M. Li, W. Maass, P. M. B. Vitányi). In proving lower bounds in the theory of Computation it is sometimes useful to give an efficient description of an incompressible string with 'holes' in it. The reconstruction of the complete string is then achieved using an additional description. In such an application we aim for a contradiction where these two descriptions together have significantly smaller length than the incompressible string they describe. Formally, let $x_1 \cdots x_k$ be a binary string of length n with the x_i's $(1 \le i \le k)$ blocks of equal length c. Suppose that d of these blocks are deleted and the relative distances in between deleted blocks are known. We can describe this information by: (1) a formalization of this discussion in $O(1)$ bits, and (2) the actual values of

$$c, m, p_1, d_1, p_2, d_2, \cdots, p_m, d_m,$$

where m $(m \le d)$ is the number of "holes" in the string, and the literal representation of

$$\hat{x} = \hat{x}_1 \hat{x}_2 \cdots \hat{x}_k.$$

Here $\hat{x}_i$ is x_i if it is not deleted, and is the empty string otherwise; p_j, d_j indicates that the next p_j consecutive x_i's (of length c each) are one contiguous group followed by a gap of $d_j c$ bits long. Therefore, $k - d$ is the number of (non-empty) $\hat{x}_i$'s, with.

$$ k = \sum_{i=1}^{m} p_i + d_i \ \& \ \ d = \sum_{i=1}^{m} d_i. $$

The actual values of the parameters and $\hat{x}$ are coded in a self-delimiting manner. Then, by the convexity of the logarithm function, the total number of bits needed to describe the above information is no more than:

$$ \sum_{i=1}^{k} |\hat{x}_i| + 3d \log(k/d) + O(\log n). $$

We then proceed by showing that we can describe x by this description plus some description of the deleted x_i's, so that the total requires considerably less than n bits. Choosing x such that $K(x) \geq n$ then gives a contradiction. See [38] and [44].

Example 8 (Quantitative Estimates). Consider the conditional complexity of a string x, with x an element of a given finite set M, given some string y. Let $d(M)$ denote the number of elements in M. Then the fraction of $x \in M$ for which $K(x \mid y) < |d(M)| - m$, does not exceed 2^{-m+1}. Namely, if $K(x \mid y) < n$ then there is a program p of length at most $n - 1$ such that $F_0(p, y) = x$. Hence there cannot be more such x than there are such p, which is at most 2^n. Moreover, $d(M) \geq 2^{|d(M)|-1}$ Combining these two observations, we find that the fraction is less than $2^{|d(M)|-m}(2^{|d(M)|-1})^{-1} = 2^{-m+1}$. Hence we have shown that the conditional complexity of the majority of elements in a finite set cannot be significantly less than the complexity of the size of that set. The following Lemma says that it can not be significantly more either.

Lemma. (Kolmogorov) *Let A be a r.e. set of pairs (x, y), and let $M_y = \{x : (x, y) \in A\}$. Then, up to a constant depending only on A, $K(x \mid y) \leq |d(M_y|$.*

Proof. Let A be enumerated by a Turing machine T. Using y, modify T to T_y such that T_y enumerates all pairs (x, y) in A, without repetition. In order of enumeration we select the pth pair (x, y), and output the first element, i.e. x. Then we find $p < d(M_y)$, such that $T_y(p) = x$. Therefore, we have by the Invariance Theorem $K(x \mid y) \le K_{T_y}(x) \le |d(M_y)|$, as required. $\qquad\qquad\square$

Example. Let A be a subset of $\{0,1\}^*$. Let $A^{\le n}$ equal $\{x \in A : |x| \le n\}$. If the limit of $d(A^{\le n})/2^n$ goes to zero for n goes to infinity, then we call A *sparse*. For example, the set of all finite strings that have twice as many zeros as ones is sparse. This has as consequence that all but finitely many of these strings have short programs.

Claim. (a) (M. Sipser) If A is recursive and sparse, then for all constant c there are only finitely many x in A with $K(x) \ge |x| - c$. Using Kolmogorov's Lemma we can extend this result as follows.

(b) If A is r.e. and $d(A^{\le n})/n^{(1+\varepsilon)}2^n, \varepsilon > 0$, goes to zero for n goes to infinity, then, for all constant c, there are only finitely many x in A with $K(x) \ge |x| - c$.

(c) If A is r.e. and $d(A^{\le n}) \le p(n)$ with p a polynomial, then, for all constant $c > 0$, there are only finitely many x in A with $K(x) \ge |x|/c$.

Proof. (a) Consider the lexicographic enumeration of all elements of A. There is a constant d, such that the ith element x of A has $K(x) \le K(i) + d$. If x has length n, then the sparseness of A implies that $K(i) \le n - g(n)$, with $g(n)$ unbounded. Therefore, for each constant c, and all n, if x in A is of length n then $K(x) < n - c$, from some n onward.

(b) Fix c. Consider an enumeration of n-length elements of A. For all such x, the Lemma above in combination with the sparseness of A implies that $K(x \mid n) \le n - (1 + \varepsilon) \log n + O(1)$. Therefore, $K(x) \le n - \varepsilon \log n + O(1)$, for some other positive ε, and the right hand side of the inequality is less than $n - c$ from some n onward.

(c) Similarly to above. $\qquad\qquad\square$

Next we prove an important fact known as symmetry of information lemma, due to Kolmogorov and Levin.

Fact 6 (Symmetry of Information). *To within an additive term of* $O(\log K(x,y))$,

$$K(x,y) = K(x) + K(y \mid x).$$

In the general case it has been proved that equality up to a logarithmic error term is the best possible. From Fact 6 it follows immediately that, to within an additive term of $O(\log K(x,y))$,

$$K(x) - K(x \mid y) = K(y) - K(y \mid x).$$

Proof. ($\leq$) We can describe both x and y by giving a description of x, a description of y given x, and an indication of where to separate the two descriptions. If p is a shortest program for x, and q is a shortest program for y, then $\overline{|p|}pq$ is a description for (x,y). Thus, up to an additive constant term, $K(x,y) \leq K(x) + K(y \mid x) + 2\log(K(x))$. Obviously, $K(x) \leq K(x,y)$ up to a constant.

($\geq$) For convenience we show

$$K(y \mid x) \leq K(x,y) - K(x) + O(\log K(x,y)).$$

Assume this is not true. Then, for every constant c there are x and y such that $K(y \mid x) \geq K(x,y) - K(x) + c\log(K(x,y))$. Consider a description of y using its serial number in the recursively enumerable set $Y = \{z : K(x,z) \leq K(x,y)\}$. Given x, we cam describe y by its serial number and a description of $K(x,y)$, so $K(y \mid x) \leq \log(d(Y)) + 2\log(K(x,y))$, up to a fixed independent constant. By the contradictory assumption, $d(Y) > 2^d$ with $d = K(x,y) - K(x) + c\log(K(x,y))$. But now we can obtain a too short description for x as follows.

Each distinct pair (u,z) with $K(u,z) \leq K(x,y)$ requires a distinct description of length at most $K(x,y)$. Since there are not more than $2^{K(x,y)+1}$ such descriptions, we can reconstruct x from its serial number j in a recursively enumerable set that is small. Namely, given $K(x,y)$, consider the strings u such that $d(\{z : K(u,z) \leq K(x,y)\}) > 2^d$. Denote the recursively enumerable set of such u by X. Clearly, $x \in X$. By the upper bound on the available number of distinct descriptions, $d(X) < 2^{K(x,y)+1}/2^d$. We can now reconstruct x from $\overline{K(x,y)}\,\overline{d}j$. That is, up to an independent constant, $K(x) \leq 2\log(K(x,y)) + 2\log(d) + K(x,y) - d + 1$. For c large this implies the contradiction $K(x) < K(x)$. $\square$

Fact 7. All previous Facts are true relative to a given binary string y. for example, Fact 1 can be rewritten as: For all finite string y and all n there is an x of length n such that $K(x \mid y) \geq |x|$. And the equation for symmetry to information can be relativized with a string z:

$$K(x \mid z) - K(x \mid y \mid z) = K(y \mid z) - K(y \mid x \mid z).$$

We have introduced the basic theory of Kolmogorov complexity. We now turn to its applications.

3   How To Use It?

We show how Kolmogorov complexity can be used in various areas through illustrative examples.

3.1   Relation with Probability Theory

P.S. Laplace [32] has pointed out the following conflict between our intuition and the classical theory of probability:

> "In the game of heads and tails, if head comes up a hundred times in a row then this appears to us extraordinary, because the nearly infinite number of combinations that can arise in a hundred throws are divided in regular sequences, or those in which we observe a rule that is easy to grasp, and in irregular sequences, that are incomparably more numerous."

Yet, conventionally 100 heads are just as probable as any other sequence of heads and tails, even though we feel that it is less 'random' than some others. Nonetheless, we can vindicate our intuition formally (following P. Gács excellent exposition in [19]), using Kolmogorov complexity. Let us call a *payoff* function (or *martingale*) with respect to distribution P any nonnegative function $t(x)$ with $\sum_x P(x)t(x) \leq 1$. Suppose our favorite nonprofit casino asks 1 dollar for a game that consists of a sequence of flips of a fair coin, and claims that each outcome x has probability $P(x) = 2^{-|x|}$. To back up this claim, it must agree to pay $t(x)$ dollars on outcome x. Accordingly, we propose a payoff t_0 with respect to P_n (P restricted to sequences of n coin flips): put $t_0 = 2^{n/2}$ for all x whose even digits are 0 (head), and 0 otherwise. This bet will cost the casino $2^{50} - 1$ dollars for the outcome ($n = 100$) above. Since we must propose the payoff function beforehand, it is unlikely that we define precisely the one that detects this particular fraud. However, fraud implies regularity, and, as Laplace suggests, the number of regular bets is so small that we can afford to make all of them in advance.

> "If we seek a cause wherever we perceive symmetry, it is not
> that we regard the symmetrical event as less possible than
> the others, but, since this event ought to be the effect of a
> regular cause or that of chance, the first of these suppositions
> is more probable than the second."

Let us make this formal, using some original ideas of Solomonoff as
developed and made precise by Levin and Gács. We need the *Kraft
inequality*: For each set S of finite binary strings such that no string in
S is a proper prefix of another strings in S, we have

$$\sum_{x \in S} 2^{-|x|} \le 1.$$

We call S a *prefix-code*. With notation $K(x \mid y)$ we mean the self-
delimiting complexity of the previous section. Then for each fixed y, the
set of $K(x \mid y)$'s is the length set of a prefix-code, so Kraft's inequality
applies. For most binary strings of length n, no significantly shorter description exists, since the number of short descriptions is small. We can
sharpen this observation by, instead of counting the number of simple
sequences, measuring their probability. By Kraft's inequality, for each
fixed y,

$$\sum_{x} 2^{-K(x|y)} \le 1, \tag{1}$$

so that only a few objects can have small complexity. Conversely, Let μ
be a *computable* probability distribution, i.e., such that there is an effective procedure that, given x, computes $\mu(x)$ to any degree of accuracy.
Let $K(\mu)$ be the length of the smallest such program. Then,

$$K(x) \le -\log \mu(x) + K(\mu) + c, \tag{2}$$

with c a universal constant. Put $d(x \mid \mu) = -\log \mu(x) - K(x)$. By
(1) $t(x \mid \mu)2^{d(x|\mu)}$ is a payoff function. We now can beat any fraudu-
lent casino. We propose the payoff function $2^{-\log P_n(x) - K(x|n)}$. (We use
conditional complexity $K(x \mid n)$ because the uniform distribution P_n
depends on n.) If every other coin flip comes up heads, then $K(x \mid n) \le
(n/2) + c_0$, and hence we win $2^{t(x)} \ge c_1 2^{n/2}$ from the casino $c_0, c_1 > 0$,
even though the bet does not refer to 'heads'.

The fact that $t(x \mid \mu)$ is a payoff function, implies by Chebychev's
(first) Inequality that for any $k > 0$,

$$\mu\{x : K(x) < -\log \mu(x) - k\} < 2^{-k}. \tag{3}$$

Together, (2) and (3) say that with large probability, the complexity $K(x)$ of a random outcome x is close to its upper bound $-\log\mu(x) - K(\mu)$. If an outcome x violates any 'laws of probability', then the complexity $K(x)$ falls far below the upper bound. Indeed, a proof of some law of probability (like the law of large numbers, the law of iterated logarithm, etc.) always gives rise to some simple computable payoff function $t(x)$ taking large value on the outcomes violating the law. In general, the payoff function $t(x \mid \mu)$ is *maximal* (up to a multiplicative constant) among all payoff functions that are semicomputable (from below). Hence the quantity $d(x \mid \mu)$ constitutes a universal test of randomness — it measures the *deficiency* of randomness in the outcome x with respect to distribution μ, or the extend of justified suspicion against hypothesis μ given the outcome x.

3.2   A Priori Probability and Inductive Inference

This application stood at the cradle of Kolmogorov complexity proper. It led Solomonoff to formulate the important notion of a universal *a priori* probability, as described previously. Solomonoff's proposal [57], i.e., the solomonoff-Levin distribution we define below, is a synthesis of Occam's principle and Turing's theory of effective computability applied to inductive inference. His idea is to view a theory as a compact description of past observations together with predictions of future ones. The problem of theory formation in science is formulated as follows. The investigator observes increasingly larger initial segments of an infinite binary sequence. We can consider the infinite binary sequence as the outcome of an infinite sequence of experiments on some aspect X of Nature. To describe the underlying regularity of this sequence, the investigator tries to formulate a theory that governs X, on the basis of the outcome of past experiments. Candidate theories are identified with computer programs that compute infinite binary sequences starting with the observed initial segment.

Let $K(x)$ be the self-delimiting variant of complexity. The incomputable *Solomonoff-Levin distribution* $\mathbf{m}(x)$ can be defined as

$$\mathbf{m}(x) = \sum_{U(p)=x} 2^{-|p|},$$

with U the reference universal prefix machine, associated with the self-delimiting Kolmogorov complexity. Now $\mathbf{m}(x)$ can be interpreted as the probability that U halts with output x if we generate the input p by an indefinitely long sequence of random coin flips (U will use only the self-delimiting prefix p of this sequence as its program). See also Definition 2 in Section 2.

This distribution was conceived by Solomonoff in [56,57], but in imprecise form. The mathematically precise form, in terms of a quantity that is essentially $\mathbf{m}(x)$ was given by Levin in [67].

It can be shown that $\mathbf{m}(x) = 2^{-K(x)\pm O(1)}$. It turns out that $\mathbf{m}(x)$ has the remarkable property that the test $d(x \mid \mathbf{m})$ shows all outcomes x random with respect to it. We can interpret (2) and (3) as saying that if the real distribution is μ, then $\mu(x)$ and $m(x)$ are close to each other with large probability. Therefore, if x comes from some unknown computable distribution μ, then we can use $\mathbf{m}(x)$ as an estimate for $\mu(x)$. Accordingly, Solomonoff has called $\mathbf{m}$ 'a priori probability.' The randomness test $d(x \mid \mu)$ can be interpreted in the framework of hypothesis testing as the likelihood ratio between hypothesis μ and the fixed alternative hypothesis $\mathbf{m}$. In ordinary statistical hypothesis testing, some properties of an unknown distribution μ are taken for granted, and the role of the universal test can probably be reduced to some tests that are used in statistical practice. However, such conditions do not hold in general as is witnessed by prediction of time series in economics, pattern recognition or inductive inference (see below).

Since the a priori probability $\mathbf{m}$ is a good estimate for the actual probability, we can use the conditional a priori probability for prediction — without reference to the unknown distribution μ. For this purpose, we first define a priori probability M for the set of infinite sequences of binary digits (or, more generally, natural numbers) as in [67]. For any finite sequence $x, M(x)$ is the a priori probability that the outcome is some extension of x. Let x, y be finite sequences. Then

$$\frac{M(xy)}{M(x)} \tag{4}$$

is an estimate of the conditional probability that the next terms of the outcome will be given by y provided that the first terms are given by x. It converges to the actual conditional probability $\mu(xy)/\mu(x)$ with μ-probability 1 for any computable distribution μ [58]. Inductive inference formula (4) can be viewed as a mathematical formulation of

Occam's razor: predict by the simplest rule fitting the data. The a priori distribution M is incomputable, and the main problem of inductive inference can perhaps be stated as 'finding efficiently computable optimal approximations to M'.

To make the discussion concrete, we give a simplified Solomonoff inductive inference theory. Given the previously observed data in the form of string S over $\{0,1\}^*$, we formulate as the inference problem the prediction of the next bit in the output sequence, i.e., extrapolating the sequence S. By the Bayes's rule, for $a = 0$ or 1,

$$P(S_a \mid S) = \frac{P(S \mid S_a)P(S_a)}{P(S)},$$

where $P(S) = P(S \mid S0)P(S0) + P(S \mid S1)P(S1)$. Since $P(S \mid S_a) = 1$ for any a, we have,

$$P(S_a \mid S) = \frac{P(S_a)}{P(S0) + P(S1)},$$

In terms of inductive inference or machine learning, the final probability $P(S_a \mid S)$ is the probability of the next symbol being a given the initial sequence S. The goal of inductive inference in general is to be able to infer the underlying machinery that generated S, and hence be able to predict (extrapolate) the next symbol. Obviously we now only need the prior probabilities $P(S_a)$ to evaluate $P(S_a \mid S)$. Let K denote the self-delimiting Kolmogorov complexity. Using the Solomonoff-Levin distribution, assign $2^{-K(S_a)}$ as the prior probability to $P(S_a)$. This corresponds to the Occam's razor principle of choosing the simplest theory and has two very nice properties:

(1) $\sum_s 2^{-K(s)} \leq 1$ by Kraft's inequality. Hence this is a proper probability assignment.

(2) For any computable probability $P(x)$, there is a constant c such that $2^{-K(x)} \geq cP(x)$. In fact, $\mathbf{m}(x) = 2^{-K(x)\pm O(1)}$[20,67].

This approach is guaranteed to converge in the limit to the true solution and in fact *better than* any other method up to a constant factor. It turns out that Gold's idea of identification by enumeration [18] can also be derived from this approach [39].

The notion of complexity of equivalent theories as the length of a shortest program that computes a given string emerges forthwith, and also the invariance of this measure under changes of computers that execute them. The metaphor of Natural Law being a compressed description of observations is singularly appealing. Among others, it gives substance to the view that a Natural Law is better if it "explains" more, that is, describes more observations. On the other hand, if the sequence of observations is sufficiently random, then it is subject to no Law but its own description. This metaphorical observation was also made by Chaitin [5].

3.3 Rissanen's Minimum Description Length Principle

Scientists formulate their theories in two steps: first a scientist must, based on scientific observations or given data, formulate alternative hypotheses, and second he selects one definite hypothesis. The second step has been done by many ad hoc principles, among the most dominant, Occam's razor principle, the Maximum Likelihood principle, various ways of using Bayes's formula with different prior distributions. However, no single principle is satisfactory in all situations. But in an ideal sense, the Solomonoff approach we have discussed presents a "perfect" way of solving all induction problems using Bayes's rule with a universal prior distribution: the Solomonoff-Levin distribution m. However, due to the non-computability of the universal prior function, such a theory cannot be directly used in practice. Inspired by Solomonoff's (Kolmogorov complexity) approach, Rissanen in 1978 proposed his Minimum Description Length Principle (MDLP) [51]. Quinlan and Rivest used this principle to construct an algorithm for constructing decision trees and the result was quite satisfactory compared to existing algorithms [50]. Using MDLP, Gao and Li [17] have implemented a system (on IBM PC) which on-line recognizes/learns hand-written English and Chinese characters. The **Minimum Description Length Principle** can be intuitively stated as follows:

> *The best theory to explain a set of data is the one which minimizes the sum of*

(1) the length (encoded in binary bits) of the theory;

(2) the length (in binary bits) of data when encoded with the help of the theory.

Example (J. Kemeny). The importance of "simplicity" for inductive inference was already exploited in an elegant paper by J. Kemeny [29]. This paper clearly anticipates the ideas developed in this section, and we cite one example. We are given n points in the plane. Which polynomial fits these points best? One extreme is to put a straight line through the cluster such that the χ^2 measure is minimized. The other extreme is to simply fit an $n-1$ degree polynomial through the n points. Neither choice seems very satisfactory, and it is customary to think that the problem is not formulated precisely enough. But MDLP says that we look for the mth degree polynomial, $m \leq n - 1$, such that the description of the m-vector of coefficients of the polynomial, together with the description of the points relative to the polynomial, is minimized.

Using Kolmogorov complexity, we can derive a version of MDLP principle , and explain how and why it works [39]. From Bayes's rule,

$$P(H \mid D) = \frac{P(D \mid H)P(H)}{P(D)},$$

we need to choose the hypothesis H such that $P(H \mid D)$ is maximized, where D denotes the experimental data. Now take the negative logarithm on both sides of the Bayes's formula, we get

$$-\log P(H \mid D) = -\log P(D \mid H) - \log P(H) + \log P(D).$$

Maximizing the term $P(H \mid D)$ is equivalent to minimizing

$$-\log P(D \mid H) - \log P(H)$$

Now to get the minimum description length principle, we only need to explain above two terms properly. The term $-\log P(H)$ is straightforward. According to the Solomonoff-Levin distribution, $P(H) = 2^{-K(H)}$ where $K(H)$ is the self-delimiting Kolmogorov complexity of H. Then the term $-\log P(H)$ is precisely the *length* of a minimum length *prefixfree encoding*, or shortest program, for the hypothesis H.

We now consider the term $-\log P(D \mid H)$. Assume that P is computable. It can be shown that the universal probability distribution $\mathbf{m}(x)$ can approximate $P(x)$ in the sense that

(1) There is a constant c, such that $\mathbf{m}(x) \geq cP(x)$, and

(2) The probability of $\mathbf{m}(x) \leq kP(x)$ is at least $1 - \frac{1}{k}$.

Since $\mathbf{m}(D \mid H) = 2^{-K(D|H)\pm O(1)}$, the quantity $2^{-K(D|H)}$ is a reasonable approximation of $P(D \mid H)$, and minimizing $-\log P(D \mid H)$ can be considered as minimizing $K(D \mid H)$, i.e., finding an H such that the description length, or the Kolmogorov complexity, of D given H is minimized.

In the original Solomonoff approach, H in general is a Turing machine. In practice we must avoid such a too general approach in order to keep things computable. In different applications, the hypothesis H can mean many different things. For example, H may refer to decision trees, finite automata, Boolean formulae, or polynomials of certain degree. Thus, Rissanen suggested the following approach. First convert (or encode) H to an integer. Then we try to assign prior distribution to each integer. Assigning $2^{-K(n)}$ to integer n would be perfect but it is not computable. Jeffreys [26] suggested to assign probability $\frac{1}{n}$ to integer n. But this results an improper distribution since $\sum_{n=1}^{\infty} \frac{1}{n}$ diverges. Rissanen defined the following length function: let

$$l^*(n) = \log n + \log \log n + \log \log \log n + \cdots$$

all positive terms, and let $L(n) = l^*(n) + \log c$ where $c = 2.865064 \cdots$ It has been shown that $\sum_n 2^{-L(n)} = 1$, see[52].

3.4 Learnability in the Valiant Learning Model

This Section is rather closely related to the previous two sections. There we did not consider the number of steps involved in making an inference, or the number of examples needed to learn a concept. Solomonoff's principle shows that we learn something perfectly in the limit, but how fast this converges is not prescribed at the outset. For instance, the well-known principle of Gold [18] of inference by enumeration can be viewed as a particular case of Solomonoff's principle, cf. [39], and we note here without further explanation, and as immediately clear to those familiar with it, that the Gold paradigm of inductive inference is aimed at *precise* inference. But what if we want to learn a concept using a number of examples that is bounded a priori? Obviously if we are to *precisely* infer a law of nature, an infinite (or exponential) behavior is inherent. However, for the purpose of machine learning, it is sufficient to just *learn* such a law *approximately*: if a human child (or a computer) would recognize, with .99 *probability*, the next apple after seeing three apples, we consider that the concept of apple is learned. In 1983, Valiant [60] introduced such a learning model. Valiant emphasizes the polynomial time learnability. For simplicity and convenience, we consider the problem learning of Boolean formulae of n variables.

Following Valiant, the concept to be learned is a Boolean formula F. Those vectors v such that $F(v) = 1$ are called positive examples, the rest are negative examples of F. For any F, there are many possible boolean formulate f such that f is consistent with the concept F. Let $|f|$ denote the smallest number of symbols needed to write the representation f. The learning algorithm has available two buttons labeled POS and NEG. If POS(NEG) is pushed, a positive (negative) example is generated according to some fixed but unknown probability distribution $D^+(D^-)$ according to nature. We assume nothing about the distributions D^+ and D^- except that $\sum_{f(v)=1} D^+(v) = 1$ and $\sum_{f(v)=0} D^-(v) = 1$. Let A be a class of concepts. Then A is learnable from examples iff there exists a polynomial p and a (possibly randomized) learning algorithm L such that, for f in A and $\varepsilon > 0$, algorithm L halts in $p(n, |f|, 1/\varepsilon)$ time and examples, and outputs a formula $g \in A$ that with probability at least $1 - \varepsilon$ has the following properties: $\sum_{g(v)=0} D^+(v) < \varepsilon$ and $\sum_{g(v)=1} D^-(v) < \varepsilon$.

Many classes of concepts are shown to be learnable in Valiant's sense [3, 21, 28, 42, 53, 60, 61]. Many valiant learnable classes are NP-complete to learn in the Gold sense. (See [27] for a survey.) In [3], again by the Occam's principle, it was shown that given a set of positive and negative data, any consistent concept of size "reasonably" less than the size of data is an "approximately" correct concept. That is, if one can find a shorter representation of data, then one learns. The shorter the conjecture is, the more and better it explains with higher probability. See also [39].

3.5 Applications of Compressibility and Incompressibility

It is not surprising that some strings can be compressed arbitrary far. Easy examples are the decimal expansions for some transcendental numbers like $\pi = 3.1415\cdots$ and $e = 2.7182\cdots$. These strings can be described in $O(1)$ bits, and have therefore constant Kolmogorov complexity. A moment's reflection suggests that the set of computable numbers, i.e., the real numbers computable by Turing machines which start with a blank tape, coincides *precisely* with the set of real numbers of Kolmogorov complexity $O(1)$. A nice application of what we may call extreme compressibility of some strings is a new version of Gödel's celebrated incompleteness theorem.

3.5.1 A Version of Gödels Theorem

We expand on Fact 4 in Section 2. Recall K. Gödel's famous incompleteness result that each formal mathematical system which contains Arithmetic is either inconsistent or contains theorems which cannot be proved in the system. Y.M.Barzdin' has first formulated a new form of Gödel's incompleteness theorem in terms of simple sets. This idea was popularized by G.J. Chaitin in a sequence of later papers, e.g., [6,7]. Let us view a theorem – a true statement – together with the description of the formal system, as a description of a proof of that theorem. Just as certain numbers can be really far compressed, like π or 10^{100}, in their descriptions, in a formal mathematical system the ratio between the length of the theorems and the length of their shortest proofs can be enormous. In a sense, the argument below shows that the worst-case such ratio expressed as a function of the length of the theorem increases faster than any computable function.

In Bennett's [2] phrase: *although most numbers are random, only finitely many of them can be proved random within a given consistent axiomatic system.* If T be an axiomatized system (which is sound, i.e. all theorems provable in T are true), whose axioms and rules of inference require about k bits to describe, then T cannot be used to prove the randomness of any number much longer than k bits. If the system could prove randomness for a number much longer than k bits, then the *first* such proof (first in an unending enumeration of all proofs obtainable by repeated application of axioms and rules of inference) could be used to derive a contradiction: an approximately k-bit program to find and print out the specific random number mentioned in this proof, a number whose smallest program is by assumption considerably larger than k bits. Therefore, even though most strings are random, we will never be able to explicitly exhibit a string of reasonable size which demonstrably possesses this property. (This formulation is due to C.H. Bennett.)

Example (G.J.Chaitin). We use an approach that is slightly different from Chaitin's approach.

(a) Let axiomatized theory T be describable in k bits: $K(T) \leq k$.

(b) Assume that all true formulas in T can be proved in T.

(c) Let $S_c(x)$ be a formula in T with the meaning: "x is the lexicographically least binary string of length c with $K(x) \geq c$". Here x is a formal parameter and c an explicit constant, so $K(S_c) \leq \log c$ up to a fixed constant independent of T and c.

For each c, there exists an x such that $S_c(x) = \textbf{true}$ is a true statement by a simple counting argument. Moreover, S_c expresses that this x is unique. It is easy to see that combining the descriptions of T, S_c, we obtain a description of this x. Namely, by (b), for each candidate string y of length c, we can decide $S_c(y) = \textbf{true}$ (holds for $y = x$) or not $(S_c(y)) = \textbf{true}$ (holds for $y \neq x$), by simple enumeration of all proofs in T. We need to distinguish the descriptions of T and S_c, so we code T's description self-delimiting in not more than $2k$ bits. Hence, for some fixed constant c' independent of T and c, We find $K(x) \leq 2k + \log c + c'$, which contradicts $K(x) > c$ for all $c > c_T$, where $c_T = 3k + c'$ for another constant c'.

As Chaitin expresses it: " $\cdots$ if one has ten pounds of axioms and a twenty – pound theorem, then that theorem cannot be derived from those axioms."

Fact (Barzdin'). The previous example is based on the following more general fact. It is not too difficult to show that if $f(x) < |x|$ is a total recursive function with $\lim_{x \to \infty} f(x) = \infty$, then the set $B = \{x : K(x) \leq f(x)\}$ is *simple* in the recursive-theoretic sense of E. Post. That is, B is recursively enumerable and the complement of B is infinite but does not contain an infinite recursively enumerable subset. It then is straightforward that for every axiomatized theory F (that is consistent and sound) there are only finitely many n for which the statement "$n \notin B$" is both true and provable in F. However, from the definition of B in follows that all x with $K(x) \geq |x|$ do not belong to B, and there are infinitely many of those. Hence, if F is strong enough to express statements of the form "$n \notin B$", for instance F contains arithmetic, then infinitely many true statements can be expressed in F but are not provable. This is a version of Gödel's famous incompleteness result. It is different from Gödel's original proof in the fact that our undecidable statements are not constructive. This result is attributed to Barzdin' in Levin and Zvonkin's 1970 survey [67].

Example (Levin). Levin has tried i gauge the implications of algorithmic complexity arguments to probe depths beyond the scope of Gödel's arguments. He derives what he calls "Information Conservation Inequalities". Without going into the subtle details, Levin's argument in [33] takes the form that the information $I(\alpha : \beta)$ in a string α about a string β cannot be significantly increased by either algorithmic or probabilistic means. In other terms, any sequence that may arise in nature contains only a finite amount of information about any sequence defined mathematically. Levin says: "Our thesis contradicts the assertion of some mathematicians that the truth of any valid proposition can be verified in the course of scientific progress by means of nonformal methods (to do so by formal methods is impossible by Gödel's theorem.)" (For a continuation of this research, see [34].)

3.5.2 Computable numbers are Not Random

One can make precise the off-hand claim made in the Introduction section that the set of computable numbers coincides with the set of reals which have Kolmogorov complexity $O(1)$. (We view real number as an infinite sequence of digits.) Let $N = \{0, 1, \cdots\}$ be the set of natural numbers, let $S = \{\varepsilon, 0, 1, 01, 10, 11, \cdots\}$ be the set of finite binary strings, and let X be the set of finite binary strings. We denote by $|s|$ the length of a strung s, and by s_n the prefix if length n of a string s. (If $x \in X$ then $|x| = \infty$.) An infinite string x is *recursive* iff there is a recursive function $f : N \to S$ such that $x_n = f(n)$ for all n. It can be shown [9] that x is recursive iff there exists a constant $c > 0$ such that for all $n \in N$ we have $K(x_{1:n}) \leq K(n) + c$.

3.5.3   Weak Prime Number Theorems

Using Kolmogorov complexity, it is easy to derive a weak version of the prime number theorem. An adaptation of the proof Chaitin gives of Euclid's theorem that the number of primes is infinite [10] yields a very simple proof of a weak prime number theorem. Let $\pi(n)$ denote the number of prime numbers less than n. We prove that $\pi(n)$ is $\Omega(\log n(\log\log n)^{-1})$. Let n be a random number with $K(n) \geq \log n - O(1)$. Consider a prime factorization

$$n = p_1^{e_1} \cdot p_2^{e_2} \cdots p_m^{e_m}$$

with $p_1, p_2, \cdots$ the sequence of primes in increasing order. With $m = \pi(n)$ we can describe n by the $\pi(n)$ length vector of exponents

$$(e_1, \cdots, e_{\pi(n)}).$$

Since $p_i \geq p_1 = 2$, it holds $e_i \leq \log n$, and, bounding $K(e_i)$ by the length of self-delimiting descriptions of e_i, $K(e_i) \leq \log\log n + 2\log\log\log n$, for all $i \leq m$. Therefore, $K(n) \leq \pi(n)(\log\log n + 2\log\log\log n)$. Substituting the lower bound on $K(n)$, we obtain the claimed lower bound on $\pi(n)$ for the special sequence of random n.

Recently, P.Berman [Personal Communication] obtained the stronger result that the number of primes below n is $\Omega(n/(\log n(\log n \log n)^2))$, by an elementary Kolmogorov complexity argument. It is interesting mainly because it shows a relation between primality and prefix codes. (There are other simple elementary methods to obtain weak prime number theorems.) Recall that we identify the positive integer n with the nth binary string. Assume that we have a function $c : N \to N$ with the following property: for every two integers $m, n, c(m)$ is not a prefix of $c(n)$. Then c is called a *prefix-code*. Consider only prefix-codes c such that $c(n) = O(n^2)$. (E.g., choose $c(n) = \overline{|n|}n$, the self-delimiting description of n with binary length $|c(n)| = \log n + 2\log\log n$ bits.)

Lemma (Berman). *For an infinite subsequence of positive integers n, $C(n) = \Omega(p_n)$ where p_n is the nth prime.*

Choosing $c(n)$ as above we have $c(n) \leq n\log^2 n$. Therefore, by the Lemma, P_n is $O(n\log^2 n)$. Straightforward manipulation of the order-of-magnitude symbols then shows $\pi(n)$ is $\Omega(n/\log^2 n)$. This can be strengthened by choosing more efficient codes.

3.5.4   Characteristic Sequences of Recursively Enumerable Sets

It turns out that with Kolmogorov complexity one can quantify the distinction between r.e. sets and recursive sets. Let $x = x_1 x_2 \cdots$ be an infinite binary sequence such that the set of numbers n with $X_n = 1$ is r. e. If the complementary set $x(n) = 0$ were also r.e., then $f(n) = x_n$ would be computable, and the relative complexity $K(x_{1:n} \mid n)$ bounded ($x_{1:n}$ is the n-length prefix of x). But in the general case, when the set of 1's is r.e., $K(x_{1:n} \mid n)$ can grow unboundedly.

Theorem (Barzdin', Loveland). *For any binary sequence x with the set $M = \{n : x_n = 1$ is r. e. holds $K(x_{1:n} \mid n) \leq \log n + c_M$, where c_M is a constant dependent on M (but not dependent on n). Moreover, there are such sequences such that for any n holds $K(x_{1:n} \mid n) \geq \log n$.*

Proof. Let the number of ones in $x_{1:n}$ be $m \leq n$. Since M is r.e. we can recursively enumerate all of its elements without repetition. Given m, we know that after having enumerated m elements in M that are less or equal to n we have found them all. Since $K(m) \leq \log n + c$ for some fixed constant c, this proves the upper bound. The lower bound holds for universal sets like $K_0 = \{< x, y >: T_x$ halts on input $y\}$. See [1, 43, 67]. $\qquad\qquad\Box$

In [31] Kolmogorov gives the following interesting interpretation with respect to investigations in the foundations of mathematics. Viz., label all Diophantine equations by natural numbers. Y.V. Mateyasevich has proved that there is no general algorithm to answer the question whether the equation D_n is soluble in integers (the answer to Hilbert's 10th problem is negative). Suppose, we weaken the problem by asking for the existence of an algorithm that enables us to answer the question of the existence or nonexistence of solutions for the first n Diophantine equations with the help of some supplementary information of size related to n. The theorem above shows that this size can be as small as $\log n + O(1)$. Such information is in fact contained in the $\sim \log n$ length prefix of the mythical number Ω, that encodes the solution to the halting problem for the first n Turing machines, cf. below. We note that in the same paper Barzdin' also proved that, if we impose any recursive time bound on the decoding process, then there are recursively enumerable sets such that the characteristic sequences can only be compressed linearly. This 1968 result is the first one we know of about resource-bounded Kolmogorov complexity, a subject we do not further discuss in this paper.

3.5.5   The Number of Wisdom Ω

This Kabalistic exercise follows Chaitin [6] and Bennett[2]. A real is *normal* if each digit from 0 to 9, and each block of digits of equal length, occurs with equal asymptotic frequency. No rational number is normal to any base, and almost all irrational numbers are normal to every base. But for particular ones, like π and e, it is not known whether they are normal,although statistical evidence suggests they are. In contrast to the randomly appearing sequence of the decimal representation of π, the digit sequence of Champernowne's number $0123456789101112 13\cdots$ is very nonrandom yet provably normal [12]. Once we know the law that governs π's sequence,we can make a fortune betting at fair odds on the continuation of a given initial segment, and most gamblers would eventually win against Champernowne's number because they will discover its law.

Almost all real numbers are Kolmogorov random, which implies that no possible betting strategy, betting against fair odds on the consecutive bits, can win infinite gain. Can we exhibit a specific such number? One can define an uncomputable number $k = 0.k_1k_2\cdots$ such that $k_i = 1$ if the ith program in a fixed enumeration of programs for some fixed universal machine halts, else $k_i = 0$. By the unsolvability of the halting problem, k is noncomputable. However, by Barzdin's Theorem (before), k is *not* incompressible: each n-length prefix $k_{1:n}$ of k can be compressed to a string of length not more than $2\log n$ (since $K(k_{1:n} \mid n) \leq \log n$), from some n onwards. It is also easy to see that a gambler can still make infinite profit, by betting only on solvable cases of the halting problem, of which there are infinitely many. Chaitin [8] has found a number that is random in the strong sense needed.

Ω equals the probability that the reference self-delimiting universal Turing machine halts when its program is generated by fair coin tosses. That is $\Omega = \sum_x \mathbf{m}(x)$, with $\mathbf{m}(x)$ the Solomonoff-Levin distribution. Then, Ω is a number between 0 and 1. It is greater than 0 since some programs do halt, and it is less than one since some programs do not halt (by the Kraft inequality). It is Kolmogorov random, it is noncomputable, and no gambling scheme can make an infinite profit against it. It has the curious property that it encodes the halting problem very compactly. Namely, suppose we want to determine whether a program p halts or not. Let program p have length n. Its probability in terms of coin tosses is 2^{-n}. If we know the first n bits $\Omega_{1:n}$ of Ω,then $\Omega_{1:n} < \Omega \leq \Omega_{1:n} + 2^{-n}$ .However, dovetailing (execute phases $1, 2, \cdots$, with phase i consists of executing one step of each of the first i programs) the running of all programs sufficiently long, must yield eventually an approximation Ω' of Ω with $\Omega' > \Omega_{1:n}$. If p is not among the halted programs which contributed to Ω', then p will never halt, since otherwise its contribution would yield $\Omega > \Omega' + 2^{-n}$, which is a contradiction. (I.e., $\Omega_{1:n}$ is a short program to obtain $k_{1:m}$ with $m \approx 2^n$.)

Therefore, knowing the first 10,000 bits of Ω enables us to solve the halting of all programs of less than 10,000 bits. This includes programs looking for counter examples to Fermat's Last Theorem, Riemann's Hypothesis and most other conjectures in mathematics that can be refuted by single finite counter examples. Moreover, for all axiomatic mathematical theories which can be expressed compactly enough to be conceivably interesting to human beings, say in less than 10,000 bits, $\Omega_{10,000}$ can be used to decide for every statement in the theory whether it is true, false or independent. Finally, knowledge of Ω_n suffices to determine whether $K(x) \leq n$ for each finite binary string x. Thus, Ω is truly the number of Wisdom, and "can be known of, but not known, through human reason"[2].

Example (Chaitin). Recall that the Barzdin'-Loveland Theorem states that for all r.e. sets each n-length initial segment of their characteristic sequence has Kolmogorov complexity $O(\log n)$. Kolmogorov has remarked that this implies that the solubility of the first n diophantine equations in an effective enumeration can be decided using at most $O(\log n)$ bits extra information. Namely, given the number $m \leq n$ of soluble equations in the first n equations, we can find them all effectively in the obvious way. Chaitin observed that this is not the case if we replace the question of mere solubility by the question of whether there are finitely many or infinitely many nontrivially different solutions. Namely, no matter how many solutions we find for a given equation, by itself this can give no information on the question to be decided. It turns out that the set of indices of the diophantine equations with infinitely many different solutions is not r.e. In particular, in the characteristic sequence each initial segment of length n has Kolmogorov complexity of about n. Chaitin says that this shows that randomness is inherent not only in natural phenomena (e.g., related to quantum mechanics), but also occurs in mathematics [11]. More precisely:

Claim. There is an (exponential) Diophantine equation

$$A(n, x_1, x_2, \ldots, x_m) = 0$$

which has only finitely many solutions $x_1, x_2, \cdots, x_m$ if the nth bit of Ω is zero and which has infinitely many solutions $x_1, x_2, \cdots, x_m$ if the nth bit of Ω is one.

Proof. By dovetailing the running of all programs of the reference prefix-machine U in the obvious way we find a computable sequence of rational numbers $r_1 \leq r_2 \leq \cdots$ such that $\Omega = \lim_{n \to \infty} r_n$. The set $R = \{(n, k):$ the nth bit of r_k is a one $\}$ is a recursively enumerable (even recursive) set. The main step is to use a theorem due to J.P. Jones and Y.V. Mateyasevich (*J. Symbol. Logic* 49(1984), pp. 818-829)to the effect that "every recursively enumerable set R has a singlefold exponential Diophantine representation $A(p, y)$". That is, $A(p, y) = 0$ is an exponential Diophantine equation, and the singlefoldedness consists in the property that $p \in R$ iff there is a y such that $A(p, y) = 0$ is satisfied, and, moreover, there is only a singe such y. (Here both p and y can be multituples of integers; in our case p represents $\langle n, x_1 \rangle$, and y represents $\langle x_2, \cdots, x_m \rangle$. For technical reasons we consider as proper solutions only solutions x involving no negative integers.) Representing R this way, there is a Diophantine equation $A(n, k, x_2, \cdots, x_m) = 0$ which has exactly one solution $x_2, \cdots, x_m$ if the nth bit of the binary expansion of r_k is a one, and it has no solution $x_2, \cdots, x_m$ otherwise. Consequently, the number of different m-tuples $x_1, x_2, \cdots, x_m$ which are solutions to $A(n, x_1, x_2, \cdots, x_m) = 0$ is infinite if the nth bit of the binary expansion of Ω is a one, and this number is finite otherwise. $\quad\square$

3.6   Proving Lower Bounds Using Kolmogorov Complexity

Kolmogorov complexity has become a very powerful tool in the lower bound proofs in theoretical computer science. It was observed in [48], that static, descriptional (program size) complexity of a *single* random string can be used to obtain lower bounds on dynamic, computational (running time) complexity. The power of the static, descriptional Kolmogorov complexity in the dynamic, computational lower bound proofs rests on one single idea: There are incompressible (or Kolmogorov random) strings. In a traditional lower bound proof by counting, it usually involves all inputs (or all strings of certain length) and one shows that the lower bound has to hold for some of these ("typical") inputs. Since a particular "typical" input is *hard* to construct, the proof has to involve all the inputs. Now we understand that a "typical input" can be constructed via a Kolmogorov random string. However, as we have shown no such string can be proved to be random or "typical". In a Kolmogorov complexity proof, we choose a random string that exists. That it cannot be exhibited is no problem, since we only need existence. As a routine, the way one proves a lower bound by Kolmogorov complexity is as follows: Fix a Kolmogorov random string which we know exists. Prove the lower bound with respect to this particular *fixed* string: show that if the lower bound does not hold, then this string can be compressed. Because we are dealing with only one fixed string, the lower bound proof usually becomes quite easy and natural.

In the next sub-section, we give four examples to illustrate the basic methodology. We then give three more sophisticated examples in the next subsection.

3.6.1   Four Examples

We illustrate how Kolmogorov complexity is used to prove lower bounds by four concrete examples, with increasing difficulties of the corresponding counting arguments.

Example 1(One Tape Turing Machines, Paul).

Consider a most basic Turing machine model with only one tape, with a 2-way read/write head, which serves as both input and work tape. The input is initially put on the first n cells of the only tape. We refer a reader who is not familiar with Turing machines to [23] for a detailed definition. The following theorem was first proved by Hennie and a proof by counting, for comparison, can be found in [23], page 318. Paul [47] presented the following elegant proof. Historically, this was the first Kolmogorov complexity lower bound proof.

Theorem. *It requires $\Omega(n^2)$ steps for the above single tape TM to recognize $L = \{ww^R : w \in \{0,1\}^*\}$ (the palindromes).*

Proof [47]. Assume on the contrary, M accepts L in $o(n^2)$ time. Let $|M|$ denote the length of the description of M. Fix a Kolmogorov random string w of length n for a large enough n. Consider the computation of M on ww^R. A *crossing sequence* associated with a tape square consists of the sequence of states the finite control is in when the tape head crosses the intersquare boundary between this square and its left neighbor. If $c.s.$ is a crossing sequence, then $|c.s.|$ denotes the length of its description. consider an input of $w0^{n/3}w^R$ of length n. Divide the tape segment containing the input into three equal length segments of size $n/3$. If each crossing sequence associated with a square in the middle segment is longer than $\frac{n}{10|M|}$ then M spent $\Omega(n^2)$ time on this input. Otherwise there is a crossing sequence of length less than $\frac{n}{10|M|}$. Assume that this occurs at c_0. Now this crossing sequence requires at most $n/10$ bits to encode. W.l.o.g. assume c_0 is left of the middle. Using this crossing sequence, we re-construct w as follows. For every string $x0^{n/3}x^R$ of length n, put it on the input tape and start to simulate M. Each time when the head reaches c_0 from the left, we take the next element in the crossing sequence to skip the computation of M when the head is on the right of c_0 and resume the simulation starting from the time when the head moves back to the left (or on) c_0 again. If the simulation ends consistently, *i.e.* every time the head moves to c_0 the current status of M is consistent with that specified in the crossing sequence, then $w = x$. Otherwise, if $w \neq x$ and the crossing sequences are consistent in both computations, then M accepts a wrong input $x0^{n/3}w^R$. However this implies

$$K(w) < |c.s.| + O(\log n) < n,$$

contradicting $K(w) \geq n$. $\square$

Example 2 (Proving Languages Non-Regular, Li-Vitányi).
The classic introduction to formal language theory is [23]. An important part of formal language theory is deriving a hierarchy of language families. The main division is Chomsky hierarchy, with regular languages, context-free languages, context-sensitive languages and recursively enu-

merable languages. The common way to prove that certain languages are not regular [not context-free] is by using "pumping" lemma's, i.e., the uvw-lemma [$uvwxy$-lemma]. However, these lemma's are complicated to state and cumbersome to prove or use. In contrast, below we show how to replace such arguments by simple, intuitive and yet rigorous, Kolmogorov complexity arguments. We present some material from our paper [40]. W.l.o.g., languages are infinite sets of strings over a finite alphabet.

Regular languages coincide with the languages accepted by finite automata (FA). Another way of stating this is by the Myhill-Nerode Theorem: each regular language over alphabet V consists of the union of some equivalence classes of a right-invariant equivalence relation on V^* $(= \bigcup_{i \geq 0} V^i)$ of finite index. Let us give an example of how to use Kolmogorov complexity to prove non-regularity. We prove that $\{0^k 1^k : k \geq 1\}$ is not regular. To the contrary, suppose it is regular. Fix k with $K(k) \geq \log k$, with k large enough to derive the contradiction below. The state q of the accepting FA after processing 0^k is, up to a constant, a description of k. Namely, by running the FA, starting from state q, on a string consisting of l's, it reaches its first accepting state precisely after k l's. Hence, there is a constant c, depending only on FA, such that $\log k < c$, which is a contradiction. We generalize this observation, actually a Kolmogorov-complexity interpretation of the Myhill-Nerode Theorem, as follows. (In lexicographic order short strings precede long strings.)

Lemma (KC-Regulartay). *Let L be regular. Then for some constant c depending only on L and for each string x, if y is the nth string in the lexicographical order in $L_x = \{y : xy \in L\}$ (or in the complement of L_x) then $K(y) \leq K(n) + c$.*

Proof. Let L be a regular language. A string y such that $xy \in L$, for some x and n as in the lemma, can be described by

(a) This discussion, and a description of the FA that accepts L,

(b) The state of the FA after processing x, and the number n.   □

The KC-reularity lemma can be applied whenever the pumping lemma can be applied. It turns out that the converse of our Lemma also holds, and we have a Kolmogorov complexity *characterization* of regular languages [40]. This characterization can be used to prove the nonregularity of *all* nonregular languages, that is, it applies to situations when the normal pumping lemma(s) do(es) not apply. Further it is easier and more intuitive than pumping lemmas.

Similar general lemmas can also be proved to separate DCFL's from CFL's. Notice that there are no pumping lemmas for this purpose. Previous proofs that a CFL is not a DCFL all use *ad hoc* methods. We refer the interested readers to [40].

Example 3 (Boolean Matrix Rank, Seiferas-Yesha). This example is due to J.Seiferas and Y.Yesha [personal communication]. For all n, there is an n by n matrix over GF(2) (a matrix with zero-one entries with the usual Boolean multiplication and addition) such that every submatrix of s rows and $n - r$ columns ($r, s \leq n/4$) has at least $s/2$ linear independent rows.

Remark. Combined with the results in [4] and [66] this example implies that $TS = \Omega(n^3)$ optimal lower bound for Boolean matrix multiplication on any general random access machines, where T stands for time and S stands for space.

Proof. Fix a random sequence x of elements in GF(2) (zeros and ones) of length n^2, so $K(x) \geq |x|$. Arrange the bits of x into a matrix M, one bit per entry in, say, the rowmajor order, We claim that this matrix M satisfies the requirement. To prove this, suppose this is not true. Then consider a submatrix of M of s rows and $n - r$ columns, $r, s \leq n/4$. Suppose that there are at most $\frac{s}{2} - 1$ linearly independent row. Then $1 + \frac{s}{2}$ rows can be expressed by the linear combination of other $\frac{s}{2} - 1$ rows. Thus we can describe this submatrix using

- The $\frac{s}{2} - 1$ linear independent rows, in $(\frac{s}{2} - 1)(n - r)$ bits;

- for each of the other $\frac{s}{2} + 1$ rows, use $(\frac{s}{2} - 1)2$ bits.

Then to specify x, we only need to specify in addition to above M without the deleted columns and rows and the indices of the deleted columns and rows. When we list the indices of the deleted rows, we list the $\frac{s}{2} - 1$ linearly independent rows first. Hence we only use

$$n^2 - (n-r)s + (n-r)\log n + s \log n + (\frac{s}{2} - 1)(n - r) + (\frac{s}{2} - 1)(\frac{s}{2} + 1) < n^2$$

bits, for large n's. This contradicts to the fact $K(x) \geq |x|$. $\square$

Example 4. We prove that a $k+1$ head 1-way DFA accepts a language which cannot be accepted by a k head 1-way DFA. The problem was first raised by Rosengerg in 1965, who also claimed a wrong solution. Sudborough and later Ibarra and Kim [24] obtained a partial solution to the problem for the case of 2 heads versus 3 heads, with difficult proofs. In the 1976 Yao and Rivest [65] and Nelson [45] presented independent full solutions by counting arguments. Here we present an almost trivial proof using Kolmogorov complexity.

Proof. Let

$$L_b = \{w_1 \# \cdots \# w_b \$ w_b \# \cdots \# w_1 : w_i \in \{0,1\}^*\}.$$

as defined by Rosenberg and Yao-Rivest. Let $b = \binom{k}{2} + 1$. So L_b can be accepted by a $(k+1)$-DFA. Assume that a k-FA M also accepts L_b. Let W be a long enough Kolmogorov random string and W be equally partitioned into $w_1 w_2 \cdots w_b$. We say that the 2 w_i's in L_b are matched if there is a time such that 2 heads of M are in the 2 w_i's concurrently. Hence there is an i such that w_i is not matched. Then apparently, this w_i can be generated from $W - w_i$ and the positions of heads and states for M when a head comes in/out w_i, $K(w_i \mid W - w_i) = O(k \log n) < |w_i|/2$, contradiction. $\qquad\qquad\square$

Remark. A lower bound obtained by Kolmogorov complexity usually implies that the lower bound holds for "almost all strings". This is the case for all three examples. In this sense the lower bounds obtained by Kolmogorov complexity are usually stronger than those obtained by its counting counterpart, since it usually also implies directly the lower bounds for nondeterministic or probabilistic versions of the considered machine.

3.6.2   Three More Examples

In this section we present three more sophisticated proofs using Kolmogorov complexity. The first, due to Seiferas, greatly simplifies and earlier existing proof. The second proof would be formidable or impossible without such a tool as Kolmogorov complexity. The Kolmogorov complexity helped to obtain better results in the third example.

Example 1. It is known that linear context-free languages can be recognized on-line by a one work tape Turing machine in $O(n^2)$ time. This is due to Kasami. Gallaire using a very complicated counting argument and *de Bruijn sequences* [16] proved

Theorem.   *A multitape Turing machine requires $\Omega(N^2/\log n)$ time to on-line recognize linear context-free languages.*

The following elegant and short proof is due to Seiferas [55] who, using Kolmogorov complexity, significantly simplified Gallaire's proof.

Proof [Seiferas].

Assume M accepts L in $\Omega(n^2/\log n)$ steps on-line. Choose y such that $y = \lfloor n/2 \rfloor$ and $K(y) \geq |y|$. We specify a hard input of length n for M using y.

The idea of the proof is to construct an input prefix

$$y \# x_1 @ \cdots @ x_{i-1} @$$

such that no x_l is a reverse of a substring of y and yet each x_l is hard enough to make M run ϵn steps, where $\epsilon > 0$ does not depend on n. Now if $i = O(n/\log n)$, M is forced to run $\Omega(n^2/\log n)$ steps. Our task is to search for such x_l's. We first prove two easy lemmas.

Lemma 1. If $K(x) \geq |c|$, then, x has no repetition longer than $3 \log_2 |x|$.

Lemma 1 is Fact 3 proved in Section2.

Lemma 2. If a string has no repetition of length m, then it is uniquely determined by the set of its substrings of length $m + 1$; *i.e.*, then it is unique string with no repetition of length m and with precisely its set of substrings of length $m + 1$.

Proof of Lemma 2. Let S be the set of substrings of x of length $m + 1$. Let $a, b \in \{0, 1\}$, and $u, v, w \in \{0, 1\}^*$. The prefix of x of length $m + 1$ corresponds uniquely to the $ua \in S$ such that for no b is bu in S. For any prefix vw of x with $|w| = m$, there is a unique b such that $wb \in S$. Hence, the unique prefix of length $|vw| + 1$ is vwb. Hence the lemma by induction.   □

We continue our proof of the theorem. Assume inductively that $x_1, \cdots, x_{i-1}$ have been chosen as required, so that the input prefix

$$y \# x_1 @ \cdots @ x_{i-1} @$$

does not yet belong to L. By Lemmas 1 and 2 we let $m = 3 \log_2 n$ so that y is uniquely determined by its set of substrings of length m. We claim that, for each $i = 1, 2, \cdots, n/m$ there is an x_i of length m that is not a reverse substring of y such that appending $x_i @$ to the input requires at least $t = \epsilon n$ many additional steps by M, where $\epsilon > 0$ does not depend on n. This proves the Theorem. Assume that the claim is not true for some i. Then we can devise a short description of all the length m substrings of y, and hence, by Lemma 2, of y. Simulate M with input $y \# x_1 @ \cdots @ x_{i-1} @$ and record the following information at the time t_0 when M reads last @ sign

- this discussion,

- work tape contents within distance t of the tape heads,

- specification of M, length n, current state of M, location of M's heads.

With this information, one can easily search for all x_i's such that $|x_i| = m$ and $x_i{}^R$ is a substring of y as follows. Run M from time t_0, using above information and with input suffix $x_i @$, for t steps. By assumption, if M accepts or uses more than t steps, then x_i is a reverse substring of y. If ϵ is small in terms of M, and if n is large, then all above information adds up to fewer than $\lfloor n/2 \rfloor$ bits, a contradiction. $\square$

Example 2. Consider k-tape Turing machines, with k (work) tapes, apart from a separate input tape and (possible) a separate output tape. It has been known for over twenty years that a 1-tape Turing machine can simulate 2- (or k-) tape linear time Turing machines in $O(n^2)$ time. It was not known if this upper bound is tight. We emphasize that, for each $k \geq 1$, such k-tape Turing machines are far more powerful than the model of Example 2, where the single tape is both input tape and work tape. E.g., a 1-(work) tape Turing machine can recognize the palindromes of Example 2 in real-time $T(n) = n$ in contrast with $T(n) = \Omega(n^2)$ required in Example 2. This makes the problem very hard. In 1963, Rabin has shown that two tapes are better than one for real time computation. Aanderaa extended this result to $k + 1$ versus k. Paul, Seiferas, Simon using Kolmogorov complexity simplified this proof [48].

In 1982 Paul [49], using Kolmogorov complexity, extended this result to: on-line simulation of real-time $k + 1$-tape Turing machines by k-tape Turing machines requires $\Omega(n(\log n)^{1/(k+1)})$ time. For on-line simulation of two tapes by one tape this lower bound was improved to $\Omega(n \log n)$ in [13]. This work decreased the gap for one tape versus two tapes only slightly. But in later development w.r.t. this problem Kolmogorov complexity has been very successful. The second author, not using Kolmogorov complexity, reported in [62] a $\Omega(n^{1.5})$ lower bound on the time to simulate a single pushdown store on-line by one *oblivious* tape unit. (A Turing machine is oblivious if its head motions are the same for each input of length n.) However, using Kolmogorov complexity the technique worked also without the oblivious restriction, and yielded in quick succession [63,64], and the optimal results cited hereafter. Around 1983/1984, independently and in chronological order [1], Wolfgang Maass at UC Berkeley, the first author at Cornell and the second author at CWI Amsterdam, obtained a square lower bound on the time to simulate two tapes by one tape (deterministically), and thereby closed the gap between 1 tape versus k (w.l.o.g.2) tapes. All three relied on Kolmogorov complexity, and actually proved more in various ways.

We will show how the argument goes. For the sake of readability, we present a weaker form of the theorem and suggest only very serious readers to read this.

Theorem. *It requires* $\Omega(n^{1.5}/\log n)$ *time to deterministically simulate a 2-tape linear time TM by a 1-tape TM.*

Before we proceed to prove the theorem, we need to prove a useful lemma. Consider a 1-tape TM M. Call M's input tape head h_1 and work tape head h_2. Let x_i be a block of input of a 1-tape TM M, and

---

[1] *Historical note.* A claim for an $\Omega(n^{2-\epsilon})$ lower bound for simulation of two tapes by both one deterministic tape and one nondeterministic tape was first circulated by W.Maass in August 1983, but did not reach Li and Vitányi. Maass submitted his extended abstract containing this result to STOC by November 1983, and this did not reach the others either. The final STOC paper of May 1984 (submitted February 1984) contained the optimal $\Omega(n^2)$ lower bound for the deterministic simulation of two tapes by one tape. In M. Li: 'On 1 tape versus 2 stacks', Tech. Rept. TR-84-591, Dept. Comp. Sci., Cornell University, January 1984, the $\Omega(n^2)$ lower bound was obtained for the simulation of two pushdown stores by one deterministic tape. In: P.M.B. Vitányi, 'One queue or two pushdown stores take square time on a one-head tape unit,' Tech. Rept. CS-R8406, Centre for Mathematics and Computer Science, Amsterdam, March 1984, the $\Omega(n^2)$ lower bound was obtained for the simulation of two pushdown stores (or the simulation of *one* queue) by one deterministic tape. Maass's and Li's result were for off-line computation with one-way input, while Vitányi's result was for on-line computation. Li and Vitányi combined these and other results in [38], while Maass's result appeared in [44].

R be a tape segment on its work tape. We say that M *maps x_i into R* if h_2 never leaves tape segment R while h_1 is reading x_i. We say M maps x_i *onto* R if h_2 traverses the *entire* tape segment R while h_1 reads x_i.

We prove an intuitively straightforward lemma for one-tape machines with one-way input. The lemma states that a tape segment bordered by short c.s.'s cannot receive a lot of information without losing some.Formally:

Jamming Lemma.*Let the input string start with $x\# = x_1 x_2 \cdots x_k \#$, with the x_i's blocks of equal length C. Let R be a segment of M's storage tape and let l be an integer such that M maps each block $x_{i_1}, \cdots, x_{i_l}$ (of the x_i's) into tape segment R. The contents of the storage tape of M, at time $t_\#$ when $h_1(t_\#) = |x\#|$ and $h_1(t_\# - 1) = |x|$, can be reconstructed by using only the blocks $x_{j_1}, \cdots, x_{j_{k-1}}$ which remain from $x_1 \cdots x_k$ after deleting blocks $x_{i_1}, \cdots, x_{i_l}$, the final contents of R, the two final c.s.'s on the left and right boundaries of R, a description of M and a description of this discussion.*

Remark. Roughly speaking, if the number of missing bits $\sum_{j=1}^{l} |x_{i_j}|$ is greater than the number of added description bits $(< 3(|R| + 2|c.s.|) + O(\log |R|))$ then the Jamming Lemma implies that either $x = x_1 \cdots x_k$ is not incompressible or some information about x has been lost.

Proof of the Jamming Lemma. Let the two positions at the left boundary and the right boundary of R be l_R and r_R, respectively. We now simulate M. Put the blocks x_j of $x_{j_1} \cdots x_{j_{k-1}}$ in their correct positions on the input tape (as indicated by the h_1 values in the c.s.'s). Run M with h_2 staying to the left of R. Whenever h_2 reaches point l_R, the left boundary of R, we interrupt M and check whether the current ID matches the next ID, say ID_i, in the c.s. at l_R. Subsequently, using ID_{i+1}, we skip the input up to and including $h_1(t_{i+1})$, adjust the state of M to $M(t_{i+1})$, and continue running M. After we have finished left of R, we do the same thing right of R. At the end we have determined the appropriate contents of M's tape, apart from the contents of R, at $t_\#$ (i.e., the time when h_1 reaches $\#$). Inscribing R with its final contents from the reconstruction description gives us M's storage tape contents at time $t_\#$. Notice that although there are many unknown x_i's, they are never polled since h_1 skips over them because h_2 never goes into R. $\square$

Proof of Theorem. The witness language L is defined by:

$$L = \{x_1 @ x_2 @ \cdots @ x_k \# y_1 @ \cdots @ y_l \# (1^i, 1^j) : x_i = y_j\}. \qquad (1)$$

Obviously, L can be accepted in linear time by a 2-tape machine. Assume, by way of contradiction, that a deterministic 1-tape machine M accepts L in $T(n) \notin \Omega(n^{1.5}/\log n)$ time. We derive a contradiction by showing that some incompressible string must have a too short description.

Assume, without loss of generality, that M writes only 0's and 1's in its storage squares and that $|M| \in O(1)$ is the number of states of M. Fix a constant C and the word length n as large as needed to derive the desired contradictions below and such that the formulas in the sequel are meaningful.

First, choose an incompressible string $x \in \{0,1\}^*$ of length $|x| = n$ (i.e., $K(x) \geq n$). Let x consist of the concatenation of $k = \sqrt{n}$ substrings, $x_1, x_2, \cdots, x_k$, each substring $\sqrt{n}$ bits long. Let

$$x_1 @ x_2 @ \cdots @ x_k \#$$

be the initial input segment on M's input tape. Let time $t_\#$ be the step at which M reads $\#$. If there are more than $k/2$ of the x_i's each mapped *onto* a contiguous tape segment of size at least n/C^3, then M requires $\Omega(n^{1.5}/\log n)$ time, which is a contradiction. Therefore, there is a set X of $k/2$ x_i's such that each of them is mapped *into* some tape segment of $\leq n/C^3$ contiguous tape cells. In the remainder of the proof we restrict attention to the x_i's in this set X. Order the elements of X according to the order of the left boundaries of the tape segments *into* which they are mapped. Let x_c be the median.

The idea of the rest of proof is as follows. Our first intuition tells us: M could only somehow "copy" x_i's on its work tape, and then put y_j's on the work tape. There must be a pair of x_i and y_j that are separated by $O(n)$ distance since these blocks must occupy $O(n)$ space. Then if we ask M to check whether $x_i = y_j$, it has to spend about $\Omega(n^{1.5}/\log n)$ time. We convert this intuition into two cases: In the first case we assume that many x_i's in X are mapped (jammed) into a small tape segment R; that is, when h_1 (the input tape head) is reading them, h_2 (the storage tape head) is always in this small tape segment R.

Intuitively, some information must have been lost. We show that then, contrary to assumption, x can be compressed (by the Jamming Lemma). In the second case, we assume there is no such 'jammed' tape segment, and that the records of the x_i's in X are "spread evenly" over the storage tape. In that case, we will arrange the y_j's so that there exists a pair (x_i, y_j) such that $x_i = y_j$ and x_i and y_j are mapped into tape segments that are far apart, i.e. of distance $O(n)$. Then we complete M's input with final index $(1^i, 1^j)$ so as to force M to match x_i against y_j. Now as in example 1 of the previous subsection, either M spends too much time or we can compress x again, yielding a second contradiction and therefore $T(n) \in \Omega(n^{1.5}/\log n)$.

Case 1 (jammed). Assume there are k/C blocks $x_i \in X$ and a fixed tape segment R of length n/C^2 on the work tape such that M maps all of these x_i's into R. Let X' be the set of such blocks.

We will construct a short program which prints x. Consider the two tape segments of length $|R|$ to the left and to the right of R on the storage tape. Call them R_l and R_r, respectively. Choose positions p_l in R_l and p_r in R_r with the shortest c.s.'s in their respective tape segments. Both c.s.'s must be shorter than $\sqrt{n}/(C^2 \log n)$, for if the shortest c.s. in either tape segment is at least $\sqrt{n}/(C^2 \log n)$ long then M uses $\Omega(n^{1.5}/\log n)$ time: contradiction. Let tape segment $R'_l(R'_r)$ be the portion of $R_l(R_r)$ right (left) of $p_l(p_r)$.

Now, using the description of

- this discussion (including the text of the program below) and simulator M in $O(1)$ bits,

- the values of n, k, C, and the positions of p_l, p_r in $O(\log n)$ bits,

- $\{x_1, \cdots, x_n\} - X'$ using at most $n - n/C + O(\sqrt{n}\log n)$ bits; $O(\sqrt{n}\log n)$ bits are used for indicating indices,

- the state of M and the position of h_2 at time $t_\#$ in $O(\log n)$ bits,

- the two c.s.'s at positions p_r and p_l at time $t_\#$ in at most $2\sqrt{n}(|M|+ O(\log n))$ bits, and

- the contents at time $t_\#$ of tape segment $R'_l R R'_r$ in at most $3n/C^2 + O(\log n)$ bits, we can construct a program to check if a string y equals x by running M as follows.

Check if $|y| = |x|$. By the Jamming Lemma (using the above information as related to M's processing of the initial input segment $x_1@\cdots@x_k\#$) reconstruct the contents of M's work tape at time $t_\#$, the time h_1 gets to the first $\#$ sign. Divide y into k equal pieces and form $y1@\cdots@y_k$. Simulate M, started on the input suffix, for each i,

$$y_1@\cdots@y_k\#(1^i,1^i)$$

from time $t_\#$ onwards. By definition of L we have that M accepts for all i if and only if $y = x$.

This description of x requires not more than

$$n - \frac{n}{C} + \frac{3n}{C^2} + O(\sqrt{n}\log n) + O(\log n) \leq \gamma n$$

bits, for some positive constant $\gamma < 1$ and large enough C and n. However, this contradicts the incompressibility of $x(K(x) \geq n)$.

Case 2 (not jammed). Assume that for each fixed tape segment R, with $|R| = n/C^2$, there are at most k/C blocks $x_i \in X$ mapped into R.

Fix a tape segment of length n/C^2 into which median x_c is mapped. Call this segment R_c. So at most k/C strings x_i in set X are mapped into R_c. Therefore, for large enough $C(C > 3)$, at least $k/6$ of the x_i's in X are mapped into the tape right of R_c. Let the set of those x_i's be $X_r = \{x_{i_1}, \cdots, x_{i_{k/6}}\} \subset X$. Similarly, let $X_l = \{x_{j_1}, \cdots, x_{j_{k/6}}\} \subset X$, consist of $k/6$ strings x_i which are mapped into the tape left of R_c. Without loss of generality, assume $i_1 < i_2 < \cdots < i_{k/6}$ and $j_1 < j_2 < \cdots < j_{k/6}$.

Set $y_1 = x_{i_1}, y_2 = x_{j_1}, y_3 = x_{i_2}, y_4 = x_{j_2}$, and so forth. In general, for all integers $m, l \leq m \leq k/6$,

$$y_{2m} = x_{j_m} \quad and \quad y_{2m-1} = x_{i_m}, \tag{2}$$

We can now define an input prefix for M to be:

$$x_1@\cdots@x_k\#y_1@\cdots@y_{k/3}\#. \tag{3}$$

Claim. There exist a pair $y_{2i-1}@y_{2i}$ which is mapped to a segment of size less than $n/(4C^2)$.

Proof of Claim. If the claim is false then M uses $\Omega(n^{1.5}/\log n)$ time, a contradiction. $\qquad\qquad\square$

Now this pair y_{2i-1}, y_{2i} is mapped to a segment with distance at least n/C^3 either to x_{i_m} or to x_{j_m}. Without loss of generality, let y_{2m-1}, y_{2m} be mapped n/C^3 away from x_{i_m}. So y_{2m-1} and x_{i_m} are separated by a region R of size n/C^3. Attach the input with suffix $(1^{i_m}, 1^{2m-1})$ to get the complete input to M

$$x_1 @ \cdots @ x_k \# y_1 @ \cdots @ y_{k/3} \# (1^{i_m}, 1^{2m-1}). \tag{4}$$

So at time when M reads the second $\#$ sign, x_{i_m} is mapped into the left side of R and $y_{j_{2m-1}}$, which is equal to x_{i_m}, is mapped into the right side of R.

Determine a position p in R which has the shortest c.s. of M's computation on the input (4). If this c.s. is longer than $\sqrt{n}/(C^2 \log n)$ then M uses time $\Omega(n^{1.5}/\log n)$: contradiction. Therefore, assume it has length at most $\sqrt{n}/(C^2 \log n)$. Then again we can construct a short program P, to accept *only* x by a 'cut and paste' argument, and show that it yields too short a description of x.

Using the description of

- this discussion (including the text of the program P below) and simulator M in $O(1)$ bits,

- the values of $n, k, C = n/k$, and the position of p in $O(\log n)$ bits,

- $n - \sqrt{n}$ bits for $X - \{x_{i_m}\}$,

- $O(\log n)$ bits for the index i_m of x_{i_m} to place it correctly on the input tape, together with

- $\leq \sqrt{n}/C$ bits for the c.s. of length $\sqrt{n}/(C^2 \log n)$ at p (assuming $C > |M|$),

we can construct a program to check if a string z equals x by running M as follows.

For a candidate input string z, program P first partitions z into $z_1 @ \cdots @ z_k$ and compares the appropriate literal substrings with the literally given strings in $\{x_1, \cdots, x_k\} - \{x_{i_m}\}$. The string x_{i_m} is given in terms of the operation of M: to compare the appropriate z_{I_m} with the x_{i_m}, we simulate M. First prepare an input according to form (3) as follows. Put the elements of $\{x_1, \cdots, x_k\} - \{x_{i_m}\}$ literally into their correct places on the input tape, filling the places for x_{i_m} arbitrarily. For the y_i's in (3) substitute the appropriate substrings z_i of candidate z according to scheme (2) *i.e.*, use z_{j_m} for y_{2m} and z_{i_m} for y_{2m-1} $(1 \leq m \leq k/6)$. Note that among these are all those substrings of candidate z which have not yet been checked against the corresponding substrings of x. Adding string $(1^{i_m}, 1^{2m-1})$ completes the input to M.

Using the c.s. at point p we run M such that h_2 always stays right of p (y_{2m-1}'s side). Whenever h_2 encounters p, we check if the current ID matches the corresponding one in the c.s.. If it does then we use the next ID of the c.s. to continue. If in the course of this simulation process M rejects or there is a mismatch (that is, when h_2 gets to p, M is not in the same state or h_1's position is not as indicated in the c.s.), then $z \neq x$. Note, that it is possible for M to accept (or reject) on the left of P (x_{i_m}'s side). However, once h_2 crosses p right-to-left for the last time M does not read z_{2m-1} substituted for y_{2m-1} any more and all other z_i's in prefix (3) are 'good' ones (we have already checked them). Therefore, if the crossing sequence of IDs at p of M's computation for candidate z match those of the prescribed c.s. then we know that M accepts. By construction the outlined program P accepts the string $z = x$. Suppose P also accepts $z' \neq x$. Then the described computation of M accepts for candidate z'. We can cut and paste the two computations of M with candidate strings z and z' using the computation with z left of p and the computation with z' right of p. Then string (4) composed from x and z' according to (2) is accepted by M. Since z and z' differ in block x_{i_m} this string is not in L as defined in (1): contradiction.

The description of x requires not more than

$$n - \sqrt{n} + \frac{\sqrt{n}}{C} + O(\log n) \leq n - \gamma\sqrt{n}$$

bits for some positive $\gamma > 0$ and large enough C and n. This contradicts the incompressibility of x $(K(x) \geq n)$ again.

Case 1 and Case 2 complete the proof that $T(n) \in \Omega(n^{1.5}/\log n)$. $\square$

Example 3. We demonstrate that the Kolmogorov complexity does not only apply to lower bounds in restricted Turing machines, it also applies to lower bounds in other general models, like parallel computing models.

Fast addition or multiplication of n numbers in parallel is obviously important. In 1985 Meyer auf der Heide and Wigderson [22] proved, using Ramsey theorems, that on priority PRAM, where the concurrent read and write are allowed, ADDITION (and MULTIPLICATION) requires $\Omega(\log n)$ parallel steps. Independently, a similar lower bound on addition was obtained by Israeli and Moran [25] and Parberry [46]. All these lower bounds depend on inputs from infinite (or exponentially large) domains. However, in practice, we are often interested in small inputs of size $n^{O(1)}$ bits. It is known that addition of n numbers of $n^{1/\log\log n}$ bits each can be done in $O(\log n/\log\log n)$ time with $n^{O(1)}$ processors which is **less** than the $\Omega(\log n)$ lower bound of [22]. In [36] Kolmogorov-complexity was applied to obtain parallel lower bounds (and tradeoffs) for a large class of functions with arguments in small domains (including Addition, Multiplication ...) on priority PRAM. As a corollary, for example, we show that for numbers of $n^{O(1)}$ bits, it takes $\Omega(\log n)$ parallel steps to add them. This improved the results of [22,25,46]. We next present this rather simple proof. Independently, Paul Beame obtained similar results, but using a different partition method. A proof of the above result was given in Example 2.

Formally, a priority PRAM consists of processors $P(i)$ $i = 1, 2, \cdots,$ $n^{O(1)}$, and an infinite number of shared memory cells $C(i), i = 1, 2, \cdots$. Each step of the computation consists of three parallel phases as follows. Each processor: (1) reads from a shared memory cell, (2) performs a computation, and (3) may attempt writing into some shared memory cell. In case of *write conflicts*, the processor with the minimum index succeeds in writing.

Theorem. *Adding n integers, each of $n^{O(1)}$ bits, requires $\Omega(\log n)$ parallel steps on a priority PRAM.*

Proof[36]. Suppose that a priority PRAM M with $n^{O(1)}$ processors adds n integers in $O(\log n)$ parallel steps for infinitely many n's. The programs (maybe infinite) of M can be encoded into an oracle A. The oracle, when queried about (i, l), returns the initial section of length l of the program for $P(i)$. Fix a string $X \in \{0,1\}^{n^3}$ such that $K^A(X) \geq |X|$. Divide X equally into n parts $x_1, x_2, \cdots, x_n$. Then consider the (*fixed*) computation of M on input $(x_1, x_2, \cdots, x_n)$. We inductively define (with respect to X) a processor to be *alive* at step t in this computation if

(1) it writes the output; or

(2) it succeeds in writing something at some step $t' \geq t$ which is read at some step $t'' \geq t'$ by a processor who is alive at step t''.

An input is *useful* if it is read at some step t by a processor alive at step t. By simple induction on the step number we have: for a T step computation, the number of useful inputs and the number of processors ever alive are both $O(2^T)$.

It is not difficult to see that, given all the useful inputs and the set $ALIVE = \{(P(i), t_i) : P(i)$ was alive until step $t_i > 0\}$, we can simulate M to uniquely reconstruct the output $\sum_{i=1}^n x_i$. Since $T = o(\log n)$, we know $2^T = o(n)$. Hence there is an input x_{i_0} which is not useful. We need $O(2^T \log P) = o(n \log n)$ bits to represent $ALIVE$. To represent $\{x_i : i \neq i_0\}$ we need $n^3 - n^2 + \log n$ bits, where $\log n$ bits are needed to indicate the index i_0 of the missing input. The total number bits needed in the simulation is less than

$$J = n^3 - n^2 + O(n \log n) + O(\log n) < n^3$$

But from these J bits we can find $\sum_{i=1}^n x_i$ by simulating M using the oracle A, and then reconstruct x_{i_0} from $\sum_{i=1}^n x_i$ and $\{x_i : i \neq i_0\}$. But then

$$K^A(X) \leq J < n^3.$$

This contradicts the randomness of X. $\qquad\qquad\qquad\qquad\qquad\square$

3.6.3   Comparisons and Open Problems

In order to help the readers solve their own problems using appropriate tool, we present a problem in combinatorics together with three proofs: one by counting, one by probabilistic argument, one by Kolmogorov complexity. The problem and first two proofs are taken from a beautiful book by Erdös and Spencer [14].

A *tournament* T is a complete directed graph. *I.e.* for each pair of vertex u and v in T, exactly one of edges $(u, v), (v, u)$ is in the graph. Given a tournament T of n nodes $\{1, \cdots, n\}$, fix any standard effective coding, denote by $c(T)$, using $n(n-1)/2$ binary bits, one bit for each edge. The bit of edge (u, v) is set to 1 iff $u < v$. The next theorem and the first two proofs are from the first example in [14].

Theorem. *If $v(n)$ is the largest integer such that every tournament*

on $\{1, \cdots, n\}$ *contains a transitive subtournament on* $v(n)$ *players, then*
$v(n) \le 1 + \lceil 2 \log_2 n \rceil$.

Remark. This theorem was proved first by P.Erdös and L. Moser in 1964. P.Stearns showed by induction that $v(n) \ge 1 + \lceil \log_2 n \rceil$. Υ *Proof by Counting.* Let $v = 2 + \lceil 2 \log_2 n \rceil$. Let $\Gamma = \Gamma_n$ be the class of all tournaments on $\{1, \cdots, n\}$, $\Gamma' =$ the class of tournaments on $\{1, \cdots, n\}$ that do contain a transitive subtournament on v players. Then

$$\Gamma' = \bigcup_A \bigcup_\sigma \Gamma_{A,\sigma}$$

where $A \subseteq \{1, \cdots, n\}, |A| = v, \sigma$ is a permutation on A, and $\Gamma_{A,\sigma}$ is the set of T such that $T \mid A$ is generated by σ. If $T \in \Gamma_{A,\sigma}$, the $\binom{v}{2}$ games of $T \mid A$ are determined. Thus

$$|\Gamma_{A,\sigma}| = 2^{\binom{n}{2} - \binom{v}{2}}$$

and by elementary estimates

$$|\Gamma'| \le \sum_{A,\sigma} 2^{\binom{n}{2} - \binom{v}{2}} = \binom{n}{v} v! 2^{\binom{n}{2} - \binom{v}{2}} < 2^{\binom{n}{2}} = |\Gamma|.$$

Thus $\Gamma - \Gamma' \neq \Phi$. That is , there exists $T \in \Gamma - \Gamma'$ not containing a transitive subtournament on v players. $\square$

Proof by Probabilistic Argument. Let $v = 2 + \lceil 2 \log_2 n \rceil$. Let $\Gamma = \Gamma_n$ be the class of all tournaments on $\{1, \cdots, n\}$. Let also $A \subseteq \{1, \cdots, n\}, |A| = v, \sigma$ be a permutation on A. Let $\mathbf{T} = T_n$ be a random variable. Its values are the members of Γ where for each $T \in \Gamma, P_r(\mathbf{T} = T) = 2^{-\binom{n}{2}}$. That is, all members of Γ are equally probable values of $\mathbf{T}$. Then

$$Pr(\mathbf{T} \text{ contains a transitive subtournament on } v \text{ players})$$

$$\le \sum_A \sum_\sigma P_r(\mathbf{T} \mid A \text{ generated by } \sigma)$$

$$= \binom{n}{v} v! 2^{-\binom{v}{2}} < 1.$$

Thus some value T of $\mathbf{T}$ does not contain a transitive subtournament on v players. $\square$

Proof by Kolmogorov Complexity. Fix $T \in \Gamma_n$ such that $K(c(T) \mid n, v(n)) \geq |c(T)| = n(n-1)/2$. Suppose $v(n) = 2 + \lceil 2 \log_2 n \rceil$ and let S be the transitive tournament of $v(n)$ nodes. We effectively re-code $c(T)$ as follows in less than $|c(T)|$ bits, and hence we obtain a contradiction, by:

- List in order of dominance the index of each node in S in front of $c(T)$, using $2(\lceil \log_2 n \rceil)^2 + 2(\lceil \log_2 n \rceil) + |c(T)|$ bits;

- Delete all bits from $c(T)$ for edges in between nodes in S to save $2(\lceil \log_2 n \rceil)^2 + 2(\lceil \log_2 n \rceil) + 1$ bits.

Open Questions. We list three open questions we consider to be interesting and possibly solvable by Kolmogorov complexity.

(1) Can a k head 1-way DFA accept $\{x \mid yxz\}$, where $x, y, z \in \{0,1\}^*$? [15]

(2) For Turing machine on-line simulations, are 2 heads on one (1-dimensional) tape better than two (1-dimensional) tapes each with one head?

(3) Prove tight, or $\Omega(n^{1+\epsilon})$, lower bound for simulating two tapes by one for off-line Turing machines with an extra 2-way input tape.

4　Conclusion

The covered material represents only the onset of a potentially enormous number of applications of Kolmogorov complexity in computer sciences, mathematics, and many other areas. By the examples we have discussed, if readers get the feel of how to use this general purpose tool in their own applications, this exposition would have served its purpose.

Acknowledgments

We are grateful to Greg Chaitin, Peter Gács, Leonid Levin and Ray Solomonoff for taking lots of time to tell us about the early history of our subject, and introducing us to many exciting applications. We are deeply in debt to Juris Hartmanis and Joel Seiferas who have introduced us in various ways to Kolmogorov complexity. Additional comments were provided by Donald Loveland and Albert Meyer. P. Berman, J.Seiferas, and Y.Yesha, supplied to us (and gave us permission to use) their unpublished material about prime number theorems, and Boolean matrix rank, respectively. Comments of Charles Bennett, Peter van Emde Boas, Jan Heering, Evangelos Kranakis, Ker-I Ko, Danny Krizanc, Michiel van Lambalgen, Lambert Meertens, A.Kh. Shen', Umesh Vaszirani, and A. Verashagin are gratefully acknowledged. Th. Tsantilas and J.Rothstein supplied many useful references. The Computer Science Departments as CWI, Harvard, and York have provided great support during this work.

References

[1] Barzdin', Y.M., "Complexity of programs to determine whether natural numbers not greater than n belong to a recursively enumerable set," *Soviet Math. Dokl.* 9, pp.1251–1254(1968).

[2] Bennett,C.H., "On random and hard to describe numbers," *Mathematics Tech. Rept. RC* 7483(#32272), *IBM Watson Research Center*, Yorktown Heights (May 1979). (Also in: M.Gardner, Mathematical Games, *Scientific American*, November 1979, pp.20–34.)

[3] Blumer, A., A. Ehrenfeucht, D. Haussler, and M. Warmuth, "Classifying Learnable Geometric Concepts With the Vapnik-Chervonenkis Dimension," pp. 273–282 in *Proceedings 18th ACM Symp. on Theory of Computing* (1986).

[4] Borodin, A and S. Cook, "A time-space tradeoff for sorting on a general sequential model of computation," *12th ACM Symp. on theory of Computing* (1980).

[5] Chaitin,G.J., "On the length of programs for computing finite binary sequences: statistical considerations," *J. Assoc. Comp. Mach.* 16, pp. 145–159(1969).

[6] Chaitin, G.J. "Information-theoretic limitations of formal systems," *J. Assoc. Comp. Mach* 21, pp 403–424(1974).

[7] Chaitin, G.J., "Randomness and mathematical proof," *Scientific American* 232, pp. 47–52 (may 1975).

[8] Chaitin,G.J., "A theory of program size formally identical to information theory," *J. Assoc. Comp. Mach.* 22, pp.329–340 (1975).

[9] Chaitin, G. J., "Information-theoretic characterization of recursive infinite strings," *Theor. Comput. Sci.* 2, pp.45–48(1976).

[10] Chaitin,G.J., "Toward a mathematical definition of 'life'," pp.477–498 in *The Maximal Entropy Formalism,* ed. M. Tribus, MIT Press, Cambridge, Mass.(1979).

[11] Chaitin, G.J., *Algorithmic Information Theory,* Cambridge University Press(1987).

[12] Champernowne,D.G.,"The construction of decimals normal in the scale of ten," *J. London Math. Soc.* 8, pp. 254–260(1933).

[13] Duris, P.,Z. Galil, W. Paul, and R. Reischuk, "Two nonlinear lower bounds for on-line computations",*Information and Control* 60, pp.1–11(1984).

[14] Erdös, P. and J.Spencer, *Probabilistic methods in combinatorics,* Academic Press, New York (1974).

[15] Galil, Z. and J. Seiferas, "Time-space optimal matching," *Proc. 13th ACM Symposium on Theory of Computing*(1981).

[16] Gallaire, H., "Recognition time of context-free language by on-line Turing machines",*Information and Control* 15, pp.288–295(1969).

[17] Gao, Q. and M. Li, " An application of minimum description length principle to on-line recognition of handprinted alphanumerals," *Tech. Rept.* CS-88-13, York University, Department of Computer Science, Toronto(October 1988).

[18] Gold, E. M., "Language identification in the limit", *Information and Control* 10, pp. 447–474(1967).

[19] Gács, P. "Randomness and probability - complexity of description," pp.551–555 in *Encyclopedia Statistical Sciences,* ed. Kotz-Johnson, John Wiley and Sons(1986).

[20] Gács, P. "Lecture notes on descriptional complexity and randomness,"*Manuscript, Boston University,* Cambridge, Mass. (October 1987). (Unpublished)

[21] Haussler, D., N. Littlestone, and M. Warmuth, "Expected mistake bounds for on-line learning algorithm," *Manuscript* (1988).

[22] Heide,F. Meyer auf der and A. Wigderson, "The complexity of parallel sorting," *17th ACM Symp. on Theory of computing* (1985).

[23] Hopcroft, J.E. and J.D. Ullman, Introduction to Automata Theory, Languages, and computation, *Addison-Wesley* (1979).

[24] Ibarra, O.H. and C.E. Kim, "On 3-head versus 2 head finite automata," *Acta Informatica* 4, pp. 193–200(1975).

[25] Israeli, A. and S. Moran, *Private communication.*

[26] Jeffreys, Z., *theory of Probability,* Oxford at the Clarendon Press(1961). (Third Edition).

[27] Kearns, M., M. Li, L. Pitt, and L. Valiant, "recent results on Boolean concept Learning," *in Machine Learning,* ed T. Mitchell.(To appear).

[28] Kearns, M., M. Li, L. Pitt, and L. Valiant, "On the learnability of Boolean formulae," *Proc. 19th ACM Symp. on Theory of Computing* (1987).

[29] Kemeny, J.G., "the use of simplicity in induction," *Philos. Rev.* 62, pp. 391–408 (1953).

[30] Kolmogorov, A. N., "Three approaches to the quantitative definition of information," *Problems in Information Transmission* 1(1), pp. 1–7(1965).

[31] Kolmogorov, A. N., "Combinatorial foundations of information theory and the calculus of probabilities," *Russian Mathematical Surveys* 38, pp. 29–40(1983).

[32] Laplace, P. S., *A Philosophical Essay on Probabilities,* Dover, New York(1819). (Date is original publication).

[33] Levin,L.A., "Laws of information conservation(non-growth) and aspects of the foundation of probability theory," *Problems in information Transmission* 10, pp. 206–210(1974).

[34] Levin, L. A., "Randomness conservation inequalities; information and independence in mathematical theories," *Information and Control* 61, pp. 15–37 (1984).

[35] Li, M. and P.M.B. Vitányi, "Kolmogorov Complexity and Its Applications," *in Handbook of Theoretical Computer Science,* ed. J. van Leeuwen, North-Holland. (To appear).

[36] Li, M. and Y. Yesha, "New lower bounds for parallel computation," *18th ACM Symposium on Theory of Computing* (1986).

[37] Li, M. and P.M.B. Vitányi, "Kolmogorovskaya swozhnost 'dvadsat' let spustia," *Uspekhi Math. Nauk* 43:6, pp.129–166,(In Russian; =*Russian Mathematical Surveys*)(1988).

[38] Li, M. and P.M.B. Vitányi, "Tape versus queue and stacks: The lower bounds," *Information and Computation* 78, pp. 56–85(1988).

[39] Li, M. and P.M.B. Vitányi, "Inductive Reasoning and Kolmogorov Complexity,"*in Proc. 4th IEEE Structure in Complexity Theory Conference* (1989).

[40] Li, M. and P.M.B. Vitányi, "A new approach to formal language theory by Kolmogorov complexity," in *International Conference on Automate Languages and Programming(ICALP), Lecture Notes in Computer Science, vol.* **XXX.** Springer Verlag, Berlin(1989).

[41] Li, M. and P.M.B. Vitányi, *An Introduction to Kolmogorov Complexity and Its Applications*, Addison-Wesley(To appear). (ACM Press Series "Frontiers in Computing").

[42] Littlestone, N., "Learning quickly when irrelevant attributes abound: A new linear threshold algorithm,"*Proc. 28th IEEE Symposium on Foundations of Computer Science* (1987).

[43] Loveland, D.W., "On minimal-program complexity measures," *Proceedings ACM Symposium on Theory of Computing* (1969).

[44] Maass, W., "Combinatorial lower bound arguments for deterministic and nondeterministic Turing machines,"*Trans. Amer. Math. Soc.* 292, pp.675–693(1985).

[45] Nelson, C. G., "One-way automata on bounded languages," TR14–76, *Harvard University* (July 1976).

[46] Parberry, I., "A complexity theory of parallel computation," *Ph.D. Thesis*, Warwick University(1984).

[47] Paul, W., "Kolmogorov's complexity and lower bounds," *Proc. 2nd International Conference on Fundamentals of Computation Theory* (September 1979).

[48] Paul, W.J., J.I. Seiferas, and J. Simon, "An information Theoretic approach to time bounds for on-line computation," *J. Computer and System Sciences* 23, pp. 108–126(1981).

[49] Paul, W.J., "On-line simulation of $k + 1$ tapes by k tapes requires nonlinear time," *Information and Control,* pp. 1–8(1982).

[50] Quinlan, J. and R. Rivest,"Inferring decision trees using the minimum description length principle," *Draft, MIT Lab for Computer Science,* Cambridge, Mass. (1987).

[51] Rissanen, J., "Modeling by the shortest data description," *Automatica* 14, pp. 465–471(1978).

[52] Rissanen, J., "A universal prior for integers and estimation by minimum description length," *Annals of Statistics* 11. pp.416–431(1982).

[53] Rivest, R., "Learning Decision-Lists," Unpublished manuscript, MIT Lab. for Computer Science (December, 1986).

[54] Rogers,H,Jr.,*Theory of Recursive Functions and effective computability,* McGrawHill, New York(1967).

[55] Seiferas, J., "A simplified lower bound for context-free-language recognition," *Information and Control* 69, pp.255–260(1986).

[56] Solomonoff, R.J., "A preliminary report on a general theory of inductive inference," *Tech. Rept. ZTB–138, Zator Company,* Cambridge, Mass.(November 1960).

[57] Solomonoff, R.J., "A formal theory of inductive inference, Part 1 and Part 2," *Information and Control* 7, pp. 1–22, 224–254(1964).

[58] Solomonoff, R.J., "Complexity-based induction systems: comparisons and convergence theorems," *IEEE Transactions on Information Theory IT*-24, pp. 422–432 (1978).

[59] Sudborough, I.H., "One-way multihead writing finite automata," *Information and Control* 30, pp. 1–20,(Also FOCS 1971)(1976).

[60] Valiant, L.G., "A theory of the Learnable," *Comm. ACM* 27, pp.1134–1142(1984).

[61] Valiant, L.G., "Deductive Learning," *Phil. Trans. Royal Soc. Lond. A* 312, pp.441–446(1984).

[62] Vitányi, P.M.B., "On the simulation of many storage heads by one," *Theoretical Computer Science* 34, pp. 157–168(1984).

[63] Vitányi, P.M.B., "Square time is optimal for the simulation of a pushdown store by an oblivious one-head tape unit," *Information Processing Letters* 21, pp. 87–91(1985).

[64] Vitányi, P.M.B., "An N**1.618 lower bound on the time to simulate one queue or two pushdown stores by one tape," *Information Processing Letters* 21, pp.147–152(1985).

[65] Yao, A.C.C. and R.L. Rivest, "$k+1$ heads are better than k," *J.ACM* 25, pp 337–340, (Also in *Proc.17th FOCS*(1976), pp. 67–70.)(1978).

[66] Yesha, Y., "Time-space tradeoffs for matrix multiplication and discrete Fourier transform on any general random access computer," *J. Comput. Syst. Sciences* 29, pp. 183–197(1984).

[67] Zvonkin, A.K. and L.A. Levin,"The complexity of finite objects and the development of the concepts of information and randomness by means of the Theory of Algorithms," *Russ. Math. Surv.* 25, pp.83–124(1970).

A Tale of Gradient Projection Methods in Nonlinear Programming [*]

Xiang-Yun Gui and *Ding-Zhu Du*

Institute of Applied Mathematics,
Academia Sinica, Beijing, P.R.China

Abstract

During the past 10 years, great progress has been made on gradient projection methods in nonlinear programming. For example, the 25-year old open problem about the convergence of Rosen's method has been resolved by Du and Zhang [9] and He [18] and the degenerate problem has been resolved by Gui and Du. In this paper, we survey various results on recent developement. Some open problems will be proposed.

1   Introduction

Consider an iterative method for solving the optimization problem max $\{f(x) : x \in \Omega\}$. Suppose that x_k is the current point. We need to find a direction d_k such that searching along the direction d_k, we can find a point $x_{k+1} \in \Omega$ satisfying $f(x_{k+1}) > f(x_k)$. If x_k is not on the boundary of the feasible region Ω, then the gradient $\nabla f(x_k)$ is a candidate for d_k. If x_k is on the boundary of the region Ω, then the gradient $\nabla f(x_k)$ may point to the outside of Ω and hence does not meet our requirement. In this case, how do we choose the search direction d_k? Rosen [27,28] discovered an important technique for finding d_k in this questioned case. He projected the gradient back to the feasible region. To give an explicit description, let us first consider the linearly constrained feasible region $\Omega = \{x : a_j^T x \geq b_j, \ j = 1, \cdots, m\}$ satisfying the following condition.

---

[*]This work was supported in part by the National Science Foundation of China and the President Foundation of Academia Sinica.

Regularity Condition.   For any $x \in \Omega$, a_j, $j \in J(x)$, are linearly independent, where $J(x) = \{j \,:\, a_j^T x = b_j\}$.

Denote $J_k = J(x_k)$ and $g_k = \nabla f(x_k)$. Clearly, the projection $P_{J_k} g_k$ of the gradient g_k onto the subspace $D(J_k) = \{d \,:\, a_j^T d = 0, \ j \in J_k\}$ is always an ascending feasible direction if it is not zero. Thus, in this case, $P_{J_k} g_k$ is an appropriate candidate for d_k. The trick is how to find d_k in case $P_{J_k} g_k = 0$. Denote by A_J the matrix consisting of column vectors a_j, $j \in J$, that is, $A_J = (a_j, \ j \in J)$. Let $(u_{kj}, \ j \in J_k)^T = (A_{J_k}^T A_{J_k})^{-1} A_{J_k}^T g_k$ and $u_{kh} = \max\{u_{kj} \,:\, j \in J_k\}$. Rosen obtained the following result.

Theorem 1 (Rosen [27]). Let c_k be a non-negative number. Choose

$$
d_k = \begin{cases}
P_{J_k} g_k, & \text{if } \| P_{J_k} g_k \| > c_k u_{kh}, \\
P_{J_k \backslash \{h\}} g_k, & \text{otherwise.}
\end{cases}
$$

Then, $d_k \neq 0$ iff x is not a Kuhn-Tucker point. Furthermore, if $d_k \neq 0$ then d_k is always an ascending feasible direction.

To avoid the zig-zag phenomena, Rosen chose c_k such that $0 < c_k < \sqrt{\lambda_k}$ where λ_k is the smallest eigenvalue of the matrix $A_{J_k}^T A_{J_k}$. He also gave a convergence theorem in his paper. However it did not take very long for someone to point out that there was a mistake in his proof. Since then, the convergence problem remained open for 25 years. During this period, much efforts have been made on this problem. In this paper, we start from such efforts to give a survey on various topics about Rosen's method.

2   On the Convergence Problem

2.1   Counterexample for the Case $c_k \to 0$

Since 1971, Rosen's method has been included in many textbooks and the convergence problem has also become quite well-known. However, most books mistake Rosen's method in the form of either $c_k = 0$ or $c_k = \infty$ which is just what Rosen tried to avoid. Du, Sun and Song [3] presented a counterexample for the case $c_k = 0$. It was indicated by Du and Zhang [9] that this counterexample also works for the case $c_k \to 0$. It follows that Rosen's restriction on c_k, $0 < c_k \leq \sqrt{\lambda_k}$, is not sufficient for the convergency.

2.2 ε-Active Set Strategy

The first convergent version was found by Polak [23,24]. Comparing with the original one, Polak's version is too complicated. First of all, he used the ε-active set strategy, that is, used an index set $J_\varepsilon(x_k) = \{j : a_j^T x_k - b_j \leq \varepsilon\}$ instead of $J(x_k)$. Secondly, the version contains a special procedure, which involves computing the gradient projection several times. Polak proved the convergence of his version under the following hypothesis.

ε-Hypothesis. There exists $\varepsilon > 0$ such that for every $x \in \Omega$, a_j, $j \in J_\varepsilon(x)$ are linearly independent where $J_\varepsilon(x) = \{j : a_j^T x - b_j \leq \varepsilon\}$.

Zhang showed that

Theorem 2 (Zhang [31]). ε-Hypothesis is equivalent to the regularity condition.

Zhang's work motivated from a theorem of Yue and Han. Yue and Han utilized Polak's procedure to study the reduced graient method and obtained the following result.

Theorem 3 (Yue and Han [29]). Consider the linearly constrained set $\Omega' = \{x : Ax = b, x \geq 0\}$. where A is an $m \times n$ matrix. Denote by A^I the submatrix of A with column indices in I. Then the following two statements are equivalent.

1. (Nondegenerative Hypothesis) Every basic feasible solution of Ω' is nondegenerate.

2. There exists $\varepsilon > 0$ such that for every $x \in \Omega'$, $\mathrm{rank}(A^{I_\varepsilon(x)})=m$ where $I_\varepsilon(x) = \{i : x_i > \varepsilon\}$.

   This type of theorems are very useful for application of ε-active set strategy. Du and Sun gave a generalization of the above two theorems.

Theorem 4 (Du and Sun [5]). For any linearly constrained region Ω, there exists $\varepsilon > 0$ such that for every $x \in \Omega$ and $J \subseteq J_\varepsilon(x)$, $\mathrm{rank}(A_J)=\mathrm{rank}(A_J, b_J)$ where $b_J = (b_j, j \in J)^T$.

Du deleted Polak's ε-procedure by choosing

$$
d_k = \begin{cases}
P_{J_k} g_k, & \text{if } \| P_{J_k} g_k \| \geq u_{ks}(b_s - a_s^T x_k) \geq u_{kt}, \\
P_{J_k \setminus \{h\}} g_k, & \text{otherwise}
\end{cases}
\tag{S}
$$

where

$$
\begin{aligned}
& u_{ks}(b_s - a_s^T x_k) = \max\{u_{kj}(b_j - a_j^T x_k) \ : \ j \in J_k\}, \\
& u_{kt} = \max\{u_{kj} \ : \ j \in J_k\}, \\
& h = \begin{cases}
s, & \text{if } u_{ks}(b_s - a_s^T x_k) \geq u_{kt}, \\
t, & \text{if } u_{ks}(b_s - a_s^T x_k) < u_{kt}
\end{cases}
\end{aligned}
$$

and $J_k = J_{\varepsilon_k}(x_k)$ for a positive number ε_k determined by the following procedure (Initially, choose $\varepsilon_0 > 0$).

> **begin**
> $\varepsilon := \varepsilon_{k-1}$;
>   **while** a_j, $j \in J_\varepsilon(x_k)$, are not linearly independent
>   **do** $\varepsilon := \varepsilon/2$;
> $\varepsilon_k := \varepsilon$;
> **end**.

If Ω satisfies the regularity condition, then by Theorem 2, we have that for sufficiently large k, $\varepsilon_k = \varepsilon_{k-1}$. Using this fact, Du showed the following theorem.

Theorem 5 (Du [4]). Supposed that Ω saftisfies the regularity condition and d_k is defined by (S). Then the algorithm either stops at a Kuhn-Tucker point or generates an infinite sequence whose cluster points are all Kuhn-Tucker points.

2.3   McCormick's Line Search

McCormick [25] discovered another technique for establishing the convergency. He did not apply his technique to the gradient projection method. However, the applicability is obvious. His main idea is improving line search procedures.

To give a unified convergence theorem, Zangwill introduced a point-to-set mapping F; each iteration that find x_{k+1} from x_k is viewed as taking an element x_{k+1} from the set $F(x_k)$. (See [22] for the detail.) He showed an abstract convergence theorem under the condition that F is a close mapping, that is, $x_k \to x$, $y_k \to y$ and $y_k \in F(x_k)$ imply $y \in F(x)$. However, for a constrained optimization problem, the closeness is lost when the line search procedure is stopped by a constraint. McCormick suggested to search along a broken line, that is, if the search is stopped by reaching a new constraint then do not stop there and, keeping the constraint active, find a new direction to continue the search. Ritter [26] decompose the broken line search into several line searches and applied it to the gradient projection method. He obtained the following result.

Theorem 6 (Ritter [26]). Suppose that Ω is compact and satisfies the regularity condition. Choose $c_k = 0$ if $J_k \setminus J_{k-1} \neq \emptyset$ and $+\infty$ if $J_k \setminus J_{k-1} \neq \emptyset$. Then, the algorithm either stops at a Kuhn-Tucker point or generates an infinite sequence whose cluster points are all Kuhn-Tucker points.

2.4   Final Solution

Although several convergent versions had been found, one still wanted to know what would happen when choose c_k such that $\alpha \leq c_k \leq \sqrt{\lambda_k}$ for some constant $\alpha > 0$. In fact, all the above convergent versions use some ideas different from that of Rosen's for anti-zigzagging. Thus, it was still open whether Rosen's original idea that keeps c_k away from 0 and ∞ works or not. (Allowing $c_k \to 0$ was considered as an oversight in [27].)

Zhang [33] made the first breakthrough in this direction; he showed that if choose all c_k equal to a positive constant c, then Rosen's method is convergent in three dimensional space. Soon later, Du and Zhang [9] established a more general result that if all c_k equal a positive constant c bounded above by a number depending on a'_js, then Rosen's method is globally convergent in n-dimensional space. This is the first solution for the convergence problem. In the same paper, Du and Zhang also conjectured that the upper bound on the constant c can be deleted, that is, c can be chosen to be any positive number. This conjecture was settled through several efforts including Du [14], He [18] and Du and Zhang [10]. Finally, Du, Wu and Zhang [12] showed the following result.

Theorem 7 (Du, Wu and Zhang [12]). Suppose that Ω satisfies the regularity condition. Let α and β be two positive numbers. If c_k are chosen to satisfy that

(a) if $J_k \setminus J_{k-1} \neq \emptyset$ then $c_k \leq \beta$, and
(b) if $J_k \setminus J_{k-1} = \emptyset$ then $c_k \geq \alpha$,

then Rosen's method is globally convergent, that is, the algorithm either stops at a Kuhn-Tucker point or generates an infinite sequence whose cluster points are all Kuhn-Tucker points.

They also pointed out that condition (a) is necessary since there is a counterexample for the case $c_k \to \infty$.

2.5   A Remark on Line Search Procedures

We did not specify the line search procedure in the above convergence theorems. It does not mean that any line search procedure can be used. In fact, Theorem 6 and Theorem 7 were proved under the condition that the line search procedure has the following property.

(P) Let $\{x_k\}_{k \in K}$ be a subsequence of the sequence generated by the algorithm with the considered line search procedure such that $x_k \to x^*$ as $k \to \infty$, $k \in K$. If the cluster point x^* is not a Kuhn-Tucker point then we have

(a) $x_{k+1} - x_k \to 0$ as $k \to \infty$, $k \in K$, and
(b) there exists $k' > 0$ such that for $k > k'$ and $k \in K$,

$$J(x_{k+1}) \setminus J(x_k) \neq \emptyset$$

Almost every line search procedure appearing in literature meets the above requirement. But, the following one does not.

Inexact Line Search (ILS). Let $\{\varepsilon_k\}$ be a decreasing sequence of positive numbers approaching to zero. Choose $x_{k+1} = x_k + \lambda^* d_k$ such that $f(x_{k+1}) \geq f(x) - \varepsilon_k$ for all $x = x_k + \lambda d_k \in \Omega$ with $\lambda \geq 0$.

Hu [19] showed by a counterexample that the algorithm, which is considered in Theorem 6 or 7, with **ILS** is not globally convergent. However, the algorithm considered in Theorem 5 with **ILS** is still convergent. It is interesting to notice that **ILS** is one of line seach procedures used in Rosen's original paper [27].

3 Dealing with Degeneracy

In 1981, several authorities mentioned the degenerate problem in some international conferences. In fact, when the regularity condition is not assumed, we may face a linearly dependent set of active constaints and the multipliers u_{kj} are non-uniquely determined. If multipliers are chosen arbitrarily a dropped constraint may cause a step of length zero to be taken, and since the dropped contraint is a function of x_k the algorithm is stuck at the current iterate. Gui and Du solved the problem by establishing the following results.

Let $Jm(x)$ denote the index set of a maximally independent subset of $\{a_j : j \in J(x)\}$.

Theorem 8 (Gui and Du [17]). The following statements are equivalent.

1. x is a Kuhn-Tucker point.

2. For any $Jm(x)$,

    (a) $P_{Jm(x)} \nabla f(x) = 0$, and
    (b) $\{y : A_{Jm(x)}^T A_{J(x)} y = -A_{Jm(x)}^T \nabla f(x), \, y \geq 0\} \neq \emptyset$.

3. There exists a $Jm(x)$ such that (a) and (b) hold.

Theorem 9. (b) in Theorem 8 holds iff

$$(u_j, \, j \in J^*)^T = (A_{J^*}^T A_{J^*})^{-1} A_{J^*}^T \nabla f(x) \leq 0$$

where J^* is an index set corresponding to the basis obtained by applying the dual simplex algorithm to the following linear programming

$$\min z = \sum_{j \in J(x) \setminus Jm(x)} y_j$$
$$\text{subject to } A_{Jm(x)}^T A_{J(x)} y = A_{Jm(x)}^T \nabla f(x) \text{ and } y \geq 0.$$

Furthermore, if (b) does not hold, then there exists an index $h \in J^*$ such that $u_h = \max\{u_j : j \in J^*\} > 0$ and for each $a_j, \, j \in J(x) \setminus J^*$, when it is expressed as the linear combination of those $a_i, \, i \in J^*$, $a_j = \sum_{i \in J^*} \alpha_{ij} a_i$, we have $\alpha_{hj} \geq 0$.

4   Combining with Variable Metric Methods

Combining Rosen's method with variable metric methods such as DFP method, Goldfarb [15,16] obtained an efficient algorithm for linearly-constrained optimization problems. Since the convergence of Rosen's method was open, one also did not know whether Goldfarb's algorithm converges or not. Gui (Kwei), Lai and Uu [20] and Gui and Du [17] presented some convergent versions. However, recently, one begins to pay attention to the original form since the convergence problem on Rosen's method has been resolved. Through studying the original Goldfarb's method, Du, Wu and Zhang [12] found an interesting problem on the feasibility of the search direction. Note that at each iteration of Rosen's method, at most one constraint can be drawn from the active set, that is, it is active at x_k but not active at x_{k+1}. However, if we remove the variable metric part from Goldfarb's algorithm, the search direction is chosen in the following way.

begin

$$J := J(x_k); \ (u_j, \ j \in J)^T = (A_J^T A_J)^{-1} A_J^T g_k;$$

$$b_{hh}^{-1/2} u_h = \max\{b_{jj}^{-1/2} u_j \ : \ j \in J\}; \ (b_{jj} \text{ is the } j\text{th diagonal element}$$
$$\text{of } (A_J^T A_J)^{-1}).)$$

while $\| \ P_J g_k \ \| < 0.5 b_{hh}^{-1/2} u_{hh}$ **do** set $J := J \setminus \{h\}$ and compute

$$P_J g_k \text{ and } b_{hh}^{-1/2} u_h;$$

$$d_k := P_J g_k;$$

end.

This means that the algorithm allows more than one constraint to be drawn from the active set. Goldfarb explained that the algorithm could be more efficient in this way. Surprisingly, there is no proof for the feasibility of the search direction d_k in Goldfarb's original papers. Du, Wu and Zhang [12] pointed out that proving the feasibility of d_k may be a hard problem. In fact, they only gave a proof for the case that at most two constraints are drawn from the active set.

5 Gradient Projection Algorithm Family

5.1 Reduced Gradient Method

It was almost the same time as Rosen discovered the gradient projection method, Wolfe found the reduced gradient method. The first convergent version of reduced gradient method was given by Yue and Han [29]. They employed a clever pivot. This pivot was furthermore simplified by Du, Sun and Song [7]. Recently, Du and Du [13] presented a convergent version which does not require any special pivot.

5.2 The Family

It was questioned by Yue that what relation exists beween the gradient projection method and the reduced gradient method. Du [6] solved this problem by presenting a family. In this family,the reduced gradient method is viewed as a special case of a type of gradient projection methods.

5.3 New Type of Gradient Projection Method

Motivation from the above family, Du and Sun [5] gave an algorithm with merits of two methods. The search direction is defined to be

$$d_k = P_J g_k + B_J^T v_J$$

where $J = J_{\varepsilon_k}(x_k)$, $B_J = A_J(A_J^T A_J)^{-1}$, and $v_J = (v_j, j \in J)^T$ computed by

$$v_j = \begin{cases} u_j, & \text{if } u_j > 0, \\ b_j - a_j^T x_k, & \text{otherwise,} \end{cases}$$
$$(u_j, j \in J)^T = B_J^T g_k.$$

6 On Nonlinear Constraints

When the feasible region is constrained by nonlinear inequalities, $\Omega = \{x : h_j(x) \geq 0, \ j = 1, \cdots, m\}$, where at least one $h_j(x)$ is nonlinear Rosen [28] suggested to choose d_k in the following way.

First, project the gradient g_k into the subspace $\{y : \nabla h_j(x_k)^T y = 0, \ j \in J(x_k)\}$ where $J(x_k) = \{j : h_j(x_k) = 0\}$. Then, push the gradient projection into the feasible region.

How do we push the gradient projection into the feasible region? Polak [24], Zhang [32] and Lai [21] presented several ways. They also established the convergence by using ε-active set strategy.

Du [8] found a different approach. He chose a subspace such that the gradient projection on the subspace is always feasible. It enables us to generalize some results from linear constraints to nonlinear constraints. Recently, several results of such kind have appeared. However, we will not discuss them for the shortage of space. To end this section, we indicate that it is still unkown whether a convergent version without using ε-active set strategy exists or not.

References

[1] M. Avriel, Nonlinear Programming: Analysis and Methods, Prentice-Hall, Englewood Cliffs, N.J., 1976.

[2] M.S. Bazaraa and C.M. Shetty, Nonlinear Programming: Theory and Algorithms, John Willy & Sons, New York, 1979.

[3] D.-Z. Du, J. Sun and T.-T, Song, A counterexample for Rosen's gradient projection method, Mathematical Report, Institute of Applied Mathematics, Beijing, 1980 (in Chinese).

[4] D.-Z. Du, A modification of Rosen-Polak's algorithm, *Kexue Tongbao* 28 (1983), 301–305.

[5] D.-Z. Du and J. Sun, A new gradient projection method, *Chinese Numerical Mathematics* 4 (1983), 378–386 (in Chinese).

[6] D.-Z. Du, A family of gradient projection algorithms, *Acta Mathematicae Applicatae Sinica* (English Series) 2 (1985) 1–13.

[7] D.-Z. Du, J. Sun and T.-T. Song, Simplified finite pivoting processes in the reduced gradient algorithms, *Acta Mathematicae Applicatae Sinica* 7 (1984), 142–146 (in Chinese).

[8] D.-Z. Du, A gradient projection algorithm for convex programming with nonlinear constraints, *Acta Mathematicae Applicatae Sinica* 8 (1985), 7–16 (in Chinese).

[9] D.Z. Du and X.-S. Zhang, A convergence theorem of Rosen's gradient projection method, *Mathematical Programming*, 36 (1986), 135–144.

[10] D.-Z. Du and X.-S. Zhang, Global convergence of Rosen's gradient projection method, *Mathematical Programming* 44 (1989).

[11] D.-Z. Du and X.-Z. Zhang, Notes on a new gradient projection method, System Science and Mathematical Sciences 2 (1989).

[12] D.-Z. Du, F. Wu and X.-S. Zhang, On Rosen's gradient projection methods, Annals of Operations Research, to appear.

[13] D.-Z. Du, and X.-F. Du, A convergent reduced gradient algorithm without special pivot, to appear.

[14] D.-Z. Du, Remarks on the convergence of Rosen's gradient projection method, *Acta Mathematicae Applicatae Sinica* (English Series) 3 (1987), 270–279.

[15] D. Goldfarb, Extension of Davidon's variable metric method to maximization under linear inequality and equality constraints, *SIAM J.on Applied Mathematics* 17 (1969), 739–764.

[16] D. Goldfarb and L. Lapidus, Conjugate gradient method for nonlinear programmjing problems with linear constraints, *Industrial and Engineering Chemistry Fundamentals* 7 (1968), 142–151.

[17] X.-Y. Gui and D.-Z. Du, A superlinearly convergent method to linearly constrained optimization problem under degeneracy, *Acta Mathematicae Applicatae Sinica* (English Series) 1 (1984), 76–84.

[18] G.-Z. He, Proof of convergence of Rosen's gradient projection method, *Journal of Chengdu University of Technology and Science*, 33 (1987), 55–68 (in Chinese).

[19] X.-D. Hu, A counterexample for the convergence of Rosen's algorithm with inexact line search, to appear.

[20] H.-Y. Kwei(X.-Y. Gui), F. Wu and Y.-L. Lai, Extension of a variable metric algorithm to a linearained optimization problem – a

variation of Goldfarb's algorithm, In K.B. Haley (ed.) O.R. 1978, North-Holland Publishing Company, 1979.

[21] Y.-L. Lai, An algorithm and its convergence for nonlinear constrained convex programming, *Acta Mathematcae Applicatae Sinica* 3 (1980), 322–331 (in Chinese).

[22] D.G. Luenberger, Introduction to Linear and Nonlinear Programming, Addison-Wesley, Reading, Mass., 1973.

[23] E. Polak, On the convergence of optimization algorithms, *Revue Francaise d'Informatique et de Recherche Operationelle* 3 (1969), 17–34.

[24] E. Polak, Computational Methods in Optimization, Academic Press, New York, 1971.

[25] G. P. McCormick, Anti-zigzagging by bending, *Management Sciences* 15 (1969), 315–320.

[26] K. Ritter, Convergence and superlinear convergence of algorithms for linearly constrained minimization problems, In L.C.W. Dixon, E. Spedicado and G.P. Szego(ed.), Nonlinear Optimization: Theory and Algorithms, Part II, 1980.

[27] J.B. Rosen, The gradient projection method for nonlinear programming, Part I: linear constraints, *SIAM J.Applied Mathematics* 8 (1960), 181–217.

[28] J.B. Rosen, The gradient projection method for nonlinear programming, Part II: nonlinear constraints, *SIAM J.Applied Mathematics* 9 (1961), 514–553.

[29] M. Yue and J. Han, A new reduced gradient method, *Scientia Sinica*, 22 (1979), 1099–1113.

[30] M. Yue and J. Han, A unified approach to the feasible direction methods for nonlinear programming with linear constraints, *Acta Mathematicae Applicatae Sinica* 1 (1984), 63–73.

[31] X.-S. Zhang, An improved Rosen-Polak method, *Acta Matheaticae Applicatae Sinica* 2 (1979), 257–267 (in Chinese).

[32] X.-S. Zhang, Discussion on Polak's algorithm of nonlinear programming, *Acta Mathematicae Applicatae Sinica* 4 (1981), 1–13 (in Chinese).

[33] X.-S. Zhang, On the convergence of Rosen's gradient projection method, *Acta Mathematicae Applicatae Sinica* 8 (1985), 125–128 (in Chinese) (also, English Series 3 (1987), 280–288.).

The C^1 Closing Lemma of 2-dimensional Non-singular Endomorphisms

Lan Wen

Institute of Mathematics, Peking University

Beijing, 100871, P. R. China

Abstract

A finite sequence of ε-kernel lifts is obtained with the starting point specified. This enables us to specify a particular branch of the inverse orbit of the non-wandering point, and hence to prove the C^1 closing lemma of 2-dimensional non-singular endomorphisms.

1   Introduction

There are several proofs of C^1 closing lemma for vector fields and diffeomorphisms in the literature, see the remarkable papers [2,3,4,5,6,7]. In this paper we prove the following C^1 closing lemma for 2-dimensional non-singular endomorphisms. Let M be a 2-dimensional compact C^∞ Riemannian manifold without boundary. A C^1 map $f\colon M \to M$ is a *non-singular endomorphism* if the tangent map $T_x f$ is injective at all $x \in M$. Let $NEnd^1(M)$ be the set of 2-dimensional non-singular endomorphisms endowed with the C^1 topology.

Theorem A (the C^1 closing lemma). *Let $f\colon M \to M$ be a 2-dimensional non-singular endomorphism, and σ be a non-wandering point of f. Then for any C^1 neighborhood $\mathcal{U}$ of f in $NEnd^1(M)$, there is a $g \in \mathcal{U}$ such that σ is a periodic point of g.*

Recall that $x \in M$ is *periodic of f* if $f^n x = x$ for some $n \geq 1$, and *non-wandering of f* if, for any neighborhood U of x in M, $(f^n U) \bigcap U \neq \emptyset$ for some $n \geq 1$. Let $P(f)$ be the set of periodic points of f, and $\Omega(f)$ the set of non-wandering points of f. As in the case of vector fields and diffeomorphisms, the C^1 closing lemma implies the following general density theorem.

Theorem B. *There is a residual subset $\mathcal{B}$ of $NEnd^1(M)$ such that $P(f)$ is dense in $\Omega(f)$ for every $f \in \mathcal{B}$.*

Recall that a residual subset is a countable intersection of open and dense subsets. As $NEnd^1(M)$ is Baire, $\mathcal{B}$ is dense in $NEnd^1(M)$.

In §2 we prove some selection lemmas. In §3 we prove a linear version of the C^1 closing lemma. In §4 we define two basic perturbations. In §5 and §6 we prove Theorem A and Theorem B respectively.

This work owes much to the influence of Professor S. T. Liao. To him my gratitude is beyond description. It was his concern and encouragement throughout the preparation that made this work ever possible.

2   Selection lemmas

Let $f\colon M \to M$ be a 2-dimensional non-singular endomorphism. The *negative f-orbit* of $p \in M$ is defined as $Orb^-(p) = \bigcup_{n=1}^{\infty} f^{-n}\{p\}$, where $f^{-n}\{p\}$ denotes the preimage $(f^n)^{-1}\{p\}$. A sequence $q_0, q_1, \cdots, q_n, \cdots$ is called a *branch* of $Orb^-(p)$ if $q_0 = p$, and $fq_n = q_{n-1}$ for all $n \geq 1$. Since f is a local diffeomorphism near each $q \in Orb^-(p) - \{p\}$, $\bigcup_{n=1}^{\mu} f^{-n}\{p\}$ is finite for each integer $\mu \geq 1$ as M is compact. In this case we will call a neighborhood W of p in M a *μ-dynamical neighborhood* of p if each connected component U of $\bigcup_{n=1}^{\mu} f^{-n}(W)$ is a neighborhood of a unique point $q \in \bigcup_{n=1}^{\mu} f^{-n}\{p\}$, denoted as $U = W(q)$ and called as the W-*component at* q, and if f^n maps $W(q)$ onto W whenever $f^n(q) = p$, $n = 1, 2, \cdots, \mu$. Clearly, μ-dynamical neighborhood exists for each $\mu \geq 1$.

Fix a $\varsigma > 0$ such that the exponential map embeds $T_xM(\varsigma) = \{u \in T_xM \mid |u| \leq \varsigma\}$ into M for every $x \in M$. Sometimes we will use a special metric near a particular point defined as follows. Let σ be a point of M and U be a neighborhood of σ in M such that $U \subset T_\sigma M(\varsigma)$. Define a metric d' on U by $d'(p, q) = |u - v|$, where $p, q \in U$, $u = \exp_\sigma^{-1}(p)$, $v = \exp_\sigma^{-1}(q)$. Denote $B(p, r, d')$ the closed ball with center p and d'-radius r. The following lemma is a variation of Lemma 2.1 in [3].

Lemma 2.1. *Let σ be a non-wandering point of f. For any neighborhood U of σ in M, and any $\rho > 0$, there are two points x and y in U such that:*

(1) *$f^\psi x = y$ for some $\psi \geq 1$;*

(2) *$B(y, \rho\delta, d') \subset U$, where $\delta = d'(x, y)/2$; and*

(3) *The d'-distance between any two points of $\{x, fx, \cdots, f^{\psi-1}x, y\} \cap (B(y, \rho\delta, d'))$ is greater than δ.*

Proof. Since σ is non-wandering, there are two points $q, r \in U$ such that $f^n q = r$ for some $n \geq 1$ and that $B(r, \rho d'(q, r), d') \subset U$. Denote

$$H = \{(u, v) \mid u, v \in \{q, fq, \cdots, f^{n-1}q, r\}, f^i u = v \text{ for}$$
$$\text{some } 1 \leq i \leq n, \text{ and that } B(v, \rho d'(u, v), d') \subset U\}.$$

Then H is non-empty and finite. Let (x, y) be an element in H, say $f^\psi(x) = y$ for some $\psi \geq 1$, such that

$$d'(x, y) = \min\{d'(u, v) \mid (u, v) \in H\}, \tag{A}$$

Then x, y satisfy (1) and (2). We now prove that x, y also satisfy (3).

Suppose not. Then there are two points $w, z \in \{x, fx, \cdots, f^{\psi-1}x, y\} \cap B(y, \rho d'(x, y)/2, d')$ such that $f^j w = z$ for some $1 \leq j \leq \psi$, and that

$$d'(w, z) \leq d'(x, y)/2. \tag{B}$$

Then for any point $b \in B(z, \rho d'(w, z), d')$, we have

$$\begin{aligned}
d'(b, y) &\leq d'(b, z) + d'(z, y) \\
&< \rho d'(w, z) + \rho d'(x, y)/2 \\
&\leq \rho d'(x, y)/2 + \rho d'(x, y)/2 \\
&= \rho d'(x, y).
\end{aligned}$$

Hence $B(z, \rho d'(w, z), d') \subset B(y, \rho d'(x, y), d') \subset U$. Then

$$(w, z) \in H. \tag{C}$$

But (A), (B), (C) together gives a contradiction. This proves Lemma 2.1. $\qquad\square$

Lemma 2.2. *Let σ be a non-wandering point of f. There is a branch $\Sigma = \{\sigma_0, \sigma_1, \cdots, \sigma_n, \cdots\}$ of $Orb^-(\sigma)$ with the following property: for any $\rho > 0$ and any integer $\mu \geq 1$, if U is a μ-dynamical neighborhood of σ in M, there are two points $x, y \in U$ such that*

(1) *$f^\psi x = y$ for some $\psi > \mu$;*

(2) *$B(y, \rho\delta, d') \subset U$, where $\delta = d'(x, y)/2$;*

(3) *The distance between any two points of $\{x, fx, \cdots, f^{\psi-1}x, y\} \cap B(y, \rho\delta, d')$ is greater than δ; and that*

(4) *$f^{\psi-\mu}(x)$ is in $U(\sigma_\mu)$.*

We may call such a branch Σ a *scattering branch* of $Orb^-(\sigma)$.

    Proof. Choose a sequence

$$0 < \rho_1 < \rho_2 < \cdots < \rho_n < \cdots$$

with $\rho_n \to \infty$, and a sequence of n-dynamical neighborhoods

$$U_1 \supset U_2 \supset \cdots \supset U_n \supset \cdots$$

of σ in M with $\mathrm{diam}(U_n) \to 0$. For each $n \geq 1$, take $x_n, y_n \in U_n$, say $f^{\psi_n}(x_n) = y_n$, such that x_n, y_n satisfy (1), (2), (3) respecting U_n and ρ_n by Lemma 2.1. Note that $\psi_n > n$ since x_n can never return to U_n without touching any U_n-components of $q \in f^{-n}\{\sigma\}$. Let $z_n \in f^{-n}\{\sigma\}$ be the unique point such that

$$f^{\psi_n - n}(x_n) \in U_n(z_n),$$

and denote $Z = \{\sigma, z_1, \cdots, z_n, \cdots\}$. Z may not be a branch of $Orb^-(\sigma)$.

    As $f^{-1}\{\sigma\}$ is finite, there is a point $\sigma_1 \in f^{-1}\{\sigma\}$ such that $Orb^-(\sigma_1)$ contains infinitely many points of Z. Inductively, for each $n \geq 2$, there is a $\sigma_n \in f^{-1}\{\sigma_{n-1}\}$ such that $Orb^-(\sigma_n)$ contains infinitely many points of Z. Let

$$\Sigma = \{\sigma, \sigma_1, \sigma_2, \cdots, \sigma_n, \cdots\}.$$

Then Σ is a branch of $Orb^-(\sigma)$. We now verify that Σ is a scattering branch of $Orb^-(\sigma)$.

Let ρ be any positive number, μ be any positive integer, and U be any μ-dynamical neighborhood of σ in M. As $Orb^-(\sigma_\mu)$ contains infinitely many points of Z, there is an $i > \mu$ such that $\rho_i > \rho$, $U_i \subset U$ and that

$$z_i \in Orb^-(\sigma_\mu).$$

Let

$$x = x_i, y = y_i, \psi = \psi_i.$$

Then x, y satisfy (1), (2), (3) respecting U and ρ. Moreover, as $f^{\psi-i}(x) \in U_i(z_i)$, we have

$$f^{\psi-\mu}(x) \in U_i(f^{i-\mu}(z_i)) = U_i(\sigma_\mu) \subset U(\sigma_\mu).$$

This proves that Σ is a scattering branch of $Orb^-(\sigma)$, and hence proves Lemma 2.2. $\qquad\qquad\square$

3 The linear version of the C^1 closing lemma

Let $V_0, V_1, \cdots, V_n, \cdots$ be a sequence of 2-dimensional inner product spaces, $T_n\colon V_n \to V_0$, $n \geq 1$, be a sequence of linear isomorphisms, and let $\varepsilon > 0$.

Theorem 3.1. *Given $\{T_n\}$ and $\varepsilon > 0$, there exists a number $\rho > 2$ and an integer $\mu \geq 1$ such that: for any finite ordered set $X = \{x, x_1, \cdots, x_s, y\}$ in V_0 with the property that the distance between any two points of $X \cap B(y, \rho\sigma)$ is greater than δ, where $\delta = |x-y|/2$, there is a point $w \in (X-\{y\}) \cap (B(y, \rho\delta))$, together with $\mu+1$ points $c_0, c_1, \cdots, c_\mu$ in $B(y, \rho\delta)$, not necessarily distinct, satisfying the following two conditions* (a) *and* (b).

(a) $c_0 = w$, $c_\mu = y$; *and*

(b) $|T_n^{-1}(c_n) - T_n^{-1}(c_{n+1})| \leq \varepsilon d(T_n^{-1}(c_{n+1}), T_n^{-1}(Y))$ *for* $n = 0, 1, \cdots$, $\mu - 1$, *where* T_0 *stands for the identity,*

$$Y = (X - \{w, y\}) \bigcup (\partial B(y, \rho\delta)),$$

and d is the distance on V_0.

We will reformulate Theorem 3.1 as Theorem 3.2 below and prove Theorem 3.2 instead. (Theorem 3.1 is easier to use in §5 but Theorem 3.2 is easier to prove.)

It is well known that each linear isomorphism $T\colon W \to V$ factors as PQ where $Q\colon W \to V$ is an isometry and $P\colon V \to V$ is positive definite symmetric. Hence if B is a closed ball in W, the image $T(B)$ is a (solid) ellipse in V. We will call an ellipse E with center origin and the shorter radius 1 a *standard ellipse*, and $x+tE$ an *ellipse of type E* for any $x \in V$ and $t > 0$. If $G = x+tE$ is an ellipse and $0 < a < 1$, we will simply write $x + atE$ as aG and call it the a-kernel of G. We will say that an ellipse E *contacts a closed subset K of V* if $E \cap K \neq \emptyset$, but $(int(E)) \cap K = \emptyset$.

Let $E_1, E_2, \cdots, E_n, \cdots$ be a sequence of standard ellipses in V_0, and let $\varepsilon > 0$. Denote $E_n(x, K)$ the ellipse of type E_n which has center x and contacts K.

Theorem 3.2. *Given $\{E_n\}$ and $\varepsilon > 0$, there exists a number $\rho > 2$ and an integer $\mu \geq 1$ such that: for any finite ordered set $X = \{x, x_1, \cdots, x_s, y\}$ in V_0 with the property that the distance between any two points of $X \cap B(y, \rho\delta)$ is greater than δ, where $\delta = |x - y|/2$, there is a point $w \in (X - \{y\}) \cap (B(y, \rho\delta))$, together with $\mu+1$ points $c_0, c_1, \cdots, c_\mu$ in $B(y, \rho\delta)$, not necessarily distinct, satisfying the following two conditions (a) and (b).*

(a) $c_0 = w$, $c_\mu = y$; *and*

(b) $c_n \in \varepsilon E_n(c_{n+1}, y)$

　　for $n = 0, 1, \cdots, \mu - 1$, where E_0 stands for the unit ball, and $Y = (X - \{w, y\}) \cup (\partial B(y, \rho\delta))$.

Clearly, Theorem 3.2 implies Theorem 3.1, since the sequence $\{T_n\}$ induces a sequence of standard ellipses $\{E_n\}$, and since the condition (b) in Theorem 3.2 implies the condition (b) in Theorem 3.1.

Proof. Let $\{E_n\}$ and $\varepsilon > 0$ be given. Denote b_n the *bolicity* of E_n, namely, the ratio (the longer radius of E_n)/(the shorter radius of E_n). Let

$$\beta = [4(1 + \varepsilon)^2/\varepsilon^2],$$

where $[\cdot]$ means rounding up to the closest integer,

$$r = 2 + \varepsilon/(2(1 + \varepsilon)) \sum_{n=1}^{\beta-1} b_n,$$

$$\rho = 1 + (r/\varepsilon)[4(1 + \varepsilon)/\varepsilon],$$

r be the least integer with $(1 - \varepsilon)^{r-1} \leq \varepsilon/((1 + \varepsilon)(\rho - 1))$,

$$\theta = r[8(\rho - 1)^2],$$

$$\mu = \beta + \theta.$$

Note that ρ and μ are derived only from $\{E_n\}$ and ε.

We verify that ρ, μ satisfy Theorem 3.2. Let $X = \{x, x_1, \cdots, x_s\}$ be given such that the distance between any two points of $X \cap B(y, \rho\delta)$ is greater than δ, where $\delta = |x - y|/2$. Denote

$$Z = (X - \{y\}) \bigcup \partial B(y, \rho\delta).$$

The proof is divided into two cases.

Case 1. The longer radius of $E_n(y, Z)$ is less than or equal to $(\rho - 1)\delta$ for each $n = \beta, \beta + 1, \cdots, \mu - 1$.

In this case we follow the approach of [3]. Note that each $E_n(y, z)$ contacts some point in $(X - \{y\}) \cap B(y, (\rho - 1)\delta)$, which is a set containing at most $[8(\rho - 1)^2]$ points as the distance between any two of its points is greater than δ. There are θ of these ellipses. Hence one can find r of them, $E_n(y, Z)$, $n = n_1, n_2, \cdots, n_r$, where $\beta \leq n_1 < n_2 < \cdots < n_r \leq \mu - 1$, contacting some common point in $(X - \{y\}) \cap B(y, (\rho - 1)\delta)$. Then let w be such a common point, $c_0 = w$, and let $c_{n_i} = w + (1 - \varepsilon)^{r-i}(y - w)$ for $i = 1, 2, \cdots, r$. It is ready to see that $c_{n_i} \in \varepsilon E_{n_i}(c_{n_{i+1}}, Z) \subset \varepsilon E_{n_i}(c_{n_{i+1}}, Y)$ for $i = 1, 2, \cdots, r - 1$. For the initial point $w = c_0$, note that $|c_{n_1} - c_0| = (1 - \varepsilon)^{r-1}|y - w| \leq (1 - \varepsilon)^{r-1}(\rho - 1)\delta \leq \varepsilon\delta/(1 + \varepsilon)$, and that $E_0(c_0, Y)$ is a ball of radius no less than δ as $d(w, Y) \geq \delta$. Thus $c_{n_1} \in \dfrac{\varepsilon}{1 + \varepsilon} E_0(c_0, Y)$. It is easy to see that this implies $c_0 \in \varepsilon E_0(c_{n_1}, Y)$. Now let $c_n = c_0$ for $1 \leq n < n_1$, $c_n = c_{n_i}$ for $n_i \leq n < n_{i+1}$, $i = 1, 2, \cdots, r - 1$, and let $c_n = y$ for $n_{i_r} \leq n < \mu$. This proves Theorem 3.2 in case 1.

Case 2. For some integer ι, $\beta \leq \iota \leq \mu - 1$, the longer radius of $E_\iota(y, Z)$ is greater than $(\rho - 1)\delta$.

In this case there are two subcases to consider. Let L be the closed interval on the longer radius of $E_\iota(y, Z)$ with center y and half length $r\delta[4(1 + \varepsilon)/\varepsilon]$.

Subcase 2.1. $L + B(0, \varepsilon\delta/(1 + \varepsilon))$ contains some point p in $X - \{y\}$.

In this subcase we need only two moves in total. Let q be a point on L with $|q - p| \leq \varepsilon\delta/(1 + \varepsilon)$. Then $q \in (\varepsilon/(1 + \varepsilon))E_0(p, Y)$. Hence $p \in \varepsilon E_0(q, Y)$. For the other move, note that the radius of $E_\iota(y, Z)$ is greater than $(\rho - 1)\delta = (r\delta/\varepsilon)[4(1 + \varepsilon)/\varepsilon]$. Hence $q \in \varepsilon E_\iota(y, Z) \subset \varepsilon E_\iota(y, Y)$. Therefore we can let $w = p$, $c_0 = w$, $c_1 = c_2 = \cdots = c_\iota = q$, and let $c_{\iota+1} = \cdots = c_\mu = y$. This proves Theorem 3.2 in subcase 2.1.

Subcase 2.2. $L + B(0, \varepsilon\delta/(1 + \varepsilon))$ does not intersect $X - \{y\}$.

In this subcase we will find up to β moves to get the longer radius of $E_\iota(y, Z)$, and one more move to get y. For the main part of the rest of the proof, it will be more convenient to use the condition $c_{n+1} \in (\varepsilon/(1 + \varepsilon))E_n(c_n, Y)$ which implies $c_n \in \varepsilon E_n(c_{n+1}, Y)$. Thus we let

$$a = \varepsilon/(1 + \varepsilon).$$

For any $u \in V_0$, denote $P(u)$ the union of the two lines $P_1(u)$ and $P_2(u)$ parallel to L such that $d(u, P_1(u)) = d(u, P_2(u)) = a\delta/2$. First we disregard X and consider $P(u)$ instead.

Let $u_0 = x$, u_1 be the point on the perpendicular segment from x to L such that $|u_1 - x| = a\delta$. We define u_{n+1}, $n = 1, 2, \cdots, \beta - 1$, inductively. Note that $E_n(u_n, P(u_n))$ contacts $P(u_n)$ at two points. Let t_n be the one closer to L. Then define $u_{n+1} = u_n + a(t_n - u_n)$. Clearly, $u_{n+1} \in aE_n(u_n, P(u_n))$. Now there is an integer k, $0 \leq k \leq \beta - 1$, such that the segment from u_k to u_{k+1} hits L at some point q, since $\beta = [4/a^2]$, $d(u_{n+1}, L) = d(u_n, L) - a^2\delta/2$, and since $d(u_0, L) \leq 2\delta$. Then let $z_n = u_n$ for $0 \leq n \leq k$, and $z_n = q$ for $k + 1 \leq n \leq \beta$. Clearly, $z_{n+1} \in aE_n(z_n, P(z_n))$ for $n = 1, 2, \cdots, \beta - 1$. Note that each $E_n(z_n, P(z_n))$, $n = 1, 2, \cdots, \beta - 1$, is contained in $B(y, r\delta)$. In fact, if $v \in E_n(z_n, P(z_n))$, $1 \leq n \leq \beta - 1$, then $|v - z_n| \leq b_n a\delta/2$ as $E_n(z_n, P(z_n))$ has bolicity b_n and the shorter radius no greater than $a\delta/2$. Hence

$$|v - y| \leq |v - z_n| + |z_n - z_{n-1}| + \cdots + |z_2 - z_1| + |z_1 - y|$$
$$\leq 2\delta + \sum_{i=1}^{n} b_i a\delta/2 \leq r\delta.$$

Now we recover the set X. If there is an n, $1 \leq n \leq \beta - 1$, such that $E_n(u_n, P(u_n))$ contains some point $x' \in Z = (X - \{y\}) \bigcup \partial B(y, \rho\delta)$, then x' must be in $X - \{y\}$ as $|x' - y| \leq r\delta$. In this case we change the initial point x to x', and start the above process all over again. This will give us a sequence $z'_0, z'_1, \cdots, z'_\beta$ such that $z'_0 = x'$, $|z'_1 - z'_0| = a\delta$, $z'_\beta \in L$, $z'_{n+1} \in$

$aE_n(z'_n, P(z'_n))$ for $n = 1, 2, \cdots, \beta - 1$, and that each $E_n(z'_n, P(z'_n))$ is contained in $B(y, 2r\delta)$. Note that $d(x', L) \leq d(x, L) - a\delta/2$. Thus there are at most $[4/a]$ changes of initial points possible. Also note that $L + B(0, a\delta)$ does not intersect $X - \{y\}$ in subcase 2.2. Hence up to $[4/a]$ changes of initial points we will get a point $w \in X - \{y\}$ and a sequence $c_0, c_1, \cdots, c_\beta$ in $B(y, \rho\delta)$, such that $c_0 = w$, $|c_1 - c_0| = a\delta$, $c_\beta \in L$, $c_{n+1} \in aE_n(c_n, P(c_n))$ for $n = 1, 2, \cdots, \beta - 1$, each $E_n(c_n, P(c_n))$ is contained in $B(y, r\delta[4/a])$, and that each $E_n(c_n, P(c_n))$ does not intersect Z for $n = 1, 2, \cdots, \beta - 1$. Thus

$$c_{n+1} \in aE_n(c_n, Z) \in aE_n(c_n, Y)$$

for $n = 1, 2, \cdots, \beta - 1$. For $n = 0$, $c_1 \in aE_n(c_n, Y)$ also holds since $|c_1 - c_0| = a\delta$. As noticed before, this implies that

$$c_n \in \varepsilon E_n(c_{n+1}, Y)$$

for $n = 1, 2, \cdots, \beta - 1$.

For the last move, note that $|c_\beta - y| \leq r\delta[4/a]$, and that the longer radius of $E_\iota(y, Z)$ is greater than $(\rho - 1)\delta = r\delta[4/a]/\varepsilon$. Then

$$c_\beta \in \varepsilon E_\iota(y, Z) \subset \varepsilon E_\iota(y, Y).$$

Thus we can let $c_n = y$ for $n = \beta + 1, \beta + 2, \cdots, \mu$. This proves Theorem 3.2 in subcase 2.2, and hence completes the proof of Theorem 3.2. $\quad\square$

4 Two basic perturbation lemmas

This section is devoted to two basic perturbation lemmas, which concern the ε-kernel lifts defined in Lemma 4.1 and the local linearizations defined in Lemma 4.2. These results are known and used in [1,3,4,5,7]. We summarize them here just for convenience.

For the purpose of dealing with some singularities in a later paper, we consider in this section $C^1(M)$, the set of C^1 maps of M into itself. For simplicity we assume that M is Riemann embedded into some R^d. $C^1(M)$ has then a C^1 metric d_1 inherited from $C^1(M, R^d)$ compatible with its C^1 topology.

Recall that the number $\varsigma > 0$ is fixed so that $\exp_p$ embeds $\{u \in T_pM \mid |u| \leq \varsigma\}$ into M for each $p \in M$. The following Lemma 4.1 is essentially Theorem 6.1 in [7].

Lemma 4.1. *For any $\eta > 0$, there is an $\varepsilon > 0$ such that for any $f \in C^1(M)$, any $p \in M$, and any two vectors $v_1, v_2 \in T_pM$ with $B(v_2, |v_1 - v_2|/\varepsilon) \subset \{u \in T_pM \mid |u| \leq \varsigma\}$, there is a diffeomorphism $h = h_{p,\varepsilon,v_1,v_2} : M \to M$, called an ε-kernel lift, such that:*

(1) $h(\exp_p(v_2)) = \exp_p(v_1)$;

(2) $supp(h) \subset \exp_p(B(v_2, |v_1 - v_2|/\varepsilon))$, *here the support means the closure of the set where h differs from the identity;*

(3) $d_1(hf, f) < \eta$.

Proof. See [7,p.288]. $\qquad\qquad\square$

Next we define the local linearizations. The following lemma is essentially Lemma 1.1 in [1].

Lemma 4.2. *Let $f \in C^1(M)$, $p \in M$, $\mu \geq 1$ be given such that all terms in $\bigcup\limits_{n=1}^{\mu+1} f^{-n}\{p\}$ are distinct and are not singularities. Then for any $\eta > 0$, there is a $\lambda > 0$, and a map $f_1 \in C^1(M)$, called a local linearization of f, with the following properties (1)–(5).*
Write $\bar{V}' = \{u \in T_pM \mid |u| \leq \lambda\}$, $V' = \{u \in T_pM \mid |u| \leq \lambda/4\}$, $W = \exp_p(W')$, $V = \exp_p(V')$.

(1) *W is $(\mu + 1)$-dynamical for both f and f_1, and the W-components for f and f_1 are the same, namely, $W_f(q) = W_{f_1}(q)$ for each $q \in \bigcup\limits_{n=1}^{\mu+1} f^{-n}\{p\}$;*

(2) *$f_1 = \exp_{f(q)} \cdot (T_q f) \cdot \exp_q^{-1}$ on each $V_{f_1}(q)$ for $q \in \bigcup\limits_{n=1}^{\mu} f^{-n}\{p\}$;*

(3) *$f_1^{\mu+1} = f^{\mu+1}$ on each $W(q)$ for $q \in f^{-\mu-1}\{p\}$. In particular, $f_1 = \exp_{f(q)} \cdot (T_{f(q)} f^\mu)^{-1} \cdot \exp_p^{-1} \cdot f^{\mu+1}$ on each $V(q)$ ($V_f(q) = V_1(q)$ here) for $q \in f^{-\mu-1}\{p\}$;*

(4) *$f_1 = f$ on $M - \bigcup\{W(q) \mid q \in \bigcup\limits_{n=1}^{\mu+1} f^{-n}\{p\}\}$;*

(5) *$d_1(f_1, f) < \eta$.*

Proof. Take a C^∞ bump function $\alpha : R^m \to R$ supported on the unit ball $B(0,1)$ of R^m, such that $\alpha = 1$ on $B(0,1/4)$ and has C^1 size ≤ 2. Take a $(\mu + 1)$-dynamical neighborhood U of p such that

$$U(q) \subset \exp_q(\{u \in T_q M \mid |u| \leq \varsigma\}) \text{ for each } q \in \bigcup_{n=0}^{\mu+1} f^{-n}\{p\} \text{ and take a}$$

$\lambda > 0$ such that $W \subset int(U)$.

As noticed above, $\bigcup_{n=1}^{\mu+1} f^{-n}\{p\}$ is finite. For each $q \in \bigcup_{n=1}^{\mu} f^{-n}\{p\}$, say, $f^n q = p$, the map

$$\hat{F}_q : \exp_p^{-1}(U) \to T_q M$$
$$\hat{F}_q(u) = \exp_q^{-1} \cdot (f^n \mid_{U(q)})^{-1} \cdot \exp_p(u) +$$
$$+ \quad \alpha(u/\lambda)((T_q f^n)^{-1}(u) - \exp_q^{-1} \cdot (f^n \mid_{U(q)})^{-1} \cdot \exp_p(u))$$

is a diffeomorphism of $\exp_p^{-1}(U)$ onto $\exp_q^{-1}(U(q))$ if U and λ are small. Define

$$F_q : U \to M$$
$$F_q = \exp_q \cdot \hat{F} \cdot \exp_p^{-1} .$$

Note that $F_q = \exp_q \cdot (T_q f^n)^{-1} \cdot \exp_p^{-1}$ on V, and $F_q = (f^n \mid_{U(q)})^{-1}$ on $U - W$. Now define

$$f_1 : M \to M$$

$$f_1 = \begin{cases} F_{f(q)} \cdot F_q^{-1} & \text{on each } W(q) \text{ for } q \in \bigcup_{n=1}^{\mu} f^{-n}\{p\}, \\ F_{f(q)} \cdot f^{\mu+1} & \text{on each } W(q) \text{ for } q \in f^{-\mu-1}\{p\}, \\ f & \text{on the rest of } M. \end{cases}$$

It is easy to verify that f_1 is well defined and satisfies (1), (2), (3), (4). A straightforward computation similar to that of [1,p.303] shows that f_1 also satisfies (5) for small λ. $\qquad\square$

Roughly say, f_1 near the first $-\mu$ iterations is just Tf modulo those exp, and f_1 near the $-(\mu+1)$th iteration cancels out these linearizations. The following corollary is clear.

Corollary 4.3. *If x and $f^k x$ are both out of $\bigcup\{w(q) \mid q \in \bigcup_{n=1}^{\mu} f^{-n}\{p\}\}$, then $f^k x = f_1^k x$.* $\qquad\square$

5   The proof of Theorem A

In this section we prove the C^1 closing lemma of 2-dimensional non-singular endomorphisms.

Theorem A. *Let $f\colon M \to M$ be a 2-dimensional non-singular endomorphism, and σ be a non-wandering point of f. Then for any C^1 neighborhood $\mathcal{U}$ of f in $NEnd^1(M)$, there is a $g \in \mathcal{U}$ such that σ is a periodic point of g.*

Proof. Let $\mathcal{U}$ be any C^1 neighborhood of f in $NEnd^1(M)$. As pointed out in [2,p.28] and [5,p.967], it suffices to prove that there is a $g \in \mathcal{U}$ such that g has a periodic point near σ, since another perturbation allows us to bring a nearby periodic point onto σ. Let U be any neighborhood of σ in M. We need prove that there is a $g \in \mathcal{U}$ with a periodic point in U. Take an $\eta > 0$ such that the η-ball of f in $NEnd^1(M)$ is contained in $\mathcal{U}$. By Lemma 4.1, there is an $\varepsilon > 0$ such that

$$d_1(hf, f) < \eta/2$$

for any $f \in NEnd^1(M)$, where h is any ε-kernel lift.

We assume that σ is not periodic already of f. All terms in $Orb^-(\sigma)$ are hence distinct. Moreover, $f^{-n}\{\sigma\}$ is non-empty for each $n \geq 1$, since σ is non-wandering.

Take a scattering branch $\Sigma = \{\sigma_0, \sigma_1, \cdots, \sigma_n, \cdots\}$ of $Orb^-(\sigma)$ given by Lemma 2.2. Let ρ, μ be the two numbers guaranteed by Theorem 3.1, by treating $V_n = T_{\sigma_n}(M)$ and $T_n = T_{\sigma_n}(f^n)$.

For the f, σ, μ, there is by Lemma 4.2 a $\lambda > 0$ and a local linearization f_1 with the following properties (1)–(5). Write $W' = \{u \in T_\sigma M \mid |u| \leq \lambda\}$, $V' = \{u \in T_\sigma M \mid |u| \leq \lambda/4\}$, $W = \exp_\sigma(W')$, $V = \exp_\sigma(V')$.

(1) W is $(\mu + 1)$-dynamical for both f and f_1, and $W_f(q) = W_{f_1}(q)$
for each $q \in \displaystyle\bigcup_{n=1}^{\mu+1} f^{-n}\{\sigma\}$;

(2) $f_1 = \exp_{f(q)}(T_q f) \exp_q^{-1}$ on each $V_{f_1}(q)$ for $q \in \displaystyle\bigcup_{n=1}^{\mu} f^{-n}\{\sigma\}$;

(3) $f_1^{\mu+1} = f^{\mu+1}$ on each $W(q)$ for $q \in f^{-\mu-1}\{\sigma\}$. In particular, $f_1 = \exp_{f(q)} \cdot (T_{f(q)} f^\mu)^{-1} \cdot \exp_\sigma^{-1} \cdot f^{\mu+1}$ on each $V(q)$ for $q \in f^{-\mu-1}\{\sigma\}$;

(4) $f_1 = f$ on $M - \bigcup\{W(q) \mid q \in \bigcup_{n=1}^{\mu+1} f^{-n}\{\sigma\}\}$;

(5) $d_1(f_1, f) < \eta/2$.

Clearly, $f_1 \in NEnd^1(M)$.

By shrinking W if necessary, we assume that $V \subset W \subset U$. By Lemma 2.2, there are two points $x, y \in V$ such that:

(1) $f^\psi x = y$ for some $\psi > \mu + 1$;

(2) $B(y, \rho\delta, d') \subset V$, where $\delta = d'(x, y)/2$;

(3) The distance between any two points of $\{x, fx, \cdots, f^{\psi-1}x, y\} \cap B(y, \rho\delta, d')$ is greater than δ; and that

(4) $f^{\psi-\mu-1}(x)$ is in $V(\sigma_{\mu+1})$.

Denote $\{x, fx, \cdots, f^{\psi-1}x, y\} \cap B(y, \rho\delta, d') = \{x, x_1, \cdots, x_s, y\}$, and denote $x' = \exp_\sigma^{-1}(x)$, $x_i' = \exp_\sigma^{-1}(x_i)$ for $i = 1, 2, \cdots, s$, and $y' = \exp_\sigma^{-1}(y)$. Then $X = \{x', x_1', \cdots, x_s', y'\}$ is a finite ordered set in $T_\sigma(M)$ such that the distance between any two points of $X \cap (y', \rho\delta)$ is greater than δ, where $\delta = |x' - y'|/2$. Thus by Theorem 3.1, there is a point $w' \in (X - \{y'\}) \cap (B(y', \rho\delta))$, together with $\mu + 1$ points $c_0, c_1, \cdots, c_\mu$ in $B(y', \rho\delta)$, not necessarily distinct, satisfying the following two conditions (a) and (b).

(a) $c_0 = w'$, $c_\mu = y'$; and

(b) $|(T_{\sigma_n} f^n)^{-1}(c_n) - (T_{\sigma_n} f^n)^{-1}(c_{n+1})|$
$\leq \varepsilon d((T_{\sigma_n} f^n)^{-1}(c_{n+1}), (T_{\sigma_n} f^n)^{-1}(Y))$
for $n = 0, 1, \cdots, \mu - 1$, where $Y = (X - \{w', y'\}) \cup (\partial B(y', \rho\delta))$.

Since w' is in $X - \{y'\}$, there is an integer φ, $\mu + 1 < \varphi \leq \psi$, such that $f^\varphi w = y$ and that

$$f^{\varphi-\mu-1}(w) = f^{\psi-\mu-1}(x) \in V(\sigma_{\mu+1}),$$

where $w = \exp_\sigma(w')$.

For each σ_n, $n = 0, 1, \cdots, \mu - 1$, let h_{σ_n} be the ε-kernel lift obtained by treating in Lemma 4.1 $p = \sigma_n$, $v_1 = (T_{\sigma_n} f^n)^{-1}(c_n)$, and $v_2 = (T_{\sigma_n} f^n)^{-1}(c_{n+1})$. Define a map g by

$$g = \begin{cases} h_{\sigma_n} \cdot f_1 & \text{on } W(\sigma_{n+1}), n = 0, 1, \cdots, \mu - 1, \\ f_1 & \text{on the rest of } M. \end{cases}$$

Then $g \in NEnd^1(M)$, and $d_1(g, f_1) < \eta/2$. Hence

$$d_1(g, f) < \eta.$$

We now verify that w is periodic of g to finish the proof of Theorem A, since w is clearly in U. Let

$$z = f^{\varphi - \mu - 1}(w).$$

Then $f^{\mu+1}(z) = y$ and $z \in V(\sigma_{\mu+1})$. It suffices to verify that $g^{\varphi - \mu - 1}(w) = z$ and $g^{\mu+1}(z) = w$. By the condition (b) above, the g orbit from w to z never touches the supports of these lifts. Then $g^{\varphi - \mu - 1}(w) = f_1^{\varphi - \mu - 1}(w)$. But $f_1^{\varphi - \mu - 1}(w) = f^{\varphi - \mu - 1}(w)$ by Corollary 4.3. Therefore

$$g^{\varphi - \mu - 1}(w) = f^{\varphi - \mu - 1}(w) = z.$$

It remains to verify that $g^{\mu+1}(z) = w$. By the condition (3) above, $g(z) = f_1(z) = \exp_{\sigma_\mu} \cdot (T_{\sigma_\mu} f^\mu)^{-1} \cdot \exp_\sigma^{-1} \cdot f^{\mu+1}(z)$ since $z \in V(\sigma_{\mu+1})$. Hence

$$g(z) = \exp_{\sigma_\mu} \cdot (T_{\sigma_\mu} f^\mu)^{-1}(y')$$

since $f^{\mu+1}(z) = y$. Thus these lifts $h_{\sigma_{\mu-1}}, h_{\sigma_{\mu-2}}, \cdots, h_{\sigma_0}$ gives rise to

$$g^\mu(g(z)) = w$$

by the condition (2) above. This completes the proof of Theorem A.

$$\square$$

6   The proof of Theorem B

In this section we prove the general density theorem. The proof we present here is due to S. T. Liao for vector fields, which also works for endomorphisms. For convenience we quote the proof below since [2] is available only in Chinese edition.

Theorem B. *There is a residual subset $\mathcal{B}$ of $NEnd^1(M)$ such that $P(f)$ is dense in $\Omega(f)$ for every $f \in \mathcal{B}$.*

Proof. Take a countable basis $\{W_k\}$ of M. For each k, let

$$\mathcal{F}_k = \{f \in NEnd^1(M) \mid P(f) \cap W_k = \emptyset\},$$

and let

$$\mathcal{H}_k = \{f \in NEnd^1(M) \mid f \text{ has a hyperbolic periodic point in } W_k\}.$$

Then $\mathcal{H}_k$ is open in $NEnd^1(M)$ by Hartman theorem. If $f \in NEnd^1(M)$ has a periodic point x in W_k, with an arbitrarily small C^1 perturbation one can find a $g \in NEnd^1(M)$ such that x is a hyperbolic periodic point of g, and hence $g \in \mathcal{H}_k$. This means that $\mathcal{H}_k \cup (int\,\mathcal{F}_k)$ is open and dense in $NEnd^1(M)$. Therefore

$$\mathcal{B} = \bigcap(\mathcal{H}_k \cup (int\,\mathcal{F}_k))$$

is a residual subset of $NEnd^1(M)$. We now verify that $P(f)$ is dense in $\Omega(f)$ for every $f \in \mathcal{B}$.

Let $f \in \mathcal{B}$ and $\sigma \in \Omega(f)$ be given, and let U be a neighborhood of σ in M. Since $\{W_k\}$ is a topological basis of M, there is a W_j such that $\sigma \in W_j \subset U$. by the C^1 closing lemma Theorem B, we see that $f \notin int\,\mathcal{F}_j$. But $f \in \mathcal{H}_j \cup (int\,\mathcal{F}_j))$. Hence $f \in \mathcal{H}_j$. Then $U \cap P(f) \supset W_j \cap P(f) \neq \emptyset$. This proves that $P(f)$ is dense in $\Omega(f)$, and hence proves Theorem B. $\square$

Added in proof. See [9] for the high dimensional case.

References

[1] J. Franks, Necessary conditions for stability of diffeomorphisms, *Trans. of A. M. S.*, 158(1971), 301–308.

[2] S. T. Liao, An extension of the C^1 closing lemma, *Acta Scientiarum Naturalium Universitatis Pekinensis*, 3(1979), 1-41 (in Chinese).

[3] J. Mai, A simpler proof of C^1 closing lemma, *Scientia Sinica* (Series A), Vol XXIX, No 10 (1986), 1021–1031.

[4] J. Mai, A simpler proof of the extended C^1 closing lemma, *Chinese Science Bulletin*, 3(1989), 180–184.

[5] C. Pugh, The closing lemma, *Amer. J. Math.*, 89(1967), 905–1009.

[6] C. Pugh, An improved closing lemma and a general density theorem, *Amer. J. Math.*, 89(1967), 1010–1021.

[7] C. Pugh & C. Robinson, The C^1 closing lemma, including Hamiltonians, *Ergod. Th. and Dynam. Sys.*, 3(1983), 261–313.

[8] C. Robinson, Introduction to the closing lemma, *Lecture Notes in Math.*, 668(1978), 223–230.

[9] L. Wen, The C^1 closing lemma of non-singular endomorphisms. *Preprint.*

Non-complete Algebraic Surfaces

De-Qi Zhang
Department of Mathematics, Osaka University,
Toyonaka, Osaka 560, Japan

Abstract

In this expository article, we describe some resent results on
the classification of singular or open algebraic surfaces. We give
first the classification of del Pezzo surfaces admitting singularities
of lower multiplicity, and then try to classify regular surfaces with
quotient singularities and with numerically trivial canonical divisor. A singular version of the Noether inequality for open algebraic
surfaces is also given.

Introduction

This is an expository article on the classification of algebraic surfaces
with singularities as well as open algebraic surfaces. Excellent works
have been done by Miyanishi [3] and by Miyanishi and Tsunoda [5].
The classification of complete, smooth algebraic surfaces was done by
Italian school. A recent classification theory for algebraic varieties of
higher dimension due to S.Mori, M.Reid, Y.Kawamata, etc., indicated
that "good" minimal models of algebraic varieties have to admit, in
general, some mild singularities(cf. [6]). In the degeneration theory
of smooth surfaces, we have to admit singularities on surfaces in the
boundary. These will motivate sufficiently our study to classify singular
surface $\overline{V}$ (by the notation) with several mild singularities or the smooth
part V^0 of V which is non-complete.

We work on the complex number field C. As in the complete, smooth
case, logarithmic Kodaira dimension plays an important role in our
study (see Definitions 1 and 3 below). In the case of surfaces, logarithmic
Kodaira dimension $\overline{\kappa}(V_0)$ takes value in $\{-\infty, 0, 1, 2\}$. If $\overline{\kappa}(V_0) = -\infty$,
logarithmic del Pezzo surface $\overline{V}$ is an essential case (see Definition 5

below). A pioneering work has been done by Miyanishi and Tsunoda (cf. [5]). A classification theory was developed in [11] and [12]. In [4] and [13], we then classified such surfaces $\overline{V}$ that $\overline{V}$ has only rational double and triple singular points.

A complete, smooth surface of Kodaira dimension 0 is a K3-surface, an Enriques surface, a hyperelliptic surface or an abelian surface. The first two cases are generalized to logarithmic Enriques surfaces $\overline{V}$ which admit quotient singularities (see Definition 10 below). Such surfaces $\overline{V}$ that their canonical coverings are smooth, were classified in [14].

Now we are interested in surfaces of general type, i.e., the case $\overline{\kappa}(V^0) = 2$. An excellent work on the pluri-canonical mappings of complete, smooth surfaces of general type was done by Bombieri[1]. We intend to generalize systematically the consideration of Bombieri to include those surfaces which are not necessarily complete or smooth. In some special cases, works were done by Sakai [8] and Reider [7]. In a recent article [10], we found a singular version of the Noether inequality for non-complete algebraic surfaces which plays an important role in the complete case.

Acknowledgment. I would like to express my sincere thanks to Professor M. Miyanishi for providing me with an unpublished notes which is very helpful in the preparation of the article [12]. I also like to thank Professor S. Tsunoda for giving me useful advice.

1   Classification of Iitaka Surfaces

Let V be a nonsingular projective variety, i.e., a closed submanifold of a projective space P^M. Set $m := \dim_C V$. A summation $L = \sum_{i=1}^r r_i L_i$ is said to be a divisor of V, if $r_i \in Z$ and if L_i is a (not necessarily smooth) subvariety of codimension one in V. Let Ω_V^1 be the cotangent bundle. Then the canonical divisor K_V is, by definition, a divisor such that the line bundle $\mathcal{O}(K_V)$, defined by K_V, is isomorphic to $\Omega_V^m = \bigwedge^m \Omega_V^1$.

Definition 1. Let L be a divisor on a nonsingular projective variety V. If the complete linear system $\mid nL \mid = \phi$, i.e., if $H^0(V, \mathcal{O}(L)^{\otimes n}) = 0$, for every positive integer n, we say that the Kodaira dimension $\kappa(V, L)$ of the divisor L is equal to $-\infty$. If $\mid nL \mid \neq \phi$, we let $\Phi_{\mid nL \mid} : V \rightarrow P^N(N := \dim \mid nL \mid)$ be the rational map, which is a map into a projective space of dimension N and which is defined by $\mid nL \mid$. More precisely, let $f_1, f_2, \cdots, f_N$ be a basis of $H^0(V, \mathcal{O}(L)^{\otimes n})$, then $\Phi_{|nL|}(v) :=$ $[f_0(v) : f_1(v) : \cdots : f_N(v)]$. The Kodaira dimension $\kappa(V, L)$ is defined as Max $\dim \Phi_{|nL|}(V)$. The Kodaira dimension $\kappa(V)$ of V is defined as $\kappa(V, K_V)$, i.e., the Kodaira dimension of the canonical divisor K_V.

The following lemma is well-known.

Lemma 2. *We have $\kappa(V, L) = -\infty, 0, 1, \cdots, or \dim V$.*

Let X be a nonsingular algebraic variety, which is not necessarily complete. Then there is a nonsingular projective variety V such that the following conditions are satisfied.

(1) X is an open dense subset of V.

(2) $V - X$ is a union of several nonsingular subvarieties $D_1, \cdots, D_n$ of codimension one in V.

(3) The divisor $D = \sum_{i=1}^{n} D_i$ has only "simple normal crossings" singularities.

Definition 3. We define the logarithmic Kodaira dimension $\overline{\kappa}(X) :=$ $\kappa(V, K_V + D)$.

In fact, this definition is independent of the choice of the embedding $X \hookrightarrow V$. Since $X = V - D, X$ is complete if and only if $D = \phi$. We often confuse X with (V, D). A pair (V, D) will be called a logarithmic variety.

Definition 4. Let V be a nonsingular, rational and projective surface and let A and N be reduced and effective divisors such that $A + N$ is a reduced divisor with only normal crossings, i.e., $A + N$ is a reduced curve which might be reducible and has only "normal crossings" singularities. A pair $(V, A + N)$ is called an Iitaka surface if $-A$ is linearly equivalent to the canonical divisor K_V and if every connected component of N is the exceptional divisor of the minimal desingularization of a rational double singular point (cf. [**11; Definition 1.1 and Lemma 1.5**]).

Let $(V, A + N)$ be an Iitaka surface. We can prove that the logarithmic Kodaira dimension $\overline{\kappa}(V - (A + N)) = 0$. Note that A is either a nonsingular elliptic curve, or a nodal curve, or a loop of nonsingular rational curves. Let $f : V \to \overline{V}$ be the contraction of N. Then $-K_{\overline{V}}$ is numerically equivalent to the direct image $f_*(A)$ of A and hence $K_{\overline{V}}$ is not numerically effective. A divisor F is numerically effective (nef, for short), if, by definition, the intersection number $(F, G) \geq 0$ for every curve G on V. By the Mori theory, there is a nef divisor $\overline{H}$ and there is an extremal rational curve $\overline{E}$ on $\overline{V}$, such that $(K_{\overline{V}}, \overline{E}) < 0$ and $\overline{H}^{\perp} \cap \overline{NE}(\overline{V}) = \boldsymbol{R}_+[\overline{E}]$ (cf.[5;Lemma 8]). Here $\overline{NE}(\overline{V})$ is the closed cone generated by numerical classes of all effective 1-cycles, and $[\overline{E}]$ is the numerical class which $\overline{E}$ belongs to. We have three cases:

Case 1. $\overline{H}$ is numerically equivalent to zero.

Case 2. $\overline{H}$ is not numerically equivalent to zero, while the self intersection number $(\overline{H}^2) = 0$.

Case 3. $(\overline{H}^2) > 0$.

Suppose the case(3) occur. Then the proper transform $E := f^1(\overline{E})$ of $\overline{E}$ is a (-1)-curve, i.e., $E \cong \boldsymbol{P}^1$ (the projective space of dimension one) and $(E^2) = -1$, and the divisor $E + N$ has negative definite intersection matrix(cf. [11;Lemma 2.3]). Let $\sigma_1 : V \to V_1$ be the smooth contraction of the connected component of $E + N$ containing the (-1)-curve E, and denote by A_1 (resp. N_1) the direct image $\sigma_{1*}(A)$ (resp. $\sigma_{1*}(N)$). Then $(V_1, A_1 + N_1)$ is again an Iitaka surface (cf. [11; Remark 2.4]). Applying the Mori theory to the surface $\overline{V}_1$ obtained by contracting N_1, we have also three cases as above. If the case (3) occurs for the pair $(V_1, A_1 + N_1)$, we obtain also a smooth contraction $\sigma_2 : V_1 \to V_2$ and an Iitaka surface $(V_2, A_2 + N_2)$. Continue this process. Since the Picard numbers satisfy $+\infty > \rho(V) > \rho(V_1) > \cdots > 0$, the above process must stop at the n-th step for some $n \geq 0$. Then, for the pair $(V_n, A_n + N_n)$, only the case (1) or (2) above takes place. Thus, we have only to consider the cases (1) and (2) for the Iitaka surface $(V, A + V)$.

Suppose the case (2) occurs. Then $\overline{H} \in \boldsymbol{R}_+[\overline{E}]$ and $(\overline{E}^2) = 0$. We can show that the rational map, defined by the complete linear system $| f_*(p\overline{E}) |$ ($p >> 0$), is composed with a $\boldsymbol{P}^1$-fibration $\Phi : V \to \boldsymbol{P}^1$ whose general fiber is $\boldsymbol{P}^1$. For a divisor $\overline{F}$ on $\overline{V}$, we denote by $f^*(\overline{F})$ the total transform on V. The structure of $(V, A + V)$ is explicitly described in [11;Lemma 2.5].

Suppose the case (1) occurs. Then $\rho(\overline{V}) = 1$ and $-K_{\overline{V}}$ is an ample divisor, i.e., by definition, $|-pK_{\overline{V}}|$ $(p >> 0)$ defines a closed embedding into a projective space. Hence $\overline{V}$ is a Gorenstein logarithmic del Pezzo surface of rank one (see the definition given below). There are altogether 27 "singularity types", i.e., the types of combinations of singularities on $\overline{V}$ by [11; **Lemma 3.5, 4.2 and 4.3**]. Every Gorenstein log del Pezzo surface is also concretely constructed there.

A normal surface $\overline{V}$ is called Gorenstein if, by definition, $-K_{\overline{V}}$ is a Cartier divisor, i.e., locally defined by a single meromorphic function.

Definition 5. Let $\overline{V}$ be a normal, rational and projective surface with at worst quotient singularities, i.e., if x belongs to the singular locus $\mathrm{Sing}(\overline{V})$, then locally $(\overline{V}, x) \cong (C^2/G_x, 0)$ with a finite subgroup G_x of $GL(2, C)$. $\overline{V}$ is called a logarithmic del Pezzo surface (log del Pezzo surface, for short) if the anti-canonical divisor $-K_{\overline{V}}$ is an ample divisor. $\overline{V}$ is said to be of rank one if the Picard number $\rho(\overline{V}) = 1$.

A nonsingular log del Pezzo surface is, by definition, a del Pezzo surface. All del Pezzo surfaces are completely classified(cf. Manin [**2**]). A log del Pezzo surface $\overline{V}$ is Gorenstein if and only if $\overline{V}$ contains at worst rational double singular points. We can prove that the smooth part $V^0 := \overline{V} - (\mathrm{Sing}\overline{V})$ of a log del Pezzo surface $\overline{V}$ has logarithmic Kodaira dimension $\overline{\kappa}(V^0) = -\infty$.

2 Topologic Properties of Gorenstein Log del Pezzo surfaces

We begin with

Definition 6. Let $\overline{V}$ be a log del Pezzo surface. Denote by $V^0 = \overline{V} - (\mathrm{Sing}\ \overline{V})$ the smooth part. Let $\pi : U^0 \to V^0$ be the (topologically) universal covering. Then π is a finite morphism and U^0 is a nonsingular algebraic surface. Let $\overline{U}$ be the normalization of $\overline{V}$ in the function field $C(U^0)$. Then $\pi : U^0 \to V^0$ extends to a finite morphism $\pi : \overline{U} \to \overline{V}$. The surface $\overline{U}$ or the covering $\pi : \overline{U} \to \overline{V}$ is called the quasi-universal covering of $\overline{V}$.

Since every Gorenstein log del Pezzo surface $\overline{V}$ is concretely constructed by [11], we can consider topologic properties of $\overline{V}$. In particular, we prove the following theorem in the joint work with M.Miyanishi [4;**Theorem 1 and Lemmas 6 and 7**].

Theorem 7. *Let $\overline{V}$ be a Gorenstein log del Pezzo surface of rank one. Then the following assertions hold true.*

(1) *The smooth part $V^0 = \overline{V} - (\mathrm{Sing}\ \overline{V})$ is simply connected if and only if $\overline{V}$ is a Gorenstein, algebraic compactification of the affine plane C^2 by an irreducible curve at infinity.*

(2) *Suppose V^0 is not simply connected. Then $\overline{V} \cong P^2/G$ with a finite subgroup G of $PGL(2, C)$ if and only if the Picard number $\rho(\overline{U})$ of the quasi-universal covering $\overline{U}$ is equal to one.*

(3) *The (topological) fundamental group $\pi_1(V^0)$ is an abelian group of order ≤ 9.*

(4) *$\overline{V}$ is determined uniquely up to isomorphism by its singularity type unless the case where $\overline{V}$ contains exactly two singularities $x_i (i = 1$ and $2)$ and x_i is a rational double singular point of Dynkin type $D4$.* $\square$

In the next section, we shall classify log del Pezzo surfaces by induction. The study of Gorenstein log del Pezzo surfaces was the first approach.

3   Classification Theory of Log del Pezzo Surfaces

Let $\overline{V}$ be a log del Pezzo surfaces of rank one. Let $f : V \to \overline{V}$ be a minimal desingularization and denote by $D = f^{-1}(\mathrm{Sing}\overline{V})$ the inverse image of the singular locus. Then D is a reduced, effective divisor and consists of admissible rational rods or forks (see [5;§ 2]) for the relevant definitions). Write $D = \sum D_i$ where D_i's are irreducible components of D. Then there is a Q-divisor $D^* = \sum \alpha_i D_i$ such that $\alpha_i \in Q, 0 \leq \alpha_i < 1$ and the total transform $F^*(K_{\overline{V}})$ of the canonical divisor is numerically equivalent to the divisor $K_V + D^*$.

We often confuse $\overline{V}$ with the pair (V, D). Since $-K_{\overline{V}}$ is an ample divisor, $-(K_V + D^*)$ is a nef divisor and $(K_V + D^*)^2 > 0$. We can find a curve C on V such that $-(C, K_V + D^*)$ attains the smallest positive value. Such a curve C is called a "minimal" curve.

Suppose the complete linear system $\mid C + D + K_V \mid \neq \phi$ for some minimal curve C. Then D is decomposed into $D' + D''$ such that $C + D''$ is linearly equivalent to $-K_V$ and the pair $(V, (C + D'') + D')$ is an Iitaka surface, though $C + D''$ might not be of normal crossings (cf. [12; **Lemma 2.1**]). Therefore we are reduced to the study of Iitaka surfaces which are classified by [**11**] as we reviewed before.

In view of the above arguments, we shall assume $\mid C + D + K_V \mid = \phi$ for every minimal curve C. Then C can be taken as a (-1)-curve, unless $\overline{V}$ is a cone, contained in a projective space P^n, over a nonsingular rational curve of degree $n - 1$ in P^{n-1} (cf. [12; **Lemma 2.2**]). We shall assume further that C is a (-1)-curve. Then one of the following cases occurs (cf. [12;§4]).

Case 1. C meets exactly one component D_1 of D, and $(C, D_1) = 1$.

Case 2. C meets at least two (-2)-curve D_1 and D_2 in D.

Case 3. C meets exactly two components D_1 and D_2 of D, and $(D_2^2) \leq -3$.

Case 4. C meets exactly three components D_1, D_2 and D_3 of D, and $(D_1^2) = -2, (D_2^2 = -3, -5 \leq (D_3^2) \leq -3$.

The case (4) is treated in our unpublished notes. Suppose the case (1) or (3) occurs. Let $\sigma : V \to W$ be the blowing-down of the (-1)-curve C and denote by B the direct image $\sigma_*(D - D_1)$. Then (W, B) is again a log del Pezzo surface of rank one (cf. [**12; Lemma 4.3**]). We are reduced to considering the case (2).

In the case (2), Let $\Phi : V \to P^1$ be the P^1-fibration such that $f_1 := 2C + D_1 + D_2$ is a singular fiber. Write $D = \sum_{i=1}^{n} D_i$ with irreducible components D_i's of D. Suppose that $V - D$ is not affine-ruled. (A surface X is called affine-ruled if, by definition, X contains an open set $C \times R$ with an affine curve.) Then there are (-1)-curve $E_i(1 \leq i \leq 4)$ and there are six components of D, say $D_j(3 \leq j \leq 8)$, such that D_3 and D_4 are two disjoint cross-sections of Φ, and $f_1, f_2 :=$

$2E_1 + D_5 + D_6, f_3 := 2E_2 + D_7 + D_8$ and $f_4 := E_3 + D_9 + \cdots + D_n + E_4$ are all singular fibers of Φ (cf. [**12;Theorem 5.1**]).

This completes the classification of log del Pezzo surfaces of rank one.

4 Log del Pezzo Surfaces With Triple Points

In the doctoral dissertation [**13**], we apply the theory developed in [**12**]. Especially, we obtain the following theorem.

Theorem 8. *Let $\overline{V}$ be a log del Pezzo surface of rank one with exactly one triple and several double singular points. Then there are exactly 97 singularity types. Moreover, the following assertions hold true.*

(1) *If the smooth part $V^0 = \overline{V} - (\mathrm{Sing}\overline{V})$ is simply connected, then V^0 contains $C \times (C - \{0\})$ as a Zariski open set.*

(2) *Suppose V^0 is not simply connected. Then $\overline{V} \cong \boldsymbol{P}^2/G$ with a finite subgroup G of $PGL(2, C)$ if and only if the Picard number $\rho(\overline{U})$ of the quasi-universal covering $\overline{U}$ is equal to one.*

(3) *The (topological) fundamental group $\pi_1(V^0)$ is a finite group of order ≤ 21 which is not necessarily an abelian group. (Compare with Theorem 7.)*

(4) *There exists a $\boldsymbol{P}^1$-fibration $\Psi : V \to \boldsymbol{P}^1$ and there are components $D_1, \cdots, D_k (k \leq 3)$ such that $D - \sum_{i=1}^{k} D_i$ is contained in singular fibers of Ψ. The configuration of D and all singular fibers of Ψ are explicitly given in Appendix of [**13**].* $\square$

In view of Theorems 7 and 8, we make the following:

Conjecture 9. Let $\overline{V}$ be a log del Pezzo surface of rank one.

(1) The smooth part V^0 of $\overline{V}$ is simply connected if and only if $\overline{V}$ is a compactification of the affine plane C^2.

(2) Suppose V^0 is not simply connected. Then $\overline{V} \cong \boldsymbol{P}^2/G$ with a finite subgroup G of $PGL(2, C)$ if and only if the Picard number $\rho(\overline{U})$ of the quasi-universal covering $\overline{U}$ is equal to one.

5 Logarithmic Enriques Surfaces

We begin with

Definition 10. A projective surface $\overline{V}$ is a logarithmic Enriques surface (log Enriques surface, for short) if the following conditions are satisfied.

(1) Every singular point $x \in \mathrm{Sing}(\overline{V})$ is an isolated quotient singularity, i.e., locally, $(\overline{V}, x)$ has the same singularity as $(C^2/G_x, 0)$ with a finite subgroup G_x of $GL(2, C)$.

(2) $\overline{V}$ is a regular surface, i.e., the irregularity dim $H^1(\overline{V}, \mathcal{O}_{\overline{V}}) = 0$.

(3) There exists a positive integer N such that $NK_{\overline{V}}$ is linearly equivalent to zero, where $K_{\overline{V}}$ is the canonical divisor of $\overline{V}$.

We can prove that the smooth part $V^0 := \overline{V} - (\mathrm{Sing}\overline{V})$ of a log Enriques surface $\overline{V}$ has logarithmic Kodaira dimension $\overline{\kappa}(V_0) = 0$. By the classification theory of nonsingular projective surfaces, a nonsingular log Enriques surface is a K3-surface or an Enriques surface.

The index of $\overline{V}$ is defined as $I = \mathrm{Index}(\overline{V}) := \min\{N \in \mathbf{N}; NK_{\overline{V}} \sim 0\}$. S. Tsunoda [9;p.108] proved that $1 \leq I \leq 66$.

We consider the following problem.

Problem 11. (1) *Find all possible combinations of singularities of $\overline{V}$.*

(2) *For each $1 \leq P \leq 66$, construct concretely log Enriques surface of index P.*

We need some preparation. Let $f : V \to \overline{V}$ be a minimal desingularization and denote by $D := f^{-1}(\mathrm{Sing}\overline{V})$ the inverse image of the singular locus. Write $D = \sum D_i, D^* = \sum \alpha_i D_i$ and $f^*(K_{\overline{V}}) = K_V + D^*$ as set at the beginning of §3. We often confuse $\overline{V}$ with the pair (V, D).

Lemma 12. (1) *We have the Kodaira dimension $\kappa(V) \leq 0$ and the irregularity* dim $H^1(V, \mathcal{O}_V) = 0$. *Moreover, if $\kappa(V) = 0$, then every singular point x of $\overline{V}$ is a rational double singular point, i.e., locally, $(\overline{V}, x)$ has the same singularity as $(C^2/G_x, 0)$ with a finite subgroup G_x of $SL(2, C)$, and either V is a K3-surface or V is an Enriques surface.*

(2) *We have $I(K_V + D^*) \sim f^*(IK_{\overline{V}}) \sim 0$, unless V is an Enriques surface and $D^* = 0$.*

In view of Lemma 12, (1), we shall assume the following conditions.

(1) $\kappa(V) = -\infty$, hence V is a rational surface.

(2) Each singular point of $\overline{V}$ is not a rational double point.

Denote by V^0 the smooth part $\overline{V} - (\mathrm{Sing}\overline{V})$. Since $IK_{\overline{V}} \sim 0$, we have $IK_{V^0} \sim 0$ and hence $\mathcal{O}(K_{V^0})^{\otimes I} \cong \mathcal{O}_{V^0}$. Then we can find a $\mathbf{Z}/I\mathbf{Z}_-$ covering $\pi : \overline{U} \to \overline{V}$ satisfying the following properties, where $U^0 := \pi^{-1}(V^0)$. Indeed, the restriction $\pi|U^0$ is locally defined as $x^I = s$, where x is an variable and s is a global section of the line bundle $\mathcal{O}_{V^0}$.

(1) $\pi \mid U^0 : U^0 \to V^0$ is an etale morphism.

(2) Every singular point of $\overline{U}$ is an isolated rational double point.

(3) $K_{\overline{U}} \sim 0$.

Definition 13. $\overline{U}$ is called the canonical covering of $\overline{V}$.

Remark 14. Let $g : U \to \overline{U}$ be a minimal desingularization. Then U is an abelian surface or a K3-surface.

We now state the main results of [**14**].

Theorem 15. *Let $\overline{V}$ or synonymously (V, D) be a logarithmic Enriques surface of index 2. Then there is a logarithmic Enriques surface $\overline{W}$ or (W, B) of index 2 and with only one singular point on $\overline{W}$, such that V is obtained from W by blowing up all singular points of B, (i.e., intersection points of irreducible components of B) and then blowing down several exceptional curves of the first kind on the blown-up surface.*

*The singularity type of $\overline{V}$ is explicitly described in [**14**]. There are at most ten singularities on $\overline{V}$ and an example of a logarithmic Enriques surface with exactly ten singularities is also given in [**14**].*

Theorem 16. *Let $\overline{V}$ be a logarithmic Enriques surface such that the canonical covering is an abelian surface. Then the index $I = 3$ or 5, and the singularity type of $\overline{V}$ is explicitly given in [**14**].*

Theorem 17. *Let $\overline{V}$ be a logarithmic Enriques surface such that the index I is a prime number and the canonical covering is a K3-surface. Then the index $I \neq 2, 13$. Moreover, the singularity type of $\overline{V}$ is explicitly given in [**14**].*

Theorem 18. *Let $\overline{V}$ be a logarithmic Enriques surface such that the index I is an odd prime number and the canonical covering is a singular surface. Then the number c of all singularities on $\overline{V}$ is bounded as $\leq \mathrm{Min}\{16, 23 - I\}$. Moreover, if $c = 16$, then $I = 5$ and the singularity type of $\overline{V}$ is explicitly given in* [14].

6 Non-complete Algebraic Surfaces of General Type

Let V be a nonsingular projective surface. If V is a minimal surface of general type, we have the following inequality due to M. Noether:

$$p_g(V) \leq \tfrac{1}{2}(K_V^2) + 2,$$

where $p_g(V)$ is the geometric genus of V, K_v is the canonical divisor of V and $K_V^2 = C_1(V)^2$ with the first chern class $C_1(V)$.

We intend, in [10], to extend this inequality to a non-complete surface of general type, which is to be defined below.

Let X be a nonsingular algebraic surface, which is not necessarily complete. As in §1, there is a nonsingular completion V of X such that $D = V - X$ is a reduced effective divisor with only simple normal crossings. We shall confuse X with (V, D).

Definition 19. A pair (V, D) is said to be a logarithmic surface (log surface, for short) of general type if the logarithmic Kodaira dimension $\overline{\kappa}(V - D) = 2$.

After finitely many blowing-downs of curves on V which are exceptional curves of the first kind, we can obtain a so called minimal log surface of general type, which is defined as follows.

Definition 20. A log surface (V, D) of general type is minimal if $K_V + D$ has a decomposition into Q-divisors:

$$K_V + D = (K_V + \sum \alpha_i D_i) + \sum (1 - \alpha_i) D_i,$$

where

(1) $0 \leq \alpha_i \leq 1$ and $\alpha_i \in Q$.

(2) $K_V + \sum \alpha_i D_i$ is a numerically effective divisor of positive self intersection number,

(3) the intersection matrix of $\sum (1 - \alpha_i) D_i$ is negative definite, and

(4) $(K_V + \sum \alpha_i D_i, \sum (1 - \alpha_i) D_i) = 0$ and $\sum (1 - \alpha_i) D_i$ contains no exceptional curves of the first kind.

This decomposition is unique and called the *Zariski decomposition*. We write $\sum \alpha_i D_i$ as D^*. If $D^* = 0$, then V is a surface of general type. The connected components of $Supp(D - D^*)$ consist of twigs, rods or forks (see Miyanishi and Tsunoda [5;p. 436] for the relevant definitions), which can be contracted to normal points with quotient singularities. We denote $\dim H^0(V, \mathcal{O}(K_V + D))$ by $\bar{p}_g$.

Write $| K_V + D | = | C | + G$, where $| C |$ and G are respectively the movable part and the fixed part of $| K_V + D |$. We shall assume $C \neq 0$. Let $\Phi_{|C|} : V \to P^N$ be the rational map defined by the complete linear system $| C |$. Then $\dim \Phi_{|C|}(V) = 1$ or 2.

Now we can state the main results in the joint work [10] with Tsunoda.

Theorem 21. *Let (V, D) be a minimal log surface of general type such that* $\dim \Phi_{|C|}(V) = 1$. *Then we have:*

$$\bar{p}_g \leq \tfrac{1}{2}\{(\bar{c}_1^2) + \sqrt{(\bar{c}_1^2)^2 + 8(\bar{c}_1^2)}\} + 1 \leq \tfrac{9}{8}(\bar{c}_1^2) + 2,$$

where $(\bar{c}_1^2) = (K_V + D^*)^2$. *We have also* $\overline{P}_g < (\bar{c}_1^2) + 3$.

Theorem 22. *Let (V, D) be a minimal log surface of general type such that* $\dim \Phi_{|C|}(V) = 2$. *Then the following assertations hold, where C is replaced by a general member of* $| C |$ *if necessary.*

(1) *We have* $\bar{p}_g \leq (K_V + D^*)^2 + 2$. *Suppose* $\bar{p}_g = (K_V + D^*)^2 + 2$. *Then* $C \sim K_V + D^*$ *and* $\Phi_{|C|} : V \to P^N$ *($N := \overline{P}_g - 1$) is a birational morphism onto a normal rational surface* $\overline{W}$ *of degree $N - 1$.*

(2) *Suppose V is not a rational surface. Then* $\bar{p}_g \leq (K_V + D^*)^2 + 1$. *Moreover, if the equality holds then* $C \sim K_V + D^*, \bar{p}_g = 3$, *the arithmetic genus $p_a(C) = 2$ or 3 and $\Phi_{|C|} : V \to P^2$ is a morphism of degree two.*

(3) *Suppose V is not a ruled surface. Then $\overline{p}_g \leq \frac{1}{2}(K_V + D^*)^2 + 2$. Moreover, if the equality holds then $C \sim K_V + D^*, \frac{1}{2}(C^2) + 1 \leq P_a(C) \leq (C^2) + 1$, and either $\Phi_{|C|} : V \to \mathbf{P}^N$ $(N := \overline{p}_g - 1)$ is a birational morphism onto a surface of degree $2(N-1)$ or $\Phi_{|C|} : V \to \mathbf{P}^N$ is a morphism of degree two onto a normal rational surface $\overline{W}$ of degree $N - 1$.*

The assertion (1) of Theorem 22 has been obtained by Fujita and Sakai, independently. Sakai [8;**Theorem 6.5**] proved also the inequality $\overline{p}_g \leq \frac{1}{2}(K_V + D^*)^2 + 2$ in Theorem 22, (3) provided D is semi-stable.

In [**10**], we consider first the case where $\dim\Phi_{|C|}(V) = 1$. After proving Theorem 21, we describe precisely such pairs (V, D) that the first inequality of Theorem 21 becomes an equality. We give also an example to show that the bounds for $\overline{p}_g$ in Theorem 21 are the best possible ones. Then we consider the case where $\dim\Phi_{|C|}(V) = 2$. We prove several inequalities of the form $\dim H^0(V, \mathcal{O}(C)) \leq \alpha(C^2) + \beta$ with $\alpha, \beta \in \mathbf{Q}$. We consider also the cases where these inequalities become equalities. Then Theorem 22 is a consequence of the above inequalities and the fact $(K_V + D^*)^2 \geq (C^2)$ (cf. [**10**;**Lemma 2.9**]).

In a recent article [**15**], we give first the geometric structure of those pairs (V, D) satisfying $\overline{p}_g = \frac{1}{2}(\overline{c}_1^2) + 2$, where $(\overline{c}_1^2) := (K_V + D^*)^2$. Then we verify an inequality of the form $(\overline{c}_1^2) \geq \alpha\overline{c}_2 + \beta$, where $\overline{c}_2 := c_2(V) - e(D)$ is the Euler number of $V - D$, and $\alpha(> 0)$ and β are two rational numbers. Finally, we shall give the geometric structure of those pairs (V, D) satisfying $(\overline{c}_1^2) \geq \alpha\overline{c}_2 + \beta$.

References

[1] E. Bombieri, Canonical models of surfaces of general type, *Publ. Math. I. H. E. S.* 42(1973), 172–219.

[2] Yu. I. Manin, Cubic Forms – Algebra, Geometry, Arithmetic, *North-Holland Publishing Company – Amsterdam, London, American Elsevier Publishing Company, Inc. – New York*, 1974.

[3] M. Miyanishi, Non-complete algebraic surfaces, *Lecture Notes in Mathematics* **857**, *Berlin Heidelberg New York, Springer*, 1981.

[4] M. Miyanishi and D. -Q. Zhang, Gorenstein log del Pezzo surfaces of rank one,*J. Algebra.* 118(1988), 63–84.

[5] M. Miyanishi and S. Tsunoda, Open algebraic surfaces with Kodaira dimension $-\infty$, *Proceedings of Symposia in Pure Mathematics* 46(1987), 435–450.

[6] S. Mori, Flip theorem and the existence of minimal models for 3-folds,*J. Amer. Math. Soc.* 1(1988), 117–253.

[7] I. Reider, Vector bundles of rank 2 and linear systems on algebraic surfaces,*Ann. Math.* 127(1988), 309–316.

[8] F. Sakai, Semi-stable curves on algebraic surfaces and logarithmic pluricanonical maps, *Math. Ann.* 254(1980), 89–120.

[9] S. Tsunoda, Structure of open algebraic surfaces, *I, J. Math. Kyoto Univ.* 23(1983), 95–125.

[10] S. Tsunoda and D. -Q. Zhang, Noether's inequality for non-complete algebraic surfaces of general type,*submitted* (1989).

[11] D. -Q. Zhang, On Iitaka surfaces, *Osaka J. Math.* 24(1987), 417–460.

[12] D. -Q. Zhang, Logarithmic del Pezzo surfaces of rank one with contractible boundaries, *Osaka J. Math.* 25(1988), 461–497.

[13] D. -Q. Zhang, Logarithmic del Pezzo surfaces with rational double and triple singular points, *Tohoku Math. J.* 41(1989), 399–452.

[14] D. -Q. Zhang, Logarithmic Enriques surfaces, *submitted* (1989).

[15] D. -Q. Zhang, Noether's inequality for non-complete algebraic surfaces of general type, the part 2,*preprint* (1989).

Using the Singularity-Separating Method to Study How to Decrease the Pressure at the End of a Tube *

Xionghua Wu
Tongji University
Shanghai, P. R. China

and
Youlan Zhu
Computing Center, Academia Sinica
Beijing, P. R. China

Abstract

In this paper the singularity-separating method(S.S.M.) is used to calculate the flow field in tubes with rapidly changing cross-section -areas. How to decrease the pressure at the end of a tube is studied. The numerical results in different cases are shown. They are satisfactory.

1 Introduction

The singularity-separating method (it will be denoted by S.S.M. for short)[1] is a method which was developed in China during the 1970s. The main idea of the method is as follows: all discontinuity lines in the flow fields are considered as internal boundary lines. The difference scheme is only used in the subregions with continuous solution. One-sided difference scheme and boundary conditions are adopted on the boundaries. They are linear or nonlinear equations. The solutions on the boundaries can be obtained by solving these equations. When

---

*This work is supported in part by the AVH Foundation, SFB 123 in Heidelberg University, Chinese National Natural Science Foundation and Tongji University Science Foundation.

we solve the initial- boundary value problems of hyperbolic differential equations by using the above method, accurate numerical results can be obtained even near the discontinuity lines or boundaries. The main problem for S.S.M. is that its code is much more complicated. For solving this problem, a software for unsteady aerodynamic problems in one dimension is being worked out. Because the software has not yet been completely finished, the details about the software will be given in another paper. Here a computational example, which is met in the process of testing the software, is presented. It is the problem of tubes with rapidly changing cross-section-areas. Through solving it, we have tested our code, and have learnt more details about the problem of tubes with rapidly changing cross-section-areas.

2   Formulation of the Problem

Cartridge igniters are a sort of tubes with a contraction part and an expansion part. They are widely used in mining engineering. Because the internal diameters of the tubes are very small and the measurement of physical quantities is very difficult, numerical simulation for the flow field in the pipes is necessary. By numerical simulation, the velocity, density, pressure and other physical quantities in a pipe can be easily obtained. Some physical laws, for example, how the pressure on the closed end of tube varies with time, can be understood thoroughly. The numerical results are useful for the design and production of cartridge igniters.

The shape of a cartridge igniter is shown in Fig. 1. Its left part is a cyclindrical tube, whose diameter is $d_1 = 1.0$. From the point O to $A(OA = l_1 = 8.5)$, the tube contracts. From the point A to $B(OB = l_2 = 15.0)$, the tube is another cyclindrical tube, whose diameter is $d_2 = 0.5$. From the point B to $C(OC = l_3 = 44.0)$, the tube expands. The diameter at C is $d_3 = 4.0$. The right end is $D(OD = l_4 = 44.0)$. Now $l_3 = l_4$, $C = D$. The contraction curve or the expansion curve which links two neighboring cyclindrical tubes is a cubic curve:

$$y = (y_2 - y_1)\left(3\left(\frac{x - x_1}{x_2 - x_1}\right)^2 - 2\left(\frac{x - x_1}{x_2 - x_1}\right)\right),$$

where (x_1, y_1) and (x_2, y_2) are the coordinates of the left end $(O$ or B in Fig. 1) and the right end$(A$ or C in Fig. 1) of the curve. Because the

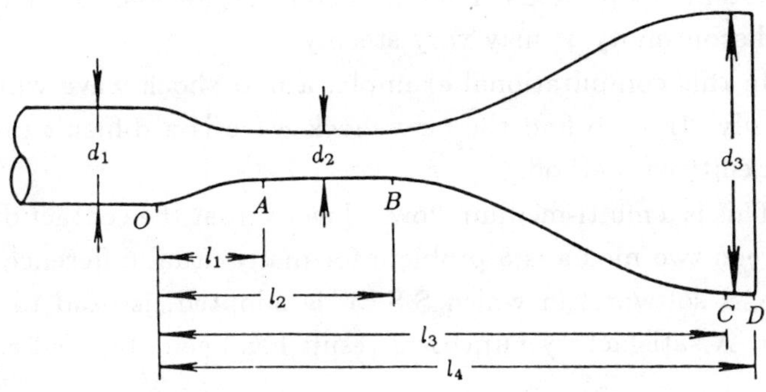

Fig. 1

diameter of the tube is a small quantity compared with the length, the problem of tube with variable cross-section-area is usually considered as a problem of unsteady flow in one dimension. The physical quantities satisfy the following system of equations:

$$(\rho A)_t + (\rho u A)_x = 0, \tag{1}$$

$$(\rho u A)_t + (\rho u^2 A + p A)_x = p A_x, \tag{2}$$

$$(e A)_t + ((e + p) u A)_x = 0, \tag{3}$$

where

$$e \equiv 0.5 \rho u^2 + \frac{p}{\gamma - 1}$$

stands for the total energy, u for the gas velocity, ρ for the density, p for the pressure, γ for the adiabatic constant, c for the sound velocity and $A(x)$ for the area of cross section at the point x.

Suppose that O is the origin of x axis. At initial time, the gas states are $\gamma_l = 1.29$, $u_l = 7.5$, $\rho_l = 9.0$, $p_l = 50.0$ at the region $x < 0$, and $\gamma_r = 1.4$, $u_r = 0.0$, $\rho_r = 1.0$, $p_r = 1.0$ at the region $x > 0$. The computing region is from $x = -20.0$ to $x = 44.0$. The boundary conditions are: at the left end u, ρ, p are equal to constants u_l, ρ_l, p_l (here, $u_l > c_l$) and at the right end the tube is closed, that is, $u = 0$.

Because of variation of cross section and reflection of discontinuities from the right side, the flow field becomes very complicated. The main difficulties of this problem are as follows:

(1) Because of variation of cross section, right hand side terms appear in the system of equations. Even if in some region the solution does not have a discontinuity, it may vary steeply.

(2) In this computational example, a new shock wave will form automatically. How to find the new shock wave is a difficult problem for the shock-fitting method.

(3) This is a multi-medium flow. How to treat the contact discontinuity between two media is a problem for many usual difference methods.

Here, a software, in which S.S.M. is adopted, is used to solve this problem. A satisfactory numerical result has been obtained easily.

3   Numerical Methods

The main idea of S.S.M. is as follows:

(1) By using the transformation:

$$\begin{cases} \xi = (x - x_i(t))/(x_{i+1}(t) - x_i(t)) + i, \\ t = t, \end{cases}$$

an arbitrary subregion whose boundaries are $x_i(t)$ and $x_{i+1}(t)$ is transformed into a strip $i \leq \xi \leq i+1$ on the ξ-t plane. The difference scheme can be used conveniently in such a strip.

(2) The system of equations (1)–(3) is rewritten in the characteristic form

$$\tilde{G}\frac{\partial \tilde{U}}{\partial t} + \tilde{\Lambda}\tilde{G}\frac{\partial \tilde{U}}{\partial x} = \tilde{F}, \tag{4}$$

where

$$\tilde{G} = \begin{pmatrix} \rho c & 0 & 1 \\ 0 & \gamma p & -\rho \\ -\rho c & 0 & 1 \end{pmatrix}, \quad \tilde{U} = \begin{pmatrix} u \\ \rho \\ p \end{pmatrix},$$

$$\tilde{\Lambda} = \begin{pmatrix} \tilde{\lambda}_1 & 0 & 0 \\ 0 & \tilde{\lambda}_2 & 0 \\ 0 & 0 & \tilde{\lambda}_3 \end{pmatrix} = \begin{pmatrix} u+c & 0 & 0 \\ 0 & u & 0 \\ 0 & 0 & u-c \end{pmatrix}$$

$$\tilde{F} = \begin{pmatrix} -\rho p u A'(x)/A(x) \\ 0 \\ -\rho p u A'(x)/A(x) \end{pmatrix}$$

The system of equations in the ξ-t coordinates is

$$G\frac{\partial U}{\partial t} + \Lambda G\frac{\partial U}{\partial \xi} = F \tag{5}$$

or

$$G_n\frac{\partial U}{\partial t} + \lambda_n G_n\frac{\partial U}{\partial \xi} = F_n, \qquad n = 1,2,3, \tag{6}$$

where

$$G(\xi,t) = \tilde{G}(x,t), \quad U(\xi,t) = \tilde{U}(x,t), \quad F(\xi,t) = \tilde{F}(x,t),$$

$$\Lambda = \begin{pmatrix} \lambda_1 & 0 & 0 \\ 0 & \lambda_2 & 0 \\ 0 & 0 & \lambda_3 \end{pmatrix},$$

$$\begin{aligned} \lambda_n &= \frac{\partial \xi}{\partial x}\tilde{\lambda}_n + \frac{\partial \xi}{\partial t} \\ &= (\tilde{\lambda}_n - (\xi - i)(v_{i+1}(t) - v_i(t)))/(x_{i+1}(t) - x_i(t)), \end{aligned}$$

$$v_i(t) = \frac{dx_i(t)}{dt}$$

is the speed of the i-th boundary line and G_n is the n-th row of G, F_n is the n-th component of F. For simplicity we omit the subscript n of λ_n, G_n and F_n in the difference equations.

In what follows we use the following notation:

$$\sigma = \frac{\lambda \Delta t}{\Delta \xi}, \qquad U_m^k = U(\xi_m, t_k).$$

For other quantities similar notations have been adopted.

If $|\sigma| < 1$, we take the following explicit scheme: At the interim level the approximation formula is

$$G_m^k U_m^{k+\frac{1}{2}} = \left(1 \mp \frac{\sigma_m^k}{2}\right) G_m^k U_m^k \pm \left(\frac{\sigma_m^k}{2}\right) G_m^k U_{m\mp 1}^k + 0.5 F_m^k \Delta t,$$

where we take the upper sign if $\sigma > 0$ or the lower sign if $\sigma < 0$. At the regular level the formula is

$$0.5((1 - B_m)(G_{m+1}^{k+\frac{1}{2}} + G_m^{k+\frac{1}{2}}) + B_m(G_m^{k+\frac{1}{2}} + G_{m-1}^{k+\frac{1}{2}}))U_m^{k+1}$$
$$= (1 - B_m)S_m^1 + B_m S_m^2,$$

where B_m is such a number that it is zero on the left boundary and one on the right boundary, the variation of B_m is smooth, and

$$
\begin{aligned}
S_m^1 \;=\; & 0.5(G_m^{k+\frac{1}{2}} + G_{m+1}^{k+\frac{1}{2}})U_{m+1}^k \\
& -0.5(0.5(\sigma_{m+1}^{k+\frac{1}{2}} + \sigma_m^{k+\frac{1}{2}}) + 1)(G_m^{k+\frac{1}{2}} + G_{m+1}^{k+\frac{1}{2}})(U_{m+1}^{k+\frac{1}{2}} - U_m^{k+\frac{1}{2}}) \\
& +0.5\Delta t(F_m^{k+\frac{1}{2}} + F_{m+1}^{k+\frac{1}{2}}), \\
S_m^2 \;=\; & 0.5(G_m^{k+\frac{1}{2}} + G_{m-1}^{k+\frac{1}{2}})U_{m-1}^k \\
& -0.5(0.5(\sigma_{m-1}^{k+\frac{1}{2}} + \sigma_m^{k+\frac{1}{2}}) - 1)(G_m^{k+\frac{1}{2}} + G_{m-1}^{k+\frac{1}{2}})(U_m^{k+\frac{1}{2}} - U_{m-1}^{k+\frac{1}{2}}) \\
& +0.5\Delta t(F_m^{k+\frac{1}{2}} + F_{m-1}^{k+\frac{1}{2}}).
\end{aligned}
$$

If $|\,\sigma\,| > 1$, we take the implicit scheme: At the interim level the approximation formula is

$$
\begin{aligned}
(1 + 0.5(\sigma_m^k + \sigma_{m+1}^k))&(G_m^k + G_{m+1}^k)U_{m+1}^{k+\frac{1}{2}} \\
+ (1 - 0.5(\sigma_m^k + \sigma_{m+1}^k))&(G_m^k + G_{m+1}^k)U_m^{k+\frac{1}{2}} \\
= (G_m^k + G_{m+1}^k)(U_m^k + U_{m+1}^k) &+ \Delta t(F_m^k + F_{m+1}^k).
\end{aligned}
$$

At the regular level the approximation formula is

$$
\begin{aligned}
(1 + 0.5(\sigma_m^{k+\frac{1}{2}} + \sigma_{m+1}^{k+\frac{1}{2}}))&(G_m^{k+\frac{1}{2}} + G_{m+1}^{k+\frac{1}{2}})U_{m+1}^{k+1} \\
+ (1 - 0.5(\sigma_m^{k+\frac{1}{2}} + \sigma_{m+1}^{k+\frac{1}{2}}))&(G_m^{k+\frac{1}{2}} + G_{m+1}^{k+\frac{1}{2}})U_m^{k+1} \\
= (G_m^{k+\frac{1}{2}} + G_{m+1}^{k+\frac{1}{2}})&(U_{m+1}^k + U_m^k) \\
- 0.5(\sigma_{m+1}^{k+\frac{1}{2}} + \sigma_m^{k+\frac{1}{2}})&(G_m^{k+\frac{1}{2}} + G_{m+1}^{k+\frac{1}{2}})(U_{m+1}^k - U_m^k) \\
+ 2\Delta t(F_m^{k+\frac{1}{2}} &+ F_{m+1}^{k+\frac{1}{2}}).
\end{aligned}
$$

The above scheme is one sided scheme on boundaries. This is a foundation for treating the values on the boundaries accurately.

(3) The unknown quantities at the inner points in the subregions are eliminated from the difference equations by using the elimination method and some relations containing only the unknowns on the boundaries are obtained. Then the unknown quantities on the boundaries can be obtained by using the boundary conditions and the relations containing only the quantities on the boundaries. The unknown quantities at the inner points can be computed by using the difference equations and the quantities at the boundary points.

(4) When the boundary line intersect with each other, we use the subroutine for solving the Riemann problem to determine the velocities and the properties of the new discontinuity lines, and the initial values of the physical quantities in each new subregion. Then a new coordinate transformation is adopted in order to turn the new subregions into strips on the ξ-t plane and the numerical computation will continue[2].

(5) How to calculate the shock wave which forms automatically as time goes on is a key to our calculation. This problem can be easily treated by the *shock-capturing* method, but it is a difficult problem for the *shock-fitting* method. In our code there is a subroutine which can be used to detect whether a characteristic line will intersect in a short time with another characteristic line of the same family. If they will intersect each other, a new internal boundary will be given along the characteristic line and it will be considered as a shock wave with vanishing strength at the very beginning. It becomes gradually a real shock wave when the characteristic lines of the same family gather to it.

(6) Because of the variation of the cross section, the physical quantities vary very quickly. For example, in the second subregion, the pressure varies from 1344 to 0.677. Such a large gradient of distributions of physical quantities brings difficulty to our computation. To overcome the difficulty, a self-adaptive mesh is adopted in our code. The number of mesh points in every subregion is adjusted by the gradient of distribution of each physical quantity and the length of subregion. The number of mesh points in a subregion will increase automatically if the gradient in it becomes large. Using such a sort of mesh the numerical results will have a high accuracy.

(7) Because the boundary between two media is treated as an internal boundary in our code, there is only one medium in each subregion. The problem of multimedium could be solved easily in our method.

4 Numerical Results

The initial data for the problem are given in section 2. The numerical results are calculated. From the numerical results, it is found that at the very beginning, a shock wave of the second family, a contact discontinuity and a shock wave of the first family appear near $x = 0$ because of the initial discontinuity decomposition. When the contact discontinuity propagates to $x = 24.4$, a new shock wave of the second family appears between the contact discontinuity and the former shock wave of the second family because of expansion of the tube. Its strength increases gradually. The pressure increases to 114 when the shock wave impinges on the closed end.

For the reason of safety, engineers want the maximum pressure on the closed end of tube to be as small as possible. Here, the pressure on the closed end of tubes with different diameters and lengths are calculated. The numerical results show that if we take $l_1 = 7.0, l_2 = 20.0, l_3 = 43.0, d_2 = 0.62, d_3 = 5.6$, the maximum pressure on the closed end will decrease to 60.94. From the numerical results it is seen that the above physical problem is very complicated. The numerical simulation is helpful to research of the problem.

Acknowledgments. The authors wish to thank prof. Dun Huang for the discussions in understanding these phenomena on this gas dynamics problem.

References

[1] Zhu Youlan, Zhong Xichang, Chen Bingmu and Zhang Zuomin, Difference Methods for Initial Boundary Value Problems and Flow Around Bodies, *Science Press*, Beijing, China, 1980; English edition, Springer, Heidelberg, 1988.

[2] Wu Xionghua and Zhu Youlan, Numerical Solution of Multimedium Flow with Various Discontinuities, *Journal of Computational Mathematics*, Vol. 1, No. 4, 303–316, 1983.

Remarks on Measure Branching Processes

Song Shiqi

Institute of Applied Mathematics, Academia Sinica,
Beijing, People's Republic of China

Abstract

Three aspects of measure branching processes are studied. The
first is about the martingale problem characterization of a measure
branching process. The second is about the canonical measures of
a measure branching process, where as the last one concerns a kind
of Dirichlet processes associated with a measure branching process.

Remark I. Let A be a Markov Pregenerator on R^n, whose domain of
definition contains the Schwatz space $S(R^n)$, such as the infinitesimal
generator of Brownian motion.

In S.Roelly-Coppoletta[5] (see also Dynkin [2], Dawson [1]) a measure
branching process associated with A is characterized as the solution of
the following martingale problem: Let $M_F(R^n)$ be the space of finite
positive measures on R^n provided with the weak convergence topology.
Let Ω be the space of all continuous mappings from $[0, \infty)$ into $M_F(R^n)$,
and $X = (X_t)$ be the coordinate process on Ω. Our martingale problem
is to find, for any $m \in M_F(R^n)$, a probability measure P^m on Ω such
that

$$
\left\{
\begin{array}{lll}
\text{(i)} & P^m(X_0 = m) = 1; & \\[2mm]
\text{(ii)} & \text{For all } f \in S, \ M^f \hat{=} \langle f, X_t \rangle - \langle f, X_0 \rangle - \displaystyle\int_0^t \langle Af, X_s \rangle ds & \\[1mm]
& \quad \text{is a martingale under } P^m; & \quad (1) \\[2mm]
\text{(iii)} & \text{For } f, g \in S, \ \langle M^f, M^g \rangle_t = \displaystyle\int_0^t \frac{r}{2} \langle fg, X_s \rangle ds & \\[1mm]
& \quad \text{under } P^m, \text{ where } r \text{ is a positive constant.} &
\end{array}
\right.
$$

(Note we use the symbol $\langle , \rangle$ to indicate both the quadratic variation of
M^f and M^g, and the integral of fg with respect to the random measure

X_s. But this would not cause any confusion, as the exact meaning of $\langle , \rangle$ will be clear in the contexts)

But we note that if G is a symmetric bilinear mapping from $S \times S$ into $C(R^n)$ such that $G(f, f)(x) \geq 0$ for every $f \in S$, the processes $\int_0^t \langle G(f, g), X_s \rangle ds$ can also be a candidate for the quadratic variation $\langle M^f, M^g \rangle$. Therefore we should have considered the following more general martingale problem : to find P^m on Ω such that

$$
\begin{cases}
\text{(i) and (ii) are satisfied and} \\
\text{(iii)}' \text{ For } f, g \in S, \ \langle M^f, M^g \rangle_t = \int_0^t \langle G(f, g), X_s \rangle ds.
\end{cases}
\tag{2}
$$

Nevertheless we will show in this paragraphs that if (2) has a solution for any $m \in M_F(R^n)$, the mapping G must take the form $G(f, g) = a(x)f(x)g(x)$ where $a(x) \in C(R^n)$. We will prove the fact only when $n = 1$. The methods we use do not depend essentially on the dimension.

Now let's consider the martingale problems (2) and suppose a solution exist for any $m \in M_F(R)$. Let $Z_t \equiv kX_{\sigma t}$ where $k \in R_+, \sigma \in R_+$ substituting Z into the formula (ii) and (iii)$'$, we get for $f \in S$,

$$
\begin{aligned}
L_t^f &\triangleq \langle kf, X_{\sigma t} \rangle - \langle kf, X_0 \rangle - \int_0^{\sigma t} \langle A(kf), X_s \rangle ds \\
&= \langle f, Z_t \rangle - \langle f, Z_0 \rangle - \sigma k \int_0^t \langle Af, X_{\sigma u} \rangle du \\
&= \langle f, Z_t \rangle - \langle f, Z_0 \rangle - \sigma \int_0^t \langle Af, Z_u \rangle du \\
&= M_{\sigma t}^{kf}
\end{aligned}
\tag{3}
$$

for $f \in S, g \in S$,

$$
\begin{aligned}
\langle L^f, L^g \rangle_t &= \langle M^{kf}, M^{kg} \rangle_{\sigma t} \\
&= \int_0^{\sigma t} \langle G(kf, kg), X_s \rangle ds \\
&= \sigma \int_0^t k^2 \langle G(f, g), X_{\sigma u} \rangle du \\
&= \sigma k \int_0^t \langle G(f, g), Z_u \rangle du.
\end{aligned}
\tag{4}
$$

Let $\sigma = \dfrac{1}{k}$. Then the law of Z_t under $P^{\frac{1}{k}m}$ solves the following martingale problem:

$$
\left\{
\begin{array}{ll}
\text{(i)} & P(Z_0 = m) = 1; \\[4pt]
\text{(ii)} & \forall f \in S, \\[4pt]
& \qquad M^f \hat{=} \langle f, Z_t \rangle - \langle f, Z_0 \rangle - \dfrac{1}{k} \displaystyle\int_0^t \langle Af, Z_u \rangle du \\[4pt]
& \qquad \text{is a martingale under } P; \\[4pt]
\text{(iii)} & \forall f \in S,\ g \in S, \\[4pt]
& \qquad \langle M^f, M^g \rangle_t = \displaystyle\int_0^t \langle G(f,g), Z_u \rangle du.
\end{array}
\right.
\tag{5}
$$

If we let $k \longrightarrow \infty$, the Aldous tightness criterion (see Roelly-Coppoletta [5] or J.Jacod Shirgaev.A.N [3]) permit us to conclude that for our propose it is sufficient to assume that $A \equiv 0$ identically.

Hence our hypothesis reduces to the following: For any $m \in M_F(R)$, there exist a P^m on Ω such that

$$
\left\{
\begin{array}{ll}
\text{(i)} & P^m(X_0 = m) = 1; \\[4pt]
\text{(ii)} & \forall f \in S,\ M^f \hat{=} \langle f, X_t \rangle - \langle f, X_0 \rangle \\[4pt]
& \qquad \text{is a } P^m\text{-martingale;} \\[4pt]
\text{(iii)} & \forall f \in S,\ g \in S, \\[4pt]
& \qquad \langle M^f, M^g \rangle_t = \displaystyle\int_0^t \langle G(f,g), X_s \rangle ds.
\end{array}
\right.
\tag{6}
$$

Let $x \in R$ and $U_\varepsilon(x)$ be the ball $\{ y \in R, |y - x| < \varepsilon \}$. Suppose $f \in S, f \geq 0$, and $f(y) = 0$ for $y \in U_\varepsilon(x)$. Let $m \in M_F(R)$ be such that support $m \subset U_\varepsilon(x)$. Then

$$
E_m(\langle f, Z_t \rangle) = \langle f, m \rangle = 0
\tag{7}
$$

for all $t \geq 0$. The positivity of f implies that $\langle f, Z_t \rangle = 0\, P^m$ almost surely.

Because S^+ is total in $C(R)$ and Z_t is finite measure valued, we can conclude that $\langle f, Z_t \rangle = 0,\ P^m - a,s$ for all $f \in S$ such that $f(y) = 0$ for $y \in U_\varepsilon(x)$. Form this it results that $\langle G(f,f), X_t \rangle = 0,\ P^m - a,s$. In particular $\langle G(f,f), m \rangle = 0$. As m is an arbitrary measure subject to supp $m \subset U_\varepsilon(x)$ and $G(f,f)(.)$ is continuous, we get $G(f,f)(x) = 0$. Thus we have proved that G possesses a local property: if $f \in S, g \in S$ and $f = g$ in a neighborhood of $x, G(f,h)(x) = G(g,h)(x)$ for all $h \in S$.

Until now we have not needed extra properties of G. From now on we suppose that G is continuous in S in the sense that if $f_n \longrightarrow f,\ n \longrightarrow \infty$, in S, and $x \in R$, $\sup_{y \in U_e(x)} |G(f - f_n, f - f_n)(y)| \longrightarrow 0$, $n \longrightarrow \infty$ for some $\varepsilon > 0$. Under this extra condition we can expand $G(f, g)(x)$ in a series. In fact, let $x \in R$, let $f \in S$, $g \in S$ be such that in a neighborhood of x, they are polynomial, i.e.

$$\begin{aligned}
f(y) &= \sum_{i=0}^{n} \frac{1}{i!} f^{(i)}(x)(y - x)^i, \\
g(y) &= \sum_{i=0}^{n} \frac{1}{i!} g^{(i)}(x)(y - x)^i.
\end{aligned} \tag{8}$$

Let (φ_i) be any family of functions in S such that $\varphi_i(y) = (y - x)^i$ in a neighborhood of x. Then by the local property of G we have

$$\begin{aligned}
G(f, g)(x) &= G\left(\sum_{i=0}^{n} \frac{1}{i!} f^{(i)}(x)\varphi_i, \sum_{j=0}^{n} \frac{1}{j!} g^{(j)}(x)\varphi_j \right)(x) \\
&= \sum_{i,j=0}^{n} f^{(i)}(x) g^{(j)}(x) a_{i,j}(x)
\end{aligned} \tag{9}$$

where $a_{ij}(x) = \frac{1}{i!}\frac{1}{j!} G(\varphi_i, \varphi_j)(x)$. Obviously $a_{ij}(x)$ does not depend on the choice of φ_i. As the totality of functions f and g with required properties is dense in S, the formula (9) holds also for arbitrary elements in S, because of the continuity in S of G. Note that the constructions of $a_{ij}(x)$ are pointwise, but the same constructions show that we can paste a_{ij} up as a continuous function.

The matrices $(a_{ij}(x))$ is obviously non negative definite. Let $\xi(x), \varsigma(x)$ be sequences of functions. We denote $(\xi, \varsigma)(x) \doteq \sum a_{ij}(x)\xi_i(x)\varsigma_j(x)$. Then we have the Schwatz inequality:

$$(\xi + \varsigma, \xi + \varsigma)^{\frac{1}{2}} \geq (\xi, \xi)^{\frac{1}{2}} - (\varsigma, \varsigma)^{\frac{1}{2}}. \tag{10}$$

Let k, l be two positive numbers. Let $\Psi \in S$ be such that $\Psi(x) = 1$ for $|x| \leq l$, $\Psi(x) = 0$ for $|x| > 2l$. Let f^k be a function in S such that $\| f^k \|_\infty \leq 1$ and there exist an open set $I^k \subset [-l, l]$ such that

(i) There is a $n \in N$ independent of k such that there is $(x_1, x_2, \cdots, x_n) \in R^n$ such that $[-l, l] \subset \bigcup_{i=1}^{n}(I^k + x_i)$ where $I^k = x_i \doteq \{y : y = x + x_i, x \in I^k\}$.

(ii) $(f^k)'(x) = k$, $(f^k)^{(n)}(x) = 0$ for $x \in I^k$, $n > 1$.

(iii) $(f^k)(x) = 0$,   for $|x| > l$.

It is easy to see that such functions exist.

Using the well known $B - G - D$ inequality for continuous martingales, we obtain

$$E_m \left(\int_0^\gamma \langle G(f^k, f^k), X_s \rangle ds \right)$$

$$\leq C E_m (\sup_{s \leq \gamma} |\langle f^k, X_s \rangle - \langle f^k, X_0 \rangle|^2)$$

$$\leq C E_m (\sup_{s \leq \gamma} |\langle \Psi, X_s \rangle - \langle \Psi, X_0 \rangle|^2)$$

$$\leq C E_m \left(\int_0^\gamma \langle G(\Psi, \Psi), X_s \rangle ds \right) + C \langle \Psi, m \rangle^2 \tag{11}$$

where γ is any stopping time which makes all quantities in (11) finite.

On other hand,

$$\int_0^t \langle G(f^k, f^k) \rangle ds$$

$$\geq \int_0^t ds \int_I X_s(dx) G(f^k, f^k)(x) \quad \text{where } I = I^k$$

$$= \int_0^t ds \int_I X_s(dx)(\xi + \varsigma, \xi + \varsigma)(x)$$

where $\xi(x) = (f^k(x), 0, 0, \cdots), \quad \varsigma(x) = (0, (f^k(x))', 0, \cdots)$

$$\geq \int_0^t ds \int_I X_s(dx)((\varsigma, \varsigma)^{\frac{1}{2}}(x) - (\xi, \xi)^{\frac{1}{2}}(x))^2$$

$$\geq \int_0^t ds \int_I X_s(dx)(\sqrt{a_{11}(x)}|(f^k(x))'| - \sqrt{a_{00}(x)}|f^k(x)|)^2$$

$$\geq \frac{1}{2} \int_0^t ds \int_I X_s(dx)(a_{11}(x)k^2 - 2a_{00}(x)). \tag{12}$$

From (11), (12) it results

$$E_m \int_0^\gamma ds \langle a_{11} 1_I, X_s \rangle \cdot k^2$$

$$\leq 2 E_m \left(\int_0^\gamma ds \langle G(f^k, f^k), X_s \rangle + \int_0^\gamma ds \langle a_{00}, X_s \rangle \right)$$

$$\leq 2C E_m \left(\int_0^\gamma \langle G(\Psi, \Psi), X_s \rangle ds + \int_0^\gamma ds \langle a_{00}, X_s \rangle \right)$$

$$+ C \langle \Psi, m \rangle^2. \tag{13}$$

Now applying the property i of I^k and noting that the procedure to obtain (13) is applicable for $I + x_i$ and $g(x) \doteq f(x - x_i)$ too, we have thus

$$k^2 \cdot E_m \int_0^\gamma ds \langle a_{11} 1_{[-l,l]}, X_s \rangle$$

$$\leq C \left\{ E_m \left(\int_0^\gamma \langle G(\Psi, \Psi), X_s \rangle ds + \int_0^\gamma ds \langle a_{00}, X_s \rangle \right) + \langle \Psi, m \rangle^2 \right\}. \quad (14)$$

As the right side of (14) does not depend on k and l, we must have $E_m \int_0^\gamma ds \langle a_{11}, X_s \rangle = 0$. However as γ is arbitrary in the class of stopping times which make r.h.s of (14) finite and $\langle a_{11}, X_s \rangle$ is continuous, we must have $\langle a_{11}, X_s \rangle \equiv 0, P_m - a, s$, In particular $\langle a_{11}, m \rangle = 0$. Because m is arbitrary in $M_F(R)$, $a_{11}(x)$ must be null.

Now we continue the procedure using the same idea, we can prove by induction that all $a_{ij} \equiv 0$ for $i \geq 1$ or $j \geq 1$. (Here we need use the non negativity of (a_{ij}).) We arrive thus our conclusion.

To end this section we would like to point out that if solutions of (2) exist only for $m's$ running over a subset of $M_F(R)$, G may take other forms. A simple example of it is to consider a family $(B_t^x)_{x \in R}$ of independent brownian motion on R index by $x \in R$. We suppose $B_0^x = x$. For any point measure $m = \sum_{i=1}^n \delta_{x_i}$, the process $X_t \hat{=} \sum_{i=1}^n \delta_{B_t^{x_i}}$ solve the martingale problem (2) with $G(f,g) = f'(x)g'(x)$.

Another remark is that if the martingale problem (2) has a uniqueness property, the solutions P^m must have the branching property: for $m \in M_F(R)$, $m' \in M_F(R)$, $P^{m+m'} = P^m * P^{m'}$. Using branching property, Watanabe's results [6] show that our conclusion is true also when we replace R by a space with only finite points.

Remark II. In this section we consider a measure branching process X characterized by the martingale problem (1). It is known that for any fixed t, and $\mu \in M_F(R)$ the law of X_t under P^μ is infinitely divisible and there exist two families of measures: $(\alpha_t(x, dy)) \subset M(R)$ and $(\Gamma_t(x, dv)) \subset M(M_F(R))$ such that

$$\forall f \in C^+(R), P^\mu(e^{-\langle f, X_t \rangle}) = e^{-\int_R \langle f, \alpha_t(x) \rangle + \langle 1 - e^{\pi_f}, \Gamma_t(x) \rangle \mu(dx)} \quad (15)$$

where π_f is the function on $M_F(R)$ defined by $v \longrightarrow \langle f, v \rangle$. The families α and Γ are called canonical measures associated with X_t. In this section we are going to determine the families α and Γ.

Let W_t be the semigroup on $C_b^+(R)$ determined by the relation:

$$P_\mu(e^{-\langle f, X_t \rangle}) = e^{\langle W_t f, \mu \rangle}. \quad (16)$$

By Dynkin [2] p.17 there exists a sequence of semigroup $(W^{(k)})$ such that W is the U-limit of $W^{(k)}$ as $k \longrightarrow \infty$. (see Dynkin[2] for the definition). For fixed $k, W_t^{(k)}$ is determined by the equation:

$$W_t = \int_0^t e^{-2ks} T_s \Psi_k(W_{t-s}) ds + e^{-2kt} T_t \tag{17}$$

where T_s is the semigroup generated by A and $\Psi_k(x)$ is defined as $2k^2(1 - e^{-\frac{x}{k}})$. In this formular two kinds of operations are concerned, i.e. the operations $W \longrightarrow \Psi_k(W)$ and $U \longrightarrow TU \hat{=} \int T(x, dy) U(y)$. We need to know how the canonical measures change after each operation. Let $\mathcal{A}$ be the space of operators defined in Dynkin [2] p.14. Let $U \in \mathcal{A}$. Then there is a family $(P_\mu^U)_{\mu \in M_F(R)}$ of probability measures on $M(M_F(R))$ such that $P_\mu^U(e^{-\pi_f}) = e^{-\langle Uf, \mu \rangle}$, $\mu \in M_F(R)$, $f \in C^+(R)$. we have the following equalities:

$$\begin{aligned}
\Psi_k U f(x) &= 2k^2(1 - e^{-\frac{1}{k} U f(x)}) \\
&= 2k^2(1 - e^{U f, \frac{1}{k} \delta_x}) \\
&= 2k^2 P_{1/k}^U(1 - e^{\pi_f}).
\end{aligned} \tag{18}$$

Hence we conclude that the canonical measures for $\Psi_k U$ is given by $\alpha(x) \equiv 0$ and $\Gamma(x, dv) = 2k^2 P_{\delta_x/k}^U(dv)$. Now look at TU. If α^U, Γ^U are the canonical measures of U, a direct calculation shows that the canonical measures for TU are given by $\alpha(x) = \int T(x, dy) \alpha^U(y)$.

Now apply these result onto the formula (17). If we denote by $(P_\mu^{t,(k)})_{\mu \in M(R)}$ the probability family associated with $W_t^{(k)}$, we have got for $f \in C_b^+(R)$

$$\begin{aligned}
W_t^{(k)} f(x) &= \int_0^t e^{-2ks} T_s \Psi_k(W_{t-s} f(x)) ds + e^{-2kt} T f(x) \\
&= \left\langle 1 - e^{\pi_f}, 2k^2 \int_0^t e^{-2ks} ds \int_R T_s(x, dy) P_{\frac{1}{k} \delta_y}^{t-s,(k)} \right\rangle \\
&\quad + e^{-2kt} T_t f(x).
\end{aligned} \tag{19}$$

From this formula it results that canonical measures of $W_t^{(k)}$ are given by $\alpha_t(x) = e^{-2ks} T_t(x)$ and $\Gamma_t^{(k)}(x) = 2k^2 \int_0^t e^{-2ks} ds \int_R T_s(x, dy) P_{\delta_y/k}^{t-s,(k)}$. Let $\Lambda_t^{(k)}(x, dudv)$ be the image measure on $[0, \infty] \times M_1(\overline{R})$ of $\frac{\pi_1}{1+\pi_1}$.

$\Gamma_t^{(k)}(x, dv)$ by the mapping $v \longrightarrow (\langle 1, v \rangle, \frac{1}{\langle 1, v \rangle} \cdot v)$. where $M_1(\overline{R})$ be the space of probability measure on $\overline{R}$, where $\overline{R}$ is the one point compactification of R. Note that $\Gamma_t^{(k)}(x, dv)$ does not charges the set $\{0\}$. In fact,

$$
\begin{aligned}
\Gamma_t^{(k)}(x, \{0\}) &= 2k^2 \int_0^t e^{-2ks} ds \int_R T_s(x, dy) \lim_{\lambda \to \infty} P_{\frac{1}{k} \delta_y}^{t-s, (k)} (e^{-\lambda \pi_1}) \\
&= 2k^2 \int_0^t e^{-2ks} ds \int_R T_s(x, dy) \lim_{\lambda \to \infty} e^{-\frac{1}{k} W_{t-s}^{(k)}(\lambda)(x)} \\
&\leq 2k^2 \int_0^t e^{-2ks} ds \int_R T_s(x, dy) \lim_{\lambda \to \infty} e^{-\frac{\lambda}{k} e^{-2k(t-s)}} \\
&= 0.
\end{aligned}
\tag{20}
$$

Thus $\Lambda_t^{(k)}(x, dudv)$ is well defined.

Let us show that $(\Lambda_t^{(k)}(x))$ are uniformly bounded by 1. Let us compute first of all $P_{\delta_y}^{t-s,(k)}(\pi_1)$.

$$
\begin{aligned}
P_{\delta_y}^{t-s,(k)}(\pi_1) &= -\frac{\partial}{\partial \lambda} P_{\delta_y}^{t-s,(k)}(e^{-\lambda \pi_1})_{\lambda=0} \\
&= -\frac{\partial}{\partial \lambda} e^{-W_t^{(k)}(\lambda)(y)} \big|_{\lambda=0} \\
&= e^{-W_t^{(k)}(0)(y)} \frac{\partial}{\partial \lambda} W_t^k(0)(y).
\end{aligned}
\tag{21}
$$

However $W_t^{(k)}$ satisfies:

$$
W_t f(x) - \int_0^t T \left[2k^2 \left(\frac{W_{t-s} f(x)}{k} - 1 + e^{-\frac{W_{t-s} f(x)}{k}} \right) \right] ds + T_t f(x).
\tag{22}
$$

If $f(x) \equiv \lambda$, we get $W_t^{(\lambda)}(x) \leq \lambda$. In particular $W_t^k(0) = 0$. By the uniqueness of equation (22) we can see that $W_t^k(\lambda)$ does not depend on x. Hence the equation (22) becomes:

$$
W_t(\lambda) = -\int_0^t 2k^2 \left(\frac{W_{t-s}(\lambda)}{k} - 1 + e^{-\frac{1}{k} W_{t-s}(\lambda)} \right) ds + \lambda.
\tag{23}
$$

Derivating both side of (23) at $\lambda = 0$, we get $\frac{\partial}{\partial \lambda} W_t^{(k)}(0) = 1$. Therefore $P_{\delta_y}^{t-s,(k)}(\pi_1) = 1$.

Now consider the measure $\Lambda_t^{(k)}$

$$\Lambda_t^{(k)}(x, [0, \infty] \times M_1(\overline{R}))$$

$$\leq 2k^2 \int_0^t e^{-2ks} ds \int_R T_s(x, dy) P_{\frac{1}{k}\delta_y}^{t-s,(k)}(\pi_1)$$

$$= 2k \int_0^t e^{-2ks} ds \int_R T_s(x, dy) P_{\delta_y}^{t-s,(k)}(\pi_1)$$

$$\leq 1. \tag{24}$$

Here the second equality is resulted from the branching property of $P_\mu^{t-s,(k)}$.

The space $[0, \infty] \times M_1(\overline{R})$ being compact and $(\Lambda_t^{(k)}(x))$ being uniformly bounded by 1, it is a tight family. Let $\Lambda_t(x)$ be any its limiting points. We will show that $\Lambda_t(x)$ concentrate on $(0, \infty) \times M_1(R)$. First of all we see

$$W_t f(x) = \int (1 - e^{-u\langle f, v \rangle}) \frac{u+1}{u} \Lambda_t(x, dudv). \tag{25}$$

Therefore we have

$$P_{\delta_x}\left(e^{-\lambda\langle 1, X_t \rangle}\right)$$
$$= e^{-\int (1-e^{-\lambda u})\frac{1+u}{u}\Lambda_t(x, dudv)}$$
$$= e^{-\lambda \Lambda_t(x, \{0\} \times M_1(\overline{R})) - \int_{u \neq 0}(1-e^{-\lambda u})\frac{1+u}{u}\Lambda_t(x, dudv)}$$
$$= e^{-\frac{\lambda}{1+\frac{1}{2}\Lambda_t}}. \tag{26}$$

Here the last equality is from Dynkin [2]. Now let $\lambda \longrightarrow \infty$, we see that

$$\lim P_{\delta_x}\left(e^{-\lambda\langle 1, X_t \rangle}\right) = e^{-\frac{2}{t}} > 0$$

therefore we must have $\Lambda_t(x, \{0\} \times M_1(\overline{R})) = 0$ and

$$\int \frac{1+u}{u} \Lambda_1(x, dudv) = \frac{2}{t}.$$

To prove that $\Lambda_t(x, \{\infty\} \times M_1(\overline{R})) = 0$, we turn back to $\Lambda_t^{(k)}$. Let

$\gamma \in R_+$, we have

$$\Lambda_t^{(k)}(x, [\gamma, \infty] \times M_1((\overline{R})))$$

$$= 2k^2 \int_0^t e^{-2ks} ds \int_R T_s(x, dy) P_{\frac{1}{k}\delta_y}^{t-s,(k)}\left(\frac{\pi_1}{1+\pi_1}; \pi_1\rangle \gamma\right)$$

$$\leq 2k^2 \int_0^t e^{-2ks} ds \int_R T_s(x, dy) P_{\frac{1}{k}\delta_y}^{t-s,(k)}(\pi_1)\frac{1}{1+\gamma}$$

$$\leq \frac{1}{1+\gamma}. \tag{27}$$

In taking limit as $k \longrightarrow \infty$, we see that $\Lambda_t(x, [\gamma, \infty] \times M_1(\overline{R})) \leq \frac{1}{1+\gamma}$. We obtain the desired result by letting $\gamma \longrightarrow \infty$. Finally let $\varphi^\gamma \hat{=} 1_{x \in [-\gamma, \gamma]}$. We have

$$P_{\delta_x}(e^{-\langle \varphi^\gamma, X_t \rangle}) = e^{-\int (1-e^{-u\langle \varphi, v\rangle})\frac{1+u}{u}\Lambda_t(x, dudv)}. \tag{28}$$

When $\gamma \longrightarrow \infty$, the left hand side of (28) tends to 1 because $X_t \in M_F(R)$, while the right hand side tends to

$$\exp\left\{-\int (1 - e^{-u\langle 1_{\{\Delta\}}, v\rangle})\frac{1+u}{u}\Lambda_t(x, dudv)\right\}$$

where $\{\Delta\} = \overline{R} - R$. This proves that for $\Lambda_t(x)$ almost all v, $v(\{\Delta\}) = 0$. The concentration of $\Lambda_t(x)$ on $[0, \infty] \times M_1(R)$ is thus proved.

Now we can say that the canonical measure of W_t is given by

$$\alpha(x) \equiv 0 \quad \text{and}$$
$$\int \Gamma_t(x, d\mu) F(\mu) \hat{=} \int F(u \cdot v)\frac{1+u}{u}\Lambda_t(x, dudv). \tag{29}$$

As $\alpha(x) \equiv 0$, we can translation the equation for W into an equation for $\Gamma_t(x)$. In fact we have

$$\int (1 - e^{-\pi_y - \lambda \pi_f})\Gamma_t(x, d\mu)$$

$$= W_t(g + \lambda f)$$

$$= -\int_0^t T_s[W_{t-s}^2(g + \lambda f)]ds + T_t(g + \lambda f). \tag{30}$$

If $f \equiv 1$. and if we derivate at $\lambda = 0$, we get

$$L_t(x, g) = -2\int_0^t T_s(x, [W_{t-s}(g) \cdot L_{t-s}(\cdot, g)])ds + 1 \tag{31}$$

where $L_t(x,g) \triangleq \langle e^{\pi_y} \pi_1, \Gamma_t(x) \rangle$. Obviously $\Gamma_t(x)$ is uniquely determined by $L_t(x,g), g \in C^+(R)$.

Remark III. The third remark is very short. Let $f \in \mathcal{D}(A)$, then

$$\int_0^t \langle Af, X_s \rangle ds = \langle f, X_t \rangle - \langle f, X_0 \rangle - M_t^f.$$

Since

$$\langle M^f, M^f \rangle_t = \int_0^t \frac{\gamma}{2} \langle f^2, X_s \rangle ds,$$

if $\| f_n - f \|_\infty \longrightarrow 0$ as $n \longrightarrow \infty$, $\langle f_n, X_t \rangle - \langle f_n, X_0 \rangle - M_t^{f_n}$ converges in probability to a process V_t^f. It is easy to see that V_t^f does not depend on the choice of (f_n) and it is a process with null quadratic variation. Therefore we proved that for any $f \in C_b(R), \langle f, X_t \rangle$ is a Dirichlet process.

References

[1] Dawson, D.A.(1975) Stochastic evolution equations. J. Multi var Anal. 3, 1–52.

[2] Dynkin, E.B(1988) Superprocesses and their linear additive functionals. Transact. Amer Math. Soc. to appear.

[3] Jacod, J and Shirgaev, A.N. *Limit Theorems for Stochastic Processes*, Springer-Verlag 1987.

[4] Kallenberg, O. *Random Measures*, Academic Press 1983.

[5] Roelly-Coppoletta, S(1986) A criterion of convergence of measure-valued processes: application to measure pranching processes. Stochastics 17, 43–65.

[6] Watanabe, S.(1969) On two dimensional Markov processes with branching property. Trans. Math. Soc. February 1969. 447–446.

Yang-Mills Functional on P_3

Shuguang Wang[*]
Oxford University

Abstract

This is an exercise designed to understand elementary aspects of Yang-Mills theory, especially, the aspects in stable bundles and the Penrose twistor map. Let P_3 denote the 3-dimensional complex projective space, and consider a nontrivial rank-2 complex vector bundle E on P_3 such that $c_1(E) = 0$ and $c_2(E) = 1$, on which the Yang-Mills functional YM can be defined as usual. The main purpose is to describe the minimum submanifold M_{P_3} of YM in various ways: firstly thanks to the complex structure on P_3, the description of M_{P_3} can be converted to that of the moduli space of certain stable bundles on P_3, the latter being known due to the previous works by Hartshorne and so on. Then through the Penrose map, it is related to the pull-back bundles of instatons over S^4.

Topologically there are two rank-2 complex vector bundles on P_3 with Chern classes $c_1 = 0$ and $c_2 = 1$. They have the Atiyah-Ress invariants $\alpha = 0, 1$ respectively. Although both of them admit holomorphic structures, only the one with $\alpha = 0$ can be given a stable holomorphic structure (see below). Let us fix such a vector bundle $\pi : E \to P_3$ and a hermitian metric h on E throughout this section. P_3 is endowed with the standard Kaehler metric ω. We then consider the Yang-Mills functional on the set of unitary connections on E,

$$YM : \mathcal{A} \to \mathcal{R}, \; A \to \int_{P_3} |F|^2 * 1$$

where F is the curvature of a connection A; and denote the minimum submanifold of YM by $\mathcal{M}$. We also let $\mathcal{G}$ be the unitary gauge group and $M_{P_3} = \mathcal{M}/\mathcal{G}$.

---

[*]The author is financially supported by K. C. Wong Education Foundation Ltd.

Let $\Omega_c^2(E)$ be the complexification of $\Omega^2(E) = \Gamma^\infty(E \otimes \Lambda^2 T^* M)$. Using the Kaehler form (metric) ω on P_3 we obtain an orthogonal decomposition

$$\Omega_c^2(E) = \Omega^{0,2}(E) + \Omega^{2,0}(E) + \{\omega^\perp\} + \{\omega\}.$$

Correspondingly we have for the curvature F of a connection A

$$F = F_{0,2} + F_{2,0} + F^\perp + \hat{F}.$$

Since E can be reduced to a $SU(2)$-vector bundle, the curvature is traceless. Therefore as in [6; §1 of Chapter 2], we see

$$YM(A) = \|F\|^2 = 8\pi^2 + \|F_{0,2}\|^2 + \|F_{2,0}\|^2 + \|\hat{F}\|^2.$$

Hence $A \in \mathcal{M}$ if and only if $F_{0,2} = F_{2,0} = \hat{F} = 0$ (if the latter kind of connection does exist — this will be seen as we proceed), or equivalently A is a Hermitian-Einstein connection (for $c_1(E) = 0$). But E is indecomposable, this yields a natural map

$$\tilde{\Psi} : \mathcal{M} \to \{\text{stable homomorphic structures on } E\}$$

by using the metric h on E, see [12] or [13]. Descending both sides one has

$$\Psi : M_{P_3} \to \{\text{isomorphism classes of stable holo. structures on } E\}$$

The well-defindness of Ψ results from the fact that $\tilde{\Psi}(A)$ is isomorphic to $\tilde{\Psi}(A')$ iff A is $\mathcal{G}^c$-equivalent to A'. Here $\mathcal{G}^c$ is the complex gauge group which extends the action of the unitary gauge group $\mathcal{G}$. Cf [7; page5]
Description 1. The composition of Ψ with the inclusion l below is a diffeomorphism:

$$M_{P_3} \xrightarrow{\Psi} \{\text{iso. classes of stable holo. structures on } E\}$$
$$\xrightarrow{l} M(0,1) = \text{moduli space of stable holo. vector}$$
$$\text{bundles on } P_3 \text{ with } c_1 = 0 \text{ and } c_2 = 1.$$

Proof. The map $l \circ \Psi$ is injective: if $[A]$ and $[A']$ in M_{P_3} are such that $\Psi[A] = \Psi[A']$, i.e., the induced holomorphic structures $\tilde{\Psi}(A)$ and $\tilde{\Psi}(A')$ are isomorphic, then A is $\mathcal{G}$-equivalent to A' since $\tilde{\Psi}(A)$ and $\tilde{\Psi}(A')$ are stable. See [7; Corollary 9].

The map $l \circ \Psi$ is surjective: Let $E' \to P_3$ be any stable holomorphic vector bundle with $c_1(E') = 0$ and $c_2(E') = 1$. After Description 3 we will prove that E' must come from the Horrocks construction. So by [1; Proposition (1.4) of Chapter IV] one has $H^0(E'(-2)) = 0$, $H^1(E'(-2)) = 0$. Hence the Atiyah-Rees invariant is given by

$$\alpha(E') = \dim H^0(E'(-2)) + \dim H^1(E'(-2)) \bmod 2 = 0$$

See [3]. Since we have chose E so that $\alpha(E) = 0$, E' is isomorphic to E topologically [3]. This shows that the inclusion l is in fact surjective. Our errand is then reduced to prove Ψ is surjective. This follows from the well-known result in [8] or [15]. We may give the details here. Let ξ be any holomorphic structure on E. The present situation is seen in the diagram

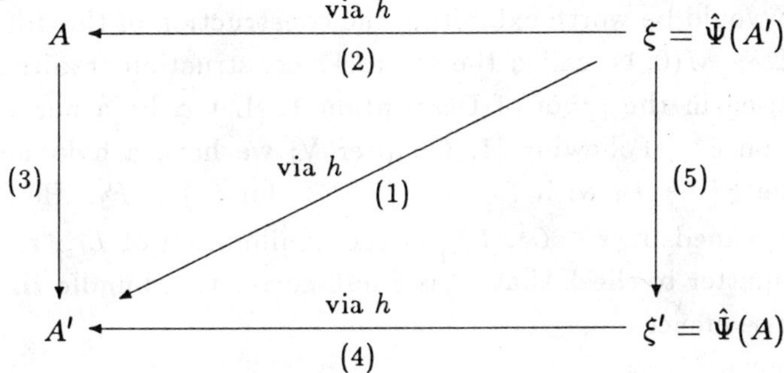

The data are explained as follows. (1) According to [8; Proposition 1] there is a metric h' on E such that ξ and h' induce a Hermitian-Einstein connection A'. (2) $A =$ connection induced by ξ and h. (3) A is $\mathcal{G}^c$-equivalent to A'. (4) ξ' is induced by A' and h. (5) That A is $\mathcal{G}^c$-equivalent to A' implies ξ is isomorphic to ξ'. To sum up $[\xi] = [\xi'] = \Psi([A'])$ and A' is Hermitian-Einstein, i.e. $[A'] \in M_{P_3}$. Hence Ψ is surjective as needed.

Remark. In the proof it is seen that if $V \to P_3$ is a complex rank-2 vector bundle with $c_1(V) = 0$, $c_2(V) = 1$ and $\alpha(V) = 1$, then V can not admit any stable holomorphic structure. So if the Yang-Mills functional has the absolute minimum, say m, then $m > 8\pi^2$. It is reasonable to make

Conjecture. The Yang-Mills functional on such V can not take its infimum $(= 8\pi^2?)$.

Compare with the case of S^6 [5; Theorem 7.11].

Remark. The same can be asked if one substitutes $c_2(V) = 2$ for $c_2(V) = 1$. In fact these are the only cases in the sense of [11; Corollary 8.4].

Both [14; page 364] and [11] characterize $M(0,1)$, which in turn gives descriptions of our M_{P_3}.

Description 2. There is a natural diffeomorphism

$$M_{P_3} \to P_5 - G(1,3)$$

where $G(1,3)$ is the Grassmannian of lines in P_3.

Denote $\mathcal{E} = \{\text{nonsingular skew forms on } C^4\}/C^*$. ($C$ is the complex number field). Description 2 is the same as

Description 3. There is a natural diffeomorphism $M_{P_3} \to \mathcal{E}$.

It would be worth exhibiting the construction of the diffeomorphism $H : \mathcal{E} \to M(0,1)$, called the Horrocks construction traditionally, which was used in the proof of Description 1. Let α be a nonsingular skew form on C^4. Following [1; Chapter V] we have a holomorphic vertor bundle $\xi^\alpha \to P_3$ with $\xi^\alpha_{(z)} = L^\alpha_{(z)}/L_{(z)}$ for $(z) \in P_3$. Here $L_{(z)}$ is the line spanned by $z \in C^4$, $L^\alpha_{(z)}$ is the annihilation of $L_{(z)}$ rel α. It is an easy matter to check that ξ^α is a null-correlation bundle viz. there is an exact sequence

$$0 \to \xi^\alpha \to T_{P_3}(-1) \to O_{P_3}(1) \to 0$$

Cf[14;§4.2]. By a conclusion there [14; Theorem 1.3.1] ξ^α is stable with $c_1 = 0$ and $c_2 = 1$. Thus $[\xi^\alpha] \in M(0,1)$ i.e. $H : \mathcal{E} \to M(0,1)$ is defined by $H([\alpha]) = [\xi^\alpha]$. We also like to give an alternative way to see why H is surjective. Let $[\xi] \in M(0,1)$, then $H^1(\xi(-2)) = 0$ by [14;Lemma 4.3.1], and there is a line on which the restriction of ξ is trivial by [4;Theorem1]. Hence [1;(2.3) on p.85] asserts that $\xi = \xi^\alpha$ for some $\alpha \in \mathcal{E}$.

Let us now consider the pull-backs of instanton bundles over S^4 via the Penrose map $P^3 \to S^4$. As is well-known, these are stable vector bundles over P^3 with extra restraints. In terms of skew forms on C^4 (note Description 3 above) this leads to

Description of instantons over S^4. There is a natural diffeomorphism between the moduli space M_{S^4} of instantons over S^4 and the set of nonsingular skew forms on C^4 (up to positive scalars) with matrix expressions in the form of

$$\begin{pmatrix} 0 & \tau & b & c \\ -\tau & 0 & -\bar{c} & \bar{b} \\ -b & \bar{c} & 0 & \rho \\ -c & -\bar{b} & -\rho & 0 \end{pmatrix}$$

where τ, ρ are positive reals, $b, c \in \mathcal{C}$ and $\rho\tau - (|b|^2 + |c|^2) > 0$. See [9].

Remark. In particular all pull-backs of instantons, which correspond to stable bundles over P_3, are minimums of the Yang-Mills functional YM on P_3 because of Description 1. There is also a direct proof in [1; 2 of Chapter IV].

Before giving the fourth description of M_{P_3} we need to review some background.

Let $V \to M$ be any vector bundle and $u : M \to M$ a diffeomorphism, suppose u is homotopic to the identity, then there is a (not canonical) diffeomorphism $\tilde{u}$ such that

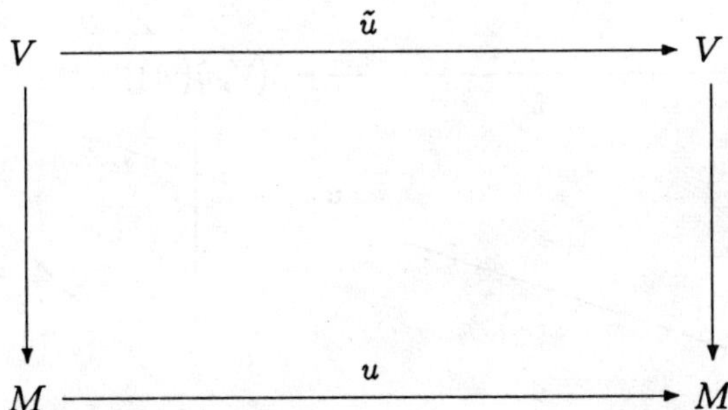

commutes and $\tilde{u}$ are fiberwise isomorphisms. In fact one has a cubic commutative diagram

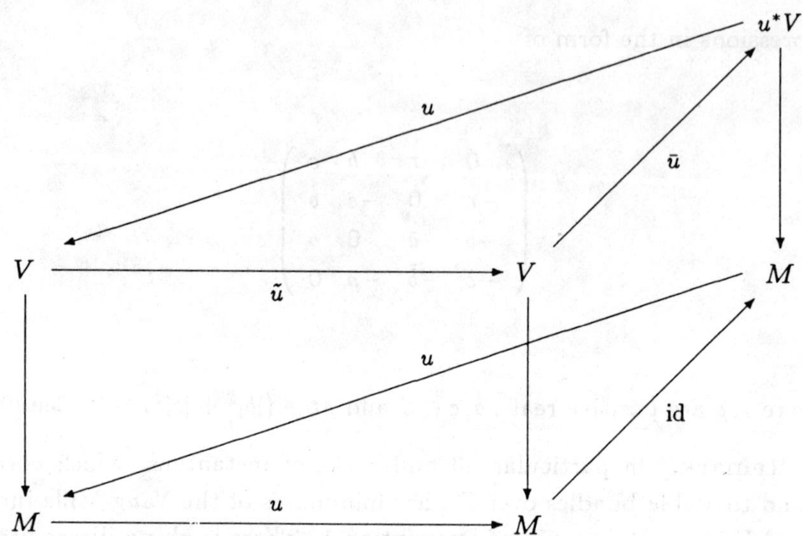

where u^*V is the pull-back of V via u, the existence of the isomorphism $\bar{u}$ is guaranteed by the classification theory (since $u \simeq \mathrm{id} : M \to M$).

Now pick any connection A on $V \to M$. One can naturally define $\hat{u}(A)$, u^*A and get a commutative diagram with connections

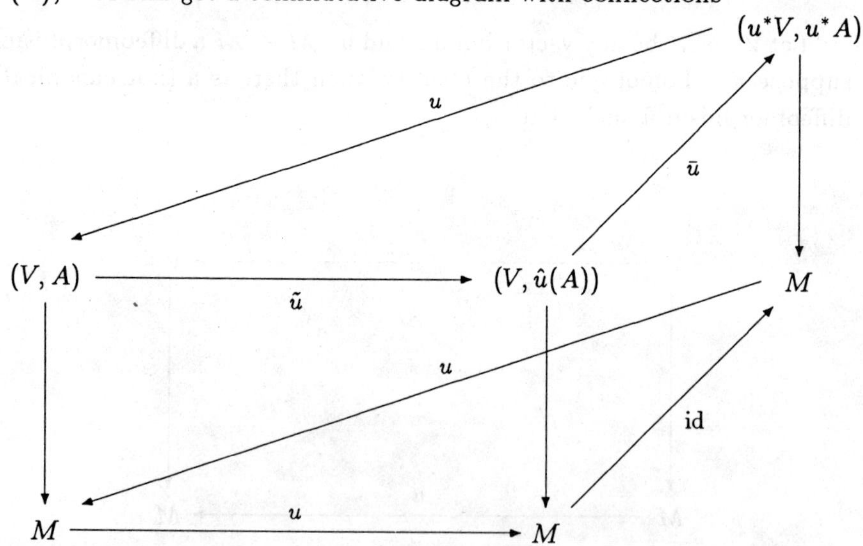

It is easy to check that $\hat{u}(A)$ is independent of choices of $\tilde{u}$ up to gauge equivalence. Furthermore if $u : M \to M$ is an isometry then the Yang-Mills functional is unchanged under the action $\hat{u}$. Essentially this is known as the enlarged gauge transformation in [5].

Back to our context. Clearly each $u \in U(4)$ induces an isometry $u : P_3 \to P_3$ which is also homotopic to the identity (due to the connectedness of $U(4)$). By the discussion above one has a mapping $\hat{u} : M_{P_3} \to M_{P_3}$ and so the entire $U(4)$ acts on M_{P_3}. We want to express this action on the skew forms (matrices) in view of Description 3. Let $u \in U(4)$. Study the following diagram

$$
\begin{array}{ccccc}
M_{P_3} \ni [A] & \xrightarrow{\;l \circ \bar{\Psi}\;} & M(0,1) \ni [\xi] & \xleftarrow{\;H\;} & \mathcal{E} \ni [\alpha] \\
\downarrow{\scriptstyle \hat{u}} & & \downarrow{\scriptstyle u^*} & & \downarrow{\scriptstyle u^*} \\
M_{P_3} \ni [\hat{u}(A)] & \xrightarrow{\;l \circ \bar{\Psi}\;} & M(0,1) \ni [u^*\xi] & \xleftarrow{\;H\;} & \mathcal{E} \ni [u^*\alpha]
\end{array}
$$

in which l, Ψ were defined in Description 1, H is the Horrocks construction, $u^*\xi$ is the pull-back of ξ and $u^*\alpha$ is the usual action of $u : \mathbb{C}^4 \to \mathbb{C}^4$ on the skew form α. The left square above commutes due to the fact that all objects are functorially associated, illustrated in the depiction

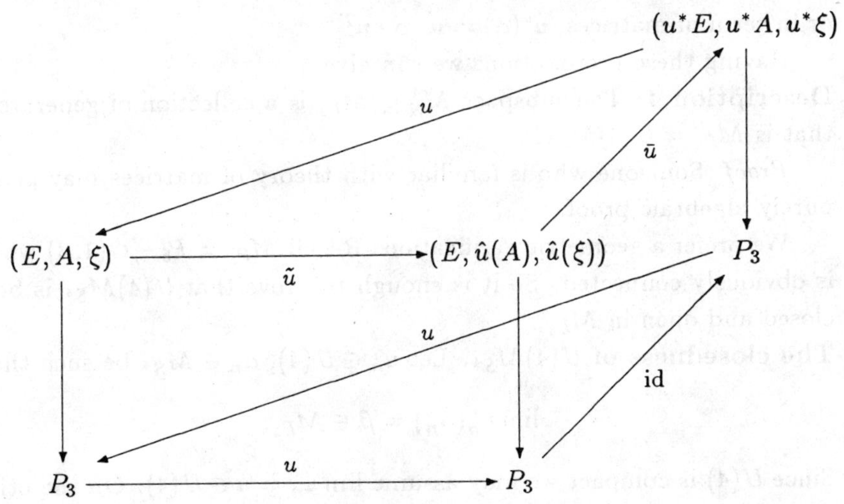

Each triple there, for example (E, A, ξ), is compatible. That means $\xi = \tilde{\Psi}(A)$, etc. Hence

$$[\hat{u}(\xi)] = l \circ \Psi([\hat{u}(A)]) = l \circ \Psi \circ \hat{u}([A]),$$
$$[u^*(\xi)] = u^*[\xi] = u^* \circ l \circ \Psi([A]).$$

But $\bar{u}$ is an isomorphism, $[u^*\xi] = [\hat{u}(\xi)]$. So

$$l \circ \Psi \circ \hat{u}([A]) = [\hat{u}(\xi)] = [u^*\xi] = u^* \circ l \circ \Psi([A]).$$

On the other hand the commutativity of the right square above follows from the isomorphism u_ℓ in

$$H(u^*(\alpha)) = \quad \xi^{u^*(\alpha)} \xrightarrow{\ u_\ell\ } u^*\xi^\alpha \ = u^* H(\alpha)$$
$$\downarrow \qquad\qquad\qquad \downarrow$$
$$P_3 \quad \xrightarrow{\ \text{id}\ } \quad P_3$$

where on each fiber u_ℓ is given by

$$\xi_{(z)}^{u^*(\alpha)} = \{L_{(z)}^{u^*(\alpha)}/L_{(z)}\} \longrightarrow \{L_{(u(z))}^{\alpha}/L_{(u(z))}\} = (u^*\xi^\alpha)_{(z)}$$
$$y + L_{(z)} \longmapsto u(y) + L_{(u(z))}.$$

In sum up, the action $U(4)$ on $\mathcal{E}$ we are interested is just the standard one i.e. for $u \in U(4)$, $\alpha \in \mathcal{E}$,

$$u^*(\alpha)(x, y) = \alpha(u(x), u(y)), \quad x, y \in C^4$$

or in term of matrices, $u^*(\alpha) = u \cdot \alpha \cdot u^T$.

Having these preparations we can give

Description 4. The subspace $M_S^4 \subseteq M_{P_3}$ is a collection of generators, that is $M_{P_3} = U(4)M_{S^4}$.

Proof. Someone who is familiar with theory of matrices may give a purely algebraic proof.

We prefer a geometric verification. Recall $M_{P_3} = P_5 - G(1, 3)$ which is obviously connected. So it is enough to prove that $U(4)M_{S^4}$ is both closed and open in M_{P_3}.

The closedness of $U(4)M_{S^4}$. Let $u_n \in U(4)$, $\alpha_n \in M_{S^4}$ be such that

$$\lim u_n^*(\alpha_n) = \beta \in M_{P_3}.$$

Since $U(4)$ is compact we may assume $\lim u_n = u \in U(4)$. On the other hand (we make no difference between u_n, α_n and their matrices)

$$|\det u_n^*(\alpha_n)| = |\det(u_n \circ \alpha_n \circ u_n^T)| = |\det u_n|^2 \cdot |\det \alpha_n| = |\det \alpha_n|$$

and

$$\lim |\det u_n^*(\alpha_n)| = |\det \beta| > 0.$$

So for large n,

$$|\det \alpha_n| \geq 1/2 \cdot |\det \beta| > 0.$$

Let

$$D = \{\alpha \in M_{S^4}; |\det \alpha| \geq 1/2 \cdot |\det \beta|\}.$$

Then $D \subseteq M_{S^4}$ is compact and $\{\alpha_n\} \subseteq D$ a.e. Hence $\{\alpha_n\}$ has a convergent subsequence $\{\alpha_N\} \to \alpha \in M_{S^4}$. Therefore

$$\beta = \lim u_N^*(\alpha_N) = u^*(\alpha) \in U(4)M_{S^4}.$$

The openness of M_{S^4}. This time we do it by counting dimensions. Let $\alpha_0 \in M_{S^4}$ be the basic instanton, i.e.

$$\alpha_0 = \begin{pmatrix} 0 & -1 & 0 & 0 \\ 1 & 0 & 0 & 0 \\ 0 & 0 & 0 & -1 \\ 0 & 0 & 1 & 0 \end{pmatrix} \text{ or } \alpha_0 \text{ with potential Im } \left(\frac{\bar{x}dx}{1 + |x|^2} \right).$$

Then $M_S^4 = SL(2, H)\{\alpha_0\}$ [1; page 24]. In other words $SL(2, H)$ is the largest subgroup of $GL(4,)$ which maps M_{S^4} into M_{S^4}.

Consider any $\alpha \in M_{S^4}$. In order to prove $U(4)M_{S^4}$ is open in M_{P_3}, it is enough to show that the transversal part of the orbit $U(4) \cdot \alpha$ has the dimensions

$$s \geq \dim M_{P_3} - \dim M_{S^4} = 10 - 5 = 5.$$

But

$$\begin{aligned} s &= \dim U(4) - \dim \text{ of the tangent part of } U(4) \cdot \alpha \\ &\geq \dim U(4) - \dim \left(U(4) \bigcap SL(2, H) \right) \\ &= \dim U(4) - \dim sp(2) = 15 - 10 = 5. \end{aligned}$$

Acknowledgment. This work is a part of author's M. Sc dissertation. The author wishes to thank his supervisor Professor Simon Donaldson for guidance and help. Thanks are also due to Professors S. S. Chern and M. F. Atiyah for suggesting author's study in Oxford.

References

[1] Atiyah, M.F., *Geometry of Yang-Mills fields. Pisa*, 1979.

[2] Atiyah, M.F., Drinfeld, V.G., Hitchin, N.J., Manin, Yu.I., Construction of instantons. *Phys. Lett.* 65(1978) 185–187.

[3] Atiyah, M.F., Rees, E., Vector bundles over projective 3-space. *Inven. Math.* 35 (1978) 131–151.

[4] Barth, W., Some properties of stable rank-2 vector bundles on P_n. *Math.Ann.* 226(1977) 125–150.

[5] Bourguignon, J.P., Lawson, H.B.Jr., Stability and isolation phenomena for Yang-Mills fields. *Comm.Math.Phys.* 79(1981) 189–230.

[6] Donaldson, S.K., *D.Phil.Thesis.* Oxford 1982.

[7] Donaldson, S.K., Anti-self dual Yang-Mills connections over complex algebraic surface and stable vector bundles. *Proc.London Math.Soc.* (3) 50(1985) 1–25.

[8] Donaldson, S.K., Infinite determinates, stable bundles and curvature.

[9] Drifeld, V.K., Manin, Yu.I., A description of instantons *Comm. Math.Phys.* 63(1978) 177–192.

[10] Hartshorne, R., Stable vector bundles of rank-2 on P^3. *Math.Ann.* 238(1978) 229–280.

[11] Hartshorne, R., Stable vector bundles and instantons. *Cmm. Math. Phys.* 59(1978) 1–15.

[12] Kobayashi, S., Curvature and stability of vector bundles. *Proc. Japan Acad. Ser.A. Math. Sci.* 58(1982) 158–162.

[13] Lubke, M., Stability of Einstein-Hermitian vector bundles. *Manusripta Math.* 42(1983) 245–275.

[14] Okonek, C., Schneider, M., Sprindler, H., *Vector bundles over projective space.* Birkhause Boston, 1981.

[15] Uhlenbeck,K.K., Yau, S.T., On the existence of Hermitian-Yang-Mills connections in stable vector bundles. *Comm.Pure and Appl.* 39(1986) 257–293.

[16] Warner, F., *Foundations of differentiable manifolds and Lie groups.* Scott, Foresman and company, 1971.

Study on Reflection of Shock Waves

Chen Shuxing
Fudan University

Abstract

Reflection of shock waves for system of conservation laws in multi-dimensional space is discussed. The paper gives the existence of solution for such problem with better smoothness than previous work.

1. Reflection of shock waves is one of the problems which occur in gas dynamics very often. For instance, when an explosive shock wave meets an obstacle, reflection of unsteady shock wave occurs, or when a shock wave caused by a flying body with supersonic speed meets another body with same speed, reflection of steady shock wave occurs. The reflection of shock wave is different from the reflection of sound wave, it is a nonlinear problem. When the place of an incident shock and the flow field on the both sides of the shock front before the reflection are given, the main problem is then to determine the place of the reflective shock front and the flow field behind the front.

The problem of shock reflection was formulated mathematically by R. Courant & K. O. Friedlichs [4]. In the fifties and early sixties many mathematicians studied quasilinear hyperbolic system with two variables systematically and proved the existence of local solution for the problem of shock reflection and many other problems. However, for the higher dimensional case, there was not essential progress in theoretical study on this subject until the eighties, although many numerical methods developed in this period. The reason is that people lacked a precise analysis for the corresponding linearized problems, because generally the method of integration along characteristics is not available any more and the classical energy estimate is not strong enough.

In 1970 H. O. Kreiss applied microlocal analysis to initial boundary problem for hyperbolic system in higher dimensions, he established the existence of such problems under the Lopatinski condition. Based on his results A. Majda discussed a initial value problem with discontinuous value for hyperbolic system of conservation laws in higher dimensions, A. Majda proved in [15] the existence of local solutions with shock front under the assumption that the planar shock front by freezing data at the origin is uniform stable. Afterwards, people obtained some further results ([10],[17],[18]). Based on these development we are going to discuss the problem of shock reflection caused by hitting a rigid wall by an arbitrary higher dimensional shock front in this paper. We will mainly discuss the reflection of unsteady shock wave, and in the last part we will briefly mention the reflection of steady shock wave.

2. In the space (t, x, y) the Euler system of unsteady inviscid incompressible flow is

$$\frac{\partial}{\partial t}\begin{bmatrix} \rho \\ \rho u \\ \rho v \\ (e + \frac{1}{2}q^2) \end{bmatrix} + \frac{\partial}{\partial x}\begin{bmatrix} \rho u \\ p + \rho u^2 \\ \rho uv \\ \rho u(i + \frac{1}{2}q^2) \end{bmatrix} + \frac{\partial}{\partial y}\begin{bmatrix} \rho v \\ \rho uv \\ p + \rho v^2 \\ \rho v(i + \frac{1}{2}q^2) \end{bmatrix} = 0 \quad (1)$$

where u, v are the components of velocity, $q^2 = u^2 + v^2$, p, ρ, e, i represent pressure, density, inner energy and entholpy respectively. In the sequel we will use $U = (u, v, p, \rho)$ to represent the unknown functions, it can determine all parameters of the flow field. On the shock front, the following Rankine-Hogoniot condition should be satisfied

$$\begin{bmatrix} \rho \\ \rho u \\ \rho v \\ \rho(e + \frac{1}{2}q^2) \end{bmatrix} \psi_t - \begin{bmatrix} \rho u \\ p + \rho u^2 \\ \rho uv \\ \rho u(i + \frac{1}{2}q^2) \end{bmatrix} + \begin{bmatrix} \rho v \\ \rho uv \\ p + \rho v^2 \\ \rho v(i + \frac{1}{2}q^2) \end{bmatrix} \psi_y = 0 \quad (2)$$

where $x = \psi(t, y)$ is the equation of the shock front and $[\ ]$ represents the jump of corresponding functions across the front. On the rigid wall Σ with the equation $x = \varphi(y)$, the boundary condition is

$$u - \varphi' v = 0. \tag{3}$$

Denote the equation of the incident shock front S_1 by $x = \psi_1(t, y)$ (without restriction we assume $\varphi(y) \le \psi_1(t, y)$). At $t = 0$ S_1 is tangential to Σ at the origin: $\varphi(0) = \varphi'(0) = \psi_1(0,0) = \psi'_{1y}(0,0) = 0$.

Assume that $\psi_1'(t,y)$ and the flow field $U_\pm(t,x,y)$ on the both sides of S_1 are given, $\psi_1(t,y) \in H^{s+1}$, $U_\pm(t,x,y) \in H^s$, they satisfy (1), (2), (3) at $t < 0$, then we have:

Theorem 1. *Under the above assumptions and $s \geq 6$ there exists Y_0, T_0, functions $l(y) \in H^s([-Y_0, Y_0])$ and $\tilde{\psi}(t,y)$, $\psi(t,y) \in H^{s+1}$ $(0 < t < T$, $-Y_0 < y < Y_0)$ satisfying $\tilde{\psi}(l(y), y) = \psi(l(y), y) = \varphi(y)$, $\tilde{\psi}(0, y) = \psi_1(0, y)$, and functions*

$$U(t,x,y) \in H^s(D) \quad \text{with} \quad D = \{l(y) < t < T_0,\ -Y_0 < y < Y_0,$$
$$x < \psi(t,y)\ \},$$

$$U_1(t,x,y) \in H^s(D_1) \quad \text{with} \quad D_1 = \{0 < t < T_0,\ -Y_0 < y < Y_0,$$
$$\left.\begin{array}{l} x > \tilde{\psi}(t,y) \quad \text{if } t < l(y) \\ x < \psi(t,y) \quad \text{if } t > l(y) \end{array}\right\},$$

$$U_2(t,x,y) \in H^s(D_2) \quad \text{with} \quad D_2 = \{0 < t < l(y),\ -Y_0 < y < Y_0,$$
$$x < \tilde{\psi}(t,y)\ \}$$

satisfying $U_1(0,x,y) = U_+(0,x,y)$, $U_2(0,x,y) = U_-(0,x,y)$, these functions satisfy system (1) in corresponding domain, boundary condition (3) on $x = \varphi(y)$ and the relation (2) on $x = \psi(t,y)$ or $x = \tilde{\psi}(t,y)$ respectively (where the value of functions on the right side of the shock front is U_1, and that on the left side is U or U_2.

Remark: Freeze all data at the origin, the wall can be regarded as the plane $x = 0$ and the shock front S_1, can be regarded as a plane shock with speed $\sigma = \psi_{1t}'(0,0)$. The unknown function on the right side of S_1 is U_{+0} and the one on the left side of S_1 is U_{-0} with $u = 0$. Moreover, U_{+0} and U_{-0} satisfy the relations

$$\rho_+(u_+ - \sigma) = -\rho_-\sigma,$$
$$p_+ + \rho_+ u_+(u_+ - \sigma) = p_-,$$
$$v_+ = v_-,$$
$$\rho_+\left(e_+ + \frac{1}{2}g_+^2\right)(u_+ - \sigma) = -\rho_-\left(e_- + \frac{1}{2}v_-^2\right)\sigma.$$

From [4] we can determine the slope σ of planar reflected shock and the flow field U_{c0} by U_{+0}, U_{-0}, and σ, and near the boundary the solution does not contain any other discontinuities (such as rarefaction wave and contact discontinuity) except the incident shock and reflected shock. Now we may view our original problem on reflection of higher dimensional shock as a perturbation of reflection of planar shock without considering, hence we also don't need to consider other discontinuities. Theorem 1 shows this assumption is reasonable.

3. The first step to prove Theorem 1 is to determine $l(y)$, $\tilde{\psi}(t,y)$ and $U_i(t,x,y)$ $(i=1,2)$.

Lemma 1. *Assume $s \geq 6$, then there exist Y_0, $T_0 > 0$ and functions*

$$
\begin{aligned}
\tilde{\psi}(t,y) &\in H^{s+1} & (0 < t < T_0,\ -Y_0 < y < Y_0), \\
U_1(t,x,y) &\in H^s & (0 < t < T_0,\ -Y_0 < y < Y_0,\ x > \tilde{\psi}(t,y)), \\
U_2(t,x,y) &\in H^s & (0 < t < T_0,\ -Y_0 < y < Y_0,\ x > \tilde{\psi}(t,y))
\end{aligned}
$$

such that $\tilde{\psi}(0,y) = \psi_1(0,y)$, $U_1(0,x,y) = U_+(0,x,y)$, $U_2(0,x,y) = U_-(0,x,y)$, and

$$
U = \begin{cases} U_1(t,x,y), & x > \max(\varphi(y), \tilde{\psi}(t,y)), \\ U_2(t,x,y), & (y) < \max(\varphi(y), \tilde{\psi}(t,y)) \end{cases}
$$

satisfies (1)–(3).

Proof. Denote the domain between S_1 and Σ on $t = 0$ by Σ_-, the domain right to S_1 by Σ_+. Extend function U_- defined on Σ_- to Σ_+ as H^s function, and use the same notation U for the extended function. Since the system (1) is symmetric hyperbolic, we know that the initial boundary problem

$$
\begin{cases}
\text{system (1),} \\
u - \varphi'v = 0 & \text{on } \Sigma, \\
U = U_- & \text{on } t = 0, x > \varphi(y)
\end{cases}
\tag{4}
$$

has a unique local H^s solution, if $s \geq 6$ (see for instance [14]).

Denote the solution of (4) by $U_2(t,x,y)$, extend U_2 to $x < \varphi(y)$, then U_2 is defined in a neighborhood of the origin in $t > 0$. Now we solve the initial boundary value problem with a free boundary

$$
\begin{cases}
\text{system (1),} \\
U = U_+ & \text{on } t = 0, x > \tilde{\psi}(0,y), \\
\tilde{\psi}(0,y) = \psi_1(0,y) & \text{on } t = 0, \\
\text{condition (2)} & \text{on } x > \tilde{\psi}(t,y)
\end{cases}
\tag{5}
$$

where the functions on the both sides of the shock are U_1 and U_2.

Noticing that the shock S_1 is an extreme shock, we may reduce the problem (5) to a nonlinear initial boundary problem discussed in a quadrant as did in [16] (here we don't need to fold the domain). Restricting in $x > \varphi(y)$, the shock S_1: $x = \tilde{\psi}(t, y)$, functions U_1 and U_2 give the place of the incident shock and the flow field on both sides of the shock for $t > 0$.

Finally, $l(y)$ can be determined from

$$\varphi(y) = \tilde{\psi}(l(y), y). \tag{6}$$

Since $\tilde{\psi}'_t \neq 0$, (6) is solvable by implicit theorem.

Remark. We mention that functions $\tilde{\psi}(t, y)$, U_1 and U_2 are independent of the extension of U_- to Σ_+ and the extension of U_2 to $x < \varphi(y)$ In fact, the shock S_1 at the origin is an extreme shock, then the characteristic cone determined by U_{-0} and issued from the origin is placed in the left side of S_1 by continuity this property holds in a neighborhood of the origin. This means that the solution in the domain confined by $\{t = 0\}$, $\{x = \varphi(y)\}$ and S_1 can be uniquely determined by the data on $\{t = 0\}$ and $\{x = \varphi(y)\}$. Then we know from energy method that $\tilde{\psi}$ and U_- are independent of extension.

4. The next problem is to determine the reflected shock S: $x = \psi(t, y)$ issued from τ: $t = l(y)$, $x = \varphi(y)$ and the solution of (1) in between S and Σ. First, let us rewrite the system (1) as

$$
\begin{pmatrix} \rho & & & \\ & \rho & & \\ & & a^{-2}\rho^{-1} & \\ & & & 1 \end{pmatrix} \frac{\partial}{\partial t} \begin{pmatrix} u \\ v \\ p \\ s \end{pmatrix}
$$

$$
+ \begin{pmatrix} \rho u & & 1 & \\ & \rho u & & \\ 1 & & a^{-2}\rho^{-1}u & \\ & & & u \end{pmatrix} \frac{\partial}{\partial x} \begin{pmatrix} u \\ v \\ p \\ s \end{pmatrix}
$$

$$
+ \begin{pmatrix} \rho v & & & \\ & \rho v & 1 & \\ & 1 & a^{-2}\rho^{-1} & \\ & & & v \end{pmatrix} \frac{\partial}{\partial y} \begin{pmatrix} u \\ v \\ p \\ s \end{pmatrix}
$$

$$
= 0 \tag{7}
$$

where s represents entropy, which is a given function of p and ρ. By introducing a transform involving the unknown function

$$
\begin{aligned}
x_1 &= (t - l(y)) \frac{x - \varphi(y)}{\psi(t, y) - \varphi(y)}, \\
x_2 &= (t - l(y)) \frac{\psi(t, y) - x}{\psi(t, y) - \varphi(y)}, \\
y &= y
\end{aligned}
\tag{8}
$$

the original problem is reduced to a nonlinear Goursat problem with two fixed boundary:

$$
\begin{aligned}
&A(U, \psi) \frac{\partial U}{\partial x_1} + B(U, \psi) \frac{\partial U}{\partial x_2} + Q(U, \psi) \frac{\partial U}{\partial y} = 0, \quad x_1 > 0, x_2 > 0, \\
&lU = 0, \quad x_1 = 0, \\
&F(x_1, y, U, \psi, \nabla \psi) = 0, \quad x_2 = 0, \quad \psi(0, y) = 0
\end{aligned}
\tag{9}
$$

where

$$
A = \begin{pmatrix}
\rho H & & \dfrac{\partial x_1}{\partial x} \\
& \rho H & \dfrac{\partial x_1}{\partial y} \\
\dfrac{\partial x_1}{\partial x} & \dfrac{\partial x_1}{\partial y} & a^{-2} \rho^{-1} H \\
& & & H
\end{pmatrix},
$$

$$
B = \begin{pmatrix}
\rho G & & \dfrac{\partial x_2}{\partial x} \\
& \rho G & \dfrac{\partial x_2}{\partial y} \\
\dfrac{\partial x_2}{\partial x} & \dfrac{\partial x_2}{\partial y} & a^{-2} \rho^{-1} G \\
& & & G
\end{pmatrix},
$$

$$
C = \begin{pmatrix}
\rho V & & \\
& \rho V & 1 \\
& 1 & a^{-2} \rho^{-1} V \\
& & & V
\end{pmatrix},
$$

$$
H = \frac{\partial x_1}{\partial t} + u \frac{\partial x_1}{\partial x} + v \frac{\partial x_1}{\partial y}, \quad G = \frac{\partial x_2}{\partial t} + u \frac{\partial x_2}{\partial x} + v \frac{\partial x_2}{\partial y}.
$$

Since $x_1 = 0$ is a characteristic surface, by the property of finite propagation speed for hyperbolic system we may assume that the whole problem is periodic with respect to y. Denote the period of y by Δ, set

$$\Omega = \{(x_1, x_2, y);\ x_1 > 0, x_2 > 0, y \in \Delta\}, \qquad \Omega_T = \Omega \cap \{x_1 + x_2 < T\},$$
$$\omega = \{(t, y);\ t > 0, y \in \Delta\}, \qquad \omega_T = \omega \cap \{t < T\}$$

and set $H_\lambda^k(\Omega_T)$, $H_\lambda^k(\omega_T)$ are Sobolev space with weight $t^{-\lambda}$

$$\|U\|_{H_\lambda^k} = \left(\sum_{|\alpha| \leq k} \lambda^{2(k-|\alpha|)} \|\partial^\alpha U\|_{L_{\lambda-\alpha_{x_1}-\alpha_{x_2}}^2}^2 \right)^{\frac{1}{2}},$$

$$\|f\|_{H_\lambda^k} = \left(\sum_{|\beta| \leq k} \lambda^{2(k-|\beta|)} \|\partial^\beta f\|_{L_{\lambda-\beta_t}^2}^2 \right)^{\frac{1}{2}}.$$

Consider the linearized problem

$$A\frac{\partial \delta U}{\partial x_1} + B\frac{\partial \delta U}{\partial x_2} + Q\frac{\partial \delta U}{\partial y} = f, \quad x_1 > 0, x_2 > 0,$$
$$l\delta U = 0, \quad x_1 = 0, \tag{10}$$
$$p\frac{\partial \delta\psi}{\partial t} + q\frac{\partial \delta\psi}{\partial y} + h\delta\psi + m\delta U = g, x_2 = 0;\ \delta\psi(0, y) = 0$$

we have

Lemma 2. *Assume that $U \in H^s$, $\psi \in H^{s+1}$, $s \geq 6$, then the problem* (10) *has a unique solution $\delta U \in H^s$, $\delta\psi \in H^{s+1}$, and for large λ*

$$\lambda\|\delta U\|_{H_{\lambda+\frac{1}{2}}^k(\Omega_T)}^2 + \|\delta U\ |_{x_2=0}\ \|_{H_\lambda^k(\omega_T)}^2 + \|\delta\psi\|_{H_{\lambda+1}^{k+1}(\omega_T)}^2$$

$$\leq C\left(\frac{1}{\lambda}\|f\|_{H_{\lambda-\frac{1}{2}}^k(\Omega_T)}^2 + \|g\|_{H_\lambda^k(\omega_T)}^2 \right). \tag{11}$$

Sketch of the proof.

Let us concentrate to deduce the energy estimate (11), because the existence and uniqueness of linearized problem can be deduced by a standard procedure. First we introduce a series of transformations to eliminate the degeneracy in the corner:

1) Blow up the corner $\{x_1 = x_2 = 0\}$ by a transformation

$$j: \begin{cases} t = x_1 + x_2, \\ \theta = x_1/(x_1 + x_2), \\ y = y. \end{cases} \tag{12}$$

Correspondingly, introduce a transformation of unknown functions

$$\delta U(x_1, x_2, y) \mapsto J_\lambda \delta U(t, \theta, y) = t^{-\lambda} \delta U(t\theta, t(1-\theta), y),$$
$$\delta \psi(t, y) \mapsto J_\lambda \delta \psi(t, y) = t^{-\lambda} \delta \psi(t, y). \tag{13}$$

2) Decompose the unknown functions $\widetilde{\delta U} = J_\lambda \delta U$ and $\widetilde{\delta \psi} = J_\lambda \delta \psi$ by a diadic partition of unity

$$\widetilde{\delta U}(t, \theta, y) = \Sigma u_j(t, \theta, y) = \Sigma \chi(2^j t) \widetilde{\delta U}(t, \theta, y),$$
$$\widetilde{\delta \psi}(t, y) = \Sigma \varphi_j(t, y) = \Sigma \chi(2^j t) \widetilde{\delta \psi}(t, y) \tag{14}$$

where $\chi(t)$ is a C^∞ cut off function with its support in $(\frac{1}{2}, 2)$

3) Use a dilation to change the scale of variables

$$\tilde{u}_j(t, \theta, y) = 2^{-\frac{i}{2}} u_j(2^{-j} t, \theta, y),$$
$$\tilde{\varphi}_j(t, y) = 2^{-\frac{i}{2}} \varphi_j(2^{-j} t, y). \tag{15}$$

By using these transformations the problem (10) can be deduced to a set of boundary problem on the domain $\{0 < \theta < 1, \frac{1}{2} < t < 1, y \in \Delta\}$:

$$\tilde{L}_\lambda^j \tilde{u}_j \equiv (\tilde{A}_j + \tilde{B}_j)\left(t\frac{\partial \tilde{u}_j}{\partial t} + \lambda \tilde{u}_j\right) + [(1-\theta)\tilde{A}_j - \theta \tilde{B}_j]\frac{\partial \tilde{u}_j}{\partial \theta}$$
$$+ t\tilde{Q}_j \frac{\partial \tilde{u}_j}{\partial \theta} = f_j^*,$$
$$l\tilde{u}_j = 0, \tag{16}$$
$$\tilde{F}_\lambda^j(\tilde{u}_j, \tilde{\varphi}_j) \equiv \tilde{p}_j \left(t\frac{\partial \tilde{\varphi}_j}{\partial t} + (\lambda + 1)\tilde{\varphi}_j\right)$$
$$+ t\left(\tilde{q}_j \frac{\partial \tilde{\varphi}_j}{\partial y} + \tilde{h}_j \tilde{\varphi}_j + \tilde{m}_j \tilde{u}_j\right) = g_j^*$$

where the coefficients have expressions as $\tilde{A}_j = A(2^{-j} t\theta, 2^{-j}(1-t)\theta, y)$ etc., the validity of (11) is equivalent to prove

$$\lambda \|\tilde{u}_j\|_{H^k(\tilde{\Omega}_{T_j})}^2 + \|\tilde{u}_j|_{x_2=0}\|_{H^k(\tilde{\omega}_{T_j})}^2$$
$$+ \|\tilde{\varphi}_j\|_{H^{k+1}(\tilde{\omega}_{T_j})}^2 + \lambda^2 \|\tilde{\varphi}_j\|_{H^k(\tilde{\omega}_{T_j})}^2$$
$$\leq C\left\{\frac{1}{\lambda}\|\tilde{L}_\lambda^j \tilde{u}_j\|_{H^k(\tilde{\Omega}_{T_j})}^2 + \|\tilde{F}_\lambda^j(\tilde{u}_j, \tilde{\varphi}_j)\|_{H^k(\tilde{\omega}_{T_j})}^2\right\}, \quad k \leq 5. \tag{17}$$

For the problem (16), the boundary $\theta = 0$ is characteristic, and $\theta = 1$ is noncharacteristic, the conditions on $\theta = 1$ satisfies Kreiss-Majda's condition. Besides, viewing that the support of $\tilde{u}_j$ is contained in $t > \frac{1}{2}$, we don't need to worry about the degeneracy of the coefficients at $t = 0$.

In order to obtain (17) for $k = 0$, we take $\eta(\theta)$ as a C^∞ $([0,1])$ function, such that $\eta = 1$ for $0 \leq \theta \leq \frac{1}{3}$ and $\operatorname{supp}\eta \subset [0, \frac{2}{3}]$, then from (17) we may deduce two boundary value problem for $w_1 = \eta\tilde{u}_j$ and $w_2 = (1 - \eta)\tilde{u}_j$. Since w_1 is a solution of a initial boundary problem of symmetric hyperbolic system with uniform characteristic boundary, satisfying admissible boundary conditions, then the estimate for w_1 holds (see [5]). As for $w_2, \tilde{\varphi}_j$ they are the solution of a coupled initial boundary value problem mentioned in [15], hence satisfies (17). Combining these estimates we obtain (17) for $\tilde{u}_j, \tilde{\varphi}_j$ which leads (11). Here we remark that in the process of using the method in [16], we only need to require the coefficients is in H with $s > [\frac{n}{2}] + 3$ rather than $2[\frac{n}{2}] + 7$ as showed in [6].

We prove (11) for the case $k > 0$ by induction. Denote $\partial_1 = \theta(1 - \theta)\partial_\theta$, $\partial_2 = \partial t$, $\partial_3 = \partial_y$ as tangential operators of domain $\{0 < \theta < 1, t > 0, y \in \Delta\}$ and act ∂_i on the both sides of (16). Then by estimating the commutator $[L_\lambda^j, \partial_i]$ we may obtain the estimate for $\partial_i\tilde{u}_j$ as (17). Returning to the coordinate system (x_1, x_2, y), we have

$$\lambda\|V\delta U\|^2_{H^k_{\lambda+\frac{1}{2}}(\Omega_T)} + \|\delta U\mid_{x_2=0}\|^2_{H^{k+1}_\lambda(W_T)} + \|\delta\psi\|^2_{H^{k+2}_{\lambda+1}(W_T)},$$

$$\leq C\left(\frac{1}{\lambda}\|f\|^2_{H^k_{\lambda-\frac{1}{2}}(\Omega_T)} + \|g\|^2_{H^k_\lambda(W_T)}\right) \tag{18}$$

where V represents any tangential operator. Now the remaining problem is to explain that $\|D\delta U\|_{H^k_{\lambda+\frac{1}{2}}(\Omega_T)}$ can be estimated by

$$K = \|\delta U\|_{H^k_{\lambda+\frac{1}{2}}(\Omega_T)} + \|V\delta U\|_{H^k_{\lambda+\frac{1}{2}}(\Omega_T)}$$

where D is an arbitrary first order operator.

By virtue of $H = 0$ on the boundary $x_1 = 0$, the rank of A is 2, hence we may obtain estimates for

$$\left\|\rho H\frac{\partial\delta u}{\partial x_1} + \frac{\partial x_1}{\partial x}\frac{\partial\delta p}{\partial x_1}\right\|_{H^k_{\lambda+\frac{1}{2}}(\Omega_T)}$$

and

$$\left\|\frac{\partial x_1}{\partial x}\frac{\partial\delta u}{\partial x_1} + \frac{\partial x_1}{\partial y}\frac{\partial\delta v}{\partial x_1} + a^{-2}p^{-1}H\frac{\partial\delta p}{\partial x_1}\right\|_{H^k_{\lambda+\frac{1}{2}}(\Omega_T)}.$$

Thus to obtain the estimate of all components of $\frac{\partial \delta U}{\partial x_1}$ we must find estimate of other two linear combinations of the components of $\frac{\partial U}{\partial x_1}$. Let $\nu \cdot \nabla = H \frac{\partial}{\partial x_1} + G \frac{\partial}{\partial x_2} + v \frac{\partial}{\partial y}$, the first two equations in (11) can be written as

$$\rho \nu \cdot \nabla \delta u + \left(\frac{\partial x_1}{\partial x} \frac{\partial}{\partial x_1} + \frac{\partial x_2}{\partial x} \frac{\partial}{\partial x_2} \right) \delta p = 0$$

$$\rho \nu \cdot \nabla \delta v + \left(\frac{\partial x_1}{\partial y} \frac{\partial}{\partial x_1} + \frac{\partial x_2}{\partial y} \frac{\partial}{\partial x_2} + \frac{\partial}{\partial y} \right) \delta p = 0$$

from which

$$\rho \nu \cdot \nabla \left(\left(\frac{\partial x_1}{\partial y} \frac{\partial}{\partial x_1} + \frac{\partial x_2}{\partial y} \frac{\partial}{\partial x_2} + \frac{\partial}{\partial y} \right) \delta u - \left(\frac{\partial x_1}{\partial x} \frac{\partial}{\partial x_1} + \frac{\partial x_2}{\partial x} \frac{\partial}{\partial x_2} \right) \delta v \right) = F \tag{19}$$

where F is a linear combination of components of δU and $V \delta U$ on the other hand, the last equation in (11) is

$$\rho \nu \cdot \nabla \delta s = 0. \tag{20}$$

By using (19) and (20) the estimates $\| \delta s \|_{H^k_{\lambda + \frac{1}{2}}(\Omega_T)}$ and

$$\left\| \left(\frac{\partial x_1}{\partial y} \frac{\partial}{\partial x_1} + \frac{\partial x_2}{\partial y} \frac{\partial}{\partial x_2} + \frac{\partial}{\partial y} \right) \delta u - \left(\frac{\partial x_1}{\partial x} \frac{\partial}{\partial x_1} + \frac{\partial x_2}{\partial x} \frac{\partial}{\partial x_2} \right) \delta v \right\|_{H^k_{\lambda + \frac{1}{2}}(\Omega_T)}$$

can be given by a lemma on first order equations. The lemma is:

Lemma 3. *Assume that $\vec{\nu}$ is a H^s vector field, tangential to $x_1 = 0$ on $x_1 = 0$, w satisfies*

$$\nu \cdot \nabla w = F + O(w), \tag{21}$$

$F \in H^k_\lambda(\Omega_T)$, $w \mid_{x_2 = 0} \in H^k_\lambda(\omega_T)$, *then for $k \leq s$*

$$\| W \|_{H^k_\lambda(\Omega_T)} \leq C (\| F \|_{H^k_\lambda(\Omega_T)} + \| w \mid_{x_2 = 0} \|_{H^k_\lambda(\omega_T)}. \tag{22}$$

Having estimates of four linearly independent combination of the components of $D \delta U$ we obtain the estimates of $\| \delta U \|_{H^{k+1}_{\lambda + \frac{1}{2}}}$.

5. Based on the result of Lemma 2 we may use Newton's iteration scheme to construct a sequence of approximate solutions as

$$\begin{cases} L(U^{(n)},\psi^{(n)})W^{(n+1)} = -L(U^{(n)},\psi^{(n)})U^{(0)}, \\ lW^{(n+1)} = 0, \\ F_{U^{(n)},\psi^{(n)}}(W^{(n+1)},\theta^{(n+1)}) = -\mathcal{F}(U^{(n)},\psi^{(n)}) + F_{U^{(n)},\psi^{(n)}}(W^{(n)},\theta^{(n)}), \\ \theta^{(n+1)}\big|_{t=0} = 0, \\ U^{(n+1)} = U^{(0)} + W^{(n+1)}, \quad \psi^{(n+1)} = \psi^{(0)} + \theta^{(n+1)}. \end{cases} \tag{23}$$

Let us choose an appropriate asymptotic solution $U^{(0)},\psi^{(0)}$ of (9) as the first term of the sequence. It means that when we substitute $U^{(0)},\psi^{(0)}$ into the left side of (9), the result is $O((x_1 + x_2)^\lambda)$ or $O(t^\lambda)$. Take λ be sufficiently large and T be sufficiently small, we can prove that (23) introduce a map $(U^{(n)},\psi^{(n)})$ to $(U^{(n+1)},\psi^{(n+1)})$ in the space

$$W^s_{\lambda,T} = \{ (U,\psi), U \in H^s_{\lambda+\frac{1}{2}}(\Omega_T), lU\big|_{x_1=0} = 0,$$
$$U\big|_{x_2=0} \in H^s_\lambda(\omega_T), \psi \in H^{s+1}_{\lambda+1}(\omega_T) \}$$

Moreover, the map is contractive according to the norm

$$\lambda\|U\|_{H^k_{\lambda+\frac{1}{2}}(\Omega_T)} + \|U\big|_{x_2=0}\|_{H^k_\lambda(\omega_T)} + \|\psi\|_{H^{k+1}_{\lambda+1}(\omega_T)}$$

with smaller k. The fact ensures the existence of a unique local solution. More details can be found in [6].

Combining the existence of the local solution of (9) and Lemma 1 we establish the result in Theorem 1.

6. Finally, let us briefly mention the reflection of steady shock wave. Here we must consider the reflection of three-dimensional shock, because only in this case the character of higher dimensional problem appears. The Euler system in this case is

$$\frac{\partial}{\partial x}\begin{bmatrix} \rho u \\ p+\rho u^2 \\ \rho uv \\ \rho uw \end{bmatrix} + \frac{\partial}{\partial y}\begin{bmatrix} \rho v \\ \rho uv \\ p+\rho v^2 \\ \rho uw \end{bmatrix} + \frac{\partial}{\partial z}\begin{bmatrix} \rho w \\ \rho uw \\ \rho vw \\ p+\rho w^2 \end{bmatrix} = 0 \tag{24}$$

where p, u, v, w, p satisfy a Bernoulli relation

$$\frac{1}{2}(u^2 + v^2 + w^2) + \frac{rp}{(r-1)p} = \text{const.} \tag{25}$$

Suppose the equation of shock S_1 is $\varphi(x, y, z) = 0$, it intersects transversally with rigid wall $\Sigma : \psi(x, y, z) = 0$. We point out that only the angle θ between S_1 and Σ is not too large, then the simple shock reflection happens (otherwise so called λ shock reflection appears). More precisely, fix a point $P \in l = S_1 \cap \Sigma$, the velocity at P ahead the shock is parallel to Σ. Denote by V the component of the velocity in the direction perpendicular to l, then V is supersonic. By V and the pressure P we can determine a shock polar (see [4]), and the simple shock reflection happens when θ is less than the critical angle determined by the shock polar.

Making a coordinate transformation we may place the origin at P, and let the x-axis be tangential to Σ and perpendicular to l. Since the form of the system (24) is invariant under any orthogonal transformation, then we may assume that for the original system (24) the x-axis has been on the place as showed above. The system is then hyperbolic with respect to x, and the variable x amounts to the variable t in the system (7). Therefore, similar to unsteady shock reflection this problem can also be solved locally, and we omit the details here.

References

[1] S. Alinhac, Existence d'ondes de rarefaction pour des ecoulements isentropiques, Sem. EDP, Ecole polytechnique, paris (1986–87)

[2] S. Alinhac, Existence d'ondes de rarefaction pour des systems quasi-lineaires hyperboliques multimensionnels (*prepublications*).

[3] R.Courant, Cauchy's problem for hyperbolic quasilinear systems of first partial differential equations in two variables, *Comm. Pure Appl. Math.*, 14(1961, 257–265).

[4] R.Courant and K.O. Friedrichs, *Supersonic flow and shock waves*, Interscience Publishers Inc., New York (1948).

[5] Chen Shucxing, On the initial-boundary value problems for quasi-linear symmetric hyperbolic system and their applications, *Chin. Ann.Math.* 1(1980), 511–521.

[6] Chen Shuxing, Smoothness of shock front solutions for system of conservation laws, *Lecture Notes is Math.* spring-Verlag, #1306(1988), 38–60.

[7] Chen Shuxing, On reflection of multidimensional shock front, (*preprint*).

[8] P. Godin, On the analytic regularity of weak solutions of analytic system of conservation laws with analytic data (*preprint*).

[9] Gu Chaohao, Li Daqian & Hou Zongyi, The Cauchy problem of quasilinear hyperbolic system with discontinuous initial values I, II, III, *Acta Math. Sinica* 4(1961), 314–323, 324–327, 5(1962), 132–143.

[10] E. Harabetian, A convergent series expansion for hyperbolic systems of conservation laws (*preprint*).

[11] H. O. Kreiss, Initial boundary value problems for hyperbolic systems, *Comm. Pure Appl. Math.*, 23(1970), 277–298.

[12] P. D. Lax, Hyperbolic system of conservation laws, *Comm. Pure Appl. Math.*, 10(1957), 537–566.

[13] Li Daqian & Yu Wenci, Some existence theorems for quasilinear hyperbolic systems of partial differential equations in two independent variables I, II, III, *Scientia Sinica* 4(1964), 529–550, 551–526, 7(1965), 1065–1067.

[14] Lin Zhenguo, Boundary value problem for system of gas dynamics with reged wall, *Scientia Sinica*, 26(1983), 226–237.

[15] A. Majda, The stability of multidimensional shock fronts, *Memoirs A. M. S.* #275(1983).

[16] A. Majda, The existence of multidimensional shock fronts, *Memoirs A. M. S.* #281(1983).

[17] A. Majda & E. Thomann, Multi-dimensional shock fronts for second order wave equations, *Comm. in PDE*, 12(1987), 777–828.

[18] G. Metivier, Interaction de deux chocs pour un systeme de deux lois de conservation, en dimension deux d'espace, *Trans. Amer. Math. Soc.* 296(1986), 431–479.

General Form of Nondegenerate Darboux Matrices of First Order for 1+1 Dimensional Unreduced Lax Pairs[*]

Zhou Zi-xiang
Institute of Mathematics, Fudan University
Shanghai, P. R. China

Abstract

In this paper, we determine all the nondegenerate Darboux matrices of first order for quite general $1 + 1$ dimensional unreduced Lax pairs, in which the potentials can be arbitrary rational functions of the spectral parameter. An explicit way to get Darboux matrix in terms of its initial value and the fundamental solution of the Lax pair is provided. Also, the permutability property of the Darboux matrices is obtained by twice Darboux transformations.

1   Introduction

Darboux matrix method is an effective method to get explicit solutions of integrable nonlinear partial differential equations (NPDEs). The main task is to construct the Darboux matrices (DMs). For a 1+1 dimensional NPDE possessing a Lax pair, if one solution of the NPDE is known, and if we can construct a DM from the solutions of the Lax pair, the problem to get special solutions of the nonlinear equation can be reduced to a linear problem to solve the Lax pair. Furthermore, we can get a series of solutions of the original NPDE by a purely algebraic algorithm so long as we know a solution of the Lax pair.

The Darboux transformations (DTs) for some specific equations have been studied since 1975 (see, for example, [6,11,13]). For the 2×2 ZS-AKNS system, DMs of first order as well as higher order in a quite

---

[*]Partially supported by ICTP, Trieste, Italy, and partially supported by the Chinese Fund of Natural Sciences and the Chinese Fund of Doctor Programmes.

general form were also known in 1984 [7,10], and they transform a solution of an equation in the system to a solution of the same equation [2,4,5,8] . As for the ZS-AKNS system of arbitrary order, the DMs in a quite general form were constructed by [3,12].

The ZS-AKNS system is

$$
\begin{cases}
\Phi_x = (\lambda J + P)\Phi, \\
\Phi_t = \displaystyle\sum_{j=0}^{n} V_{n-j}\lambda^j \Phi.
\end{cases}
\tag{1}
$$

Here J is a constant $N \times N$ diagonal matrix with mutually different diagonal entries, P is a valued in $N \times N$ off-diagonal matrices, V_j's are actually differential polynomials of P. According to [3], the following conclusion is true.

Given complex numbers $\lambda_1, \cdots, \lambda_N$, let h_i be a column solution of (1) with $\lambda = \lambda_i$, $\Lambda = \mathrm{diag}(\lambda_1, \cdots, \lambda_N)$, $H = (h_1, \cdots, h_N)$. If $\det H \neq 0$, then $G = \lambda - H\Lambda H^{-1}$ is a DM of (1).

This kind of DMs have interesting physical meanings. They are closely related to the increasing and decreasing of solitons [1,9,12].

[3] also proved that any DM in this form transforms a solution of any equation generated by (1) to a solution of the same equation. On the other hand, [12] showed that these are all the possible DMs of form $\lambda - S(x,t)$ if P is integrable with respect to x in the whole real line and G is bounded for fixed λ, t.

The natural questions are, what are the DMs for other integrable systems, and what happens for the solution which does not decay at infinity? This paper is devoted to solving these problems, and constructing all the nondegenerate DMs of first order for quite general unreduced Lax pairs.

2　Some definitions and main theorems

In this paper, m_N denoted the set of all $N \times N$ complex matrices. Ω is a simple connected domain in $\Re^2$, $\nu_1, \cdots, \nu_l$ are complex numbers. Every function is assumed to be infinitely differentiable.

An equation (or a system of equations)

$$
F(x, t, u, u_x, u_t, u_{xx}, \cdots) = 0
\tag{2}
$$

of unknowns $u = (u_1, \cdots, u_s)$ (defined in Ω) is integrable if there exist two m_N-valued functions $U[u, x, t, \lambda]$, $V[u, x, t, \lambda]$ which are differential polynomials of u, such that (2) is equivalent to

$$U_t - V_x + [U, V] = 0. \tag{3}$$

Here λ is a parameter.

(3) is the integrability condition of the linear equations

$$\begin{cases} \Phi_x = U[u, x, t, \lambda]\Phi, \\ \Phi_t = V[u, x, t, \lambda]\Phi. \end{cases} \tag{4}$$

This is called a Lax pair of (2).

For the equation (2) with Lax pair (4), a matrix $G(x, t, \lambda)$ is called a Darboux matrix if there exists $\tilde{u}$ such that for any fundamental solution Φ of (4), $\tilde{\Phi} = G\Phi$ satisfies

$$\begin{cases} \tilde{\Phi}_x = \tilde{U}\tilde{\Phi}, \\ \tilde{\Phi}_t = \tilde{V}\tilde{\Phi}, \end{cases} \tag{5}$$

Where $\tilde{U} = U[\tilde{u}, x, t, \lambda]$, $\tilde{V} = V[\tilde{u}, x, t, \lambda]$.

After the transformation,

$$\begin{cases} \tilde{U} = GUG^{-1} + G_x G^{-1}, \\ \tilde{V} = GVG^{-1} + G_t G^{-1} \end{cases} \tag{6}$$

and

$$\tilde{U}_t - \tilde{V}_x + [\tilde{U}, \tilde{V}] = 0. \tag{7}$$

Thus, $\tilde{u}$ is another solution of (2).

In this paper, we consider the Lax pair (4). Here U, V are rational functions of the spectral parameter. They are written as

$$\begin{cases} U(x, t, \lambda) = \displaystyle\sum_{j=0}^{m} U_j(x, t)\lambda^j + \sum_{k=1}^{l}\sum_{j=1}^{m_k} U_j^{(k)}(x, t)(\lambda - \nu_k)^{-j}, \\ V(x, t, \lambda) = \displaystyle\sum_{j=0}^{n} V_j(x, t)\lambda^j + \sum_{k=1}^{l}\sum_{j=1}^{n_k} V_j^{(k)}(x, t)(\lambda - \nu_k)^{-j}. \end{cases} \tag{8}$$

Most of integrable equations have Lax pairs in this form. Moreover, we do not consider reduction, i.e. the entries of $U_j, V_j, U_j^{(k)}, V_j^{(k)}$ are assumed to be independent unknowns.

In this case, the general definition of DM is equivalent to that there exist

$$\left\{\begin{array}{l} \tilde{U}(x,t,\lambda) = \displaystyle\sum_{j=0}^{m} \tilde{U}_j(x,t)\lambda^j + \sum_{k=1}^{l}\sum_{j=1}^{m_k} \tilde{U}_j^{(k)}(x,t)(\lambda - \nu_k)^{-j}, \\[4mm] \tilde{V}(x,t,\lambda) = \displaystyle\sum_{j=0}^{n} \tilde{V}_j(x,t)\lambda^j + \sum_{k=1}^{l}\sum_{j=1}^{n_k} \tilde{V}_j^{(k)}(x,t)(\lambda - \nu_k)^{-j}. \end{array}\right.$$

such that $\tilde{\Phi} = G\Phi$ satisfies

$$\left\{\begin{array}{l} \tilde{\Phi}_x = \tilde{U}\tilde{\Phi}, \\ \tilde{\Phi}_t = \tilde{V}\tilde{\Phi} \end{array}\right.$$

for any fundamental solution Φ of (4).

A nondegenerate DM of first order is the DM of form $\lambda R(x,t) - T(x,t)$ such that R is invertible and $\nu_1, \cdots, \nu_l$ are not the eigenvalues of $S = R^{-1}T$. In this case, $G = R(\lambda - S)$, and we may set $R = I$ without loss of generality since R is a trivial DM if there is no reduction. Hence, we always choose $R = I$ in this paper. Then, a DM $\lambda - S$ is nondegenerate if $\nu_1, \cdots, \nu_l$ are not the eigenvalues of S.

The following theorem provides all the possible nondegenerate DMs of first order.

Theorem 1. *If $\nu_1, \cdots, \nu_l$ are not the eigenvalues of S, then $G(x,t,\lambda) = \lambda - S(x,t)$ is a DM of (4) if and only if $S = K\Gamma K^{-1}$, where Γ is a constant matrix, and K is an m_N-valued nondegenerate solution of the integrable equations*

$$\left\{\begin{array}{l} K_x = \sum_{j=0}^{m} U_j K\Gamma^j + \sum_{k=1}^{l}\sum_{j=1}^{m_k} U_j^{(k)} K(\Gamma - \nu_k)^{-j}, \\[2mm] K_t = \sum_{j=0}^{n} V_j K\Gamma^j + \sum_{k=1}^{l}\sum_{j=1}^{n_k} V_j^{(k)} K(\Gamma - \nu_k)^{-j}. \end{array}\right. \tag{9}$$

It is required to obtain the explicit solutions of (3) and the Lax pair (4). However, it is not easy to get the explicit solution of (9). In order to derive a series of solutions of (3) by successive DTs, we hope to express the DM only through the solutions of the Lax pair. This is possible due to the following theorem.

For any open subset D of Ω, let

$$S(D) = \{H(x,t)\Lambda H(x,t)^{-1} \mid \Lambda = \mathrm{diag}(\lambda_1, \cdots, \lambda_N) \text{ is a constant diagonal matrix, } H = (h_1, \cdots, h_N) \text{ is nondegenerate,}$$
$$\text{where } h_i \text{ is a solution of (4) in } D \text{ with } \lambda = \lambda_i\},$$

then we have

Theorem 2. *If $\nu_1, \cdots, \nu_l$ are not the eigenvalues of $S(x,t)$, then $G(x,t,\lambda)$ $= \lambda - S(x,t)$ is a DM of (4) if and only if (i) $S \in S(\Omega)$, or (ii) there exist open sets $\Omega_k \subset \Omega$ $(k = 1, 2, \cdots)$ satisfying $\Omega_1 \subset \Omega_2 \subset \cdots$, $\bigcup\limits_{k=1}^{\infty} \Omega_k = \Omega$, and $S_k \in S(\Omega_k)$ such that for any point $(x,t) \in \Omega_k$, the sequences $\{S_i(x,t)\}$, $\{S_{i,x}(x,t)\}$, $\{S_{i,t}(x,t)\}$ converge to $S(x,t)$, $S_x(x,t)$, $S_t(x,t)$ respectively for $i \geq k$.*

3 Proof of the theorems

Before the proof of these two theorems, we first derive an equation which a DM should satisfy.

For $M \in m_N$ with eigenvalues different from $\nu_1, \cdots, \nu_l$, define

$$U(M) = \sum_{j=0}^{m} U_j M^j + \sum_{k=1}^{l} \sum_{j=1}^{m_k} U_j^{(k)} (M - \nu_k)^{-j}, \tag{10}$$

then we have

Lemma 1. *Under the action of the nondegenerate DM $G = \lambda - S$,*

$$GUG^{-1} + G_x G^{-1} = \sum_{j=0}^{m} \tilde{U}_j \lambda^j + \sum_{k=1}^{l} \sum_{j=1}^{m_k} \tilde{U}_j^{(k)} (\lambda - \nu_k)^{-j} - \Delta(\lambda - S)^{-1}. \tag{11}$$

Here

$$\begin{cases} \tilde{U}_j = U_j + \sum_{i=1}^{m-j} [U_{j+i}, S] S^{i-1}, \\[2mm] \tilde{U}_j^{(k)} = (S - \nu_k) U_j^{(k)} (S - \lambda_k)^{-1} - \sum_{i=1}^{m_k - j} [U_{j+i}^{(k)}, S](S - \nu_k)^{-i-1}, \\[2mm] \Delta = S_x + [S, U(S)]. \end{cases} \tag{12}$$

Therefore, G is a DM if and only if

$$\begin{cases} S_x + [S, U(S)] = 0, \\ S_t + [S, V(S)] = 0. \end{cases} \tag{13}$$

Proof.

$$GUG^{-1} + G_z G^{-1}$$
$$= (\lambda - S)U(\lambda)(\lambda - S)^{-1} - S_z(\lambda - S)^{-1}$$
$$= U(\lambda) - [S, (U(\lambda) - U(S))(\lambda - S)^{-1}] - \Delta(\lambda - S)^{-1}. \tag{14}$$

From the identities

$$(\lambda^j - S^j)(\lambda - S)^{-1} = \sum_{i=0}^{j-1} S^{j-i-1}\lambda^i,$$
$$\{(\lambda - \nu_k)^{-j} - (S - \nu_k)^{-j}\}(\lambda - S)^{-1} \tag{15}$$
$$= -\sum_{i=1}^{j}(S - \lambda_k)^{-j+i-1}(\lambda - \nu_k)^{-i},$$

we get (12) by direct calculation. $\qquad\qquad\square$

Proof of Theorem 1. Suppose K is a nondegenerate solution of (9), and let $S = K\Gamma K^{-1}$, then

$$K_z K^{-1} = U(S). \tag{16}$$

Hence

$$S_x = K_z\Gamma K^{-1} - K\Gamma K^{-1}K_z K^{-1} = [U(S), S].$$

This implies $G = \lambda - S$ is a DM by the lemma.

Conversely, suppose $\lambda - S$ is a DM, then

$$S_x = [U(S), S], \quad S_t = [V(S), S].$$

Since U_t, V_x, UV, VU are all m_N-valued rational functions of λ, we can define $U_t(S), V_x(S), (UV)(S), (VU)(S)$ as (10). Thus, the identity

$$U_t(\lambda) - V_x(\lambda) + [U(\lambda), V(\lambda)] = 0$$

implies

$$U_t(S) - V_x(S) + (UV)(S) - (VU)(S) = 0. \tag{17}$$

However,

$$(U(S))_t = \left(\sum_{j=0}^{m} U_j S^j\right)_t + \left(\sum_{k=1}^{l}\sum_{j=1}^{m_k} U_j^{(k)}(S - \nu_k)^{-j}\right)_t$$
$$= \sum_{j=0}^{m} U_{j,t} S^j + \sum_{k=1}^{l}\sum_{j=1}^{m_k} U_{j,t}^{(k)}(S - \nu_k)^{-j} + \sum_{j=0}^{m}\sum_{i=0}^{j-1} U_j S^i S_t S^{j-i-1}$$

$$-\sum_{k=1}^{l}\sum_{j=1}^{m_k}\sum_{i=0}^{j-1}U_j^{(k)}(S-\nu_k)^{-i-1}S_t(S-\nu_k)^{-j+i}$$

$$= U_t(S) + \sum_{j=0}^{m}\sum_{i=0}^{j-1}U_jS^i[V(S),S]S^{j-i-1}$$

$$-\sum_{k=1}^{l}\sum_{j=1}^{m_k}\sum_{i=0}^{j-1}U_j^{(k)}(S-\nu_k)^{-i-1}[V(S),S](S-\nu_k)^{-j+i}$$

$$= U_t(S) + \sum_{j=0}^{m}U_j[V(S),S^j] + \sum_{k=1}^{l}\sum_{j=1}^{m_k}U_j^{(k)}[V(S),(S-\nu_k)^{-j}]$$

$$= U_t(S) + \sum_{j=0}^{m}U_jV(S)S^j + \sum_{k=1}^{l}\sum_{j=1}^{m_k}U_j^{(k)}V(S)(S-\nu_k)^{-j}$$

$$-\sum_{j=0}^{m}U_jS^jV(S) - \sum_{k=1}^{l}\sum_{j=1}^{m_k}U_j^{(k)}(S-\nu_k)^{-j}V(S)$$

$$= U_t(S) + (UV)(S) - U(S)V(S),$$

likewise,

$$(V(S))_x = V_x(S) + (VU)(S) - V(S)U(S).$$

Combining these with (17), we have

$$(U(S))_t - (V(S))_x + [U(S),V(S)] = 0. \tag{18}$$

Choose any point $(x_0,t_0) \in \Omega$, (18) implies that the equation

$$\begin{cases} K_x = U(S)K, \\ K_t = V(S)K, \\ K(x_0,t_0) = I \end{cases} \tag{19}$$

has a unique solution K with $\det K \neq 0$ in Ω.

    Let

$$\Gamma = S(x_0,t_0), \quad S' = S - K\Gamma K^{-1},$$

then

$$S'_x = [U(S),S] - [U(S),K\Gamma K^{-1}] = [U(S),S'],$$
$$S'_t = [V(S),S']$$

and

$$S'(x_0,t_0) = 0.$$

Hence

$$S = K\Gamma K^{-1}$$

everywhere. (19) implies (9). This proves the theorem. □

Proof of theorem 2. First suppose $S = H\Lambda H^{-1} \in S(\Omega)$, where Λ is a constant diagonal matrix, and $H = (h_1, \cdots, h_N)$ with h_i a solution of (4) as $\lambda = \lambda_i$. Then H is a solution of (9) with $\Gamma = \Lambda$. Hence $G = \lambda - S$ is a DM of (4) by theorem 1. On the other hand, if S satisfies (ii) of this theorem, G is also a DM by the limit of (13). This proves the necessity.

Now suppose $\lambda - S$ is a DM. From Theorem 1, $S = K\Gamma K^{-1}$ where Γ is a constant matrix and K satisfies (9).

If Γ is diagonizable, $\Gamma = T\Lambda T^{-1}$ with $\Lambda = \mathrm{diag}(\lambda_1, \cdots, \lambda_N)$. Let $H = KT$. Then $S = H\Lambda H^{-1}$, and

$$\begin{cases} H_x = \displaystyle\sum_{j=1}^{m} U_j H\Lambda^j + \sum_{k=1}^{l}\sum_{j=1}^{m_k} U_j^{(k)} H(\Lambda - \nu_k)^{-j}, \\ H_t = \displaystyle\sum_{j=0}^{n} V_j H\Lambda^j + \sum_{k=1}^{l}\sum_{j=1}^{n_k} V_j^{(k)} H(\Lambda - \nu_k)^{-j}. \end{cases}$$

If we write $H = (h_1, \cdots, h_N)$, then h_i is a solution of (4) with $\lambda = \lambda_i$. Hence $S \in S(\Omega)$.

Now suppose Γ is not diagonalizable. Choose a constant matrix Θ such that $\Gamma^{(\varepsilon)} = \Gamma + \varepsilon\Theta$ is diagonalizable with eigenvalues different from ν_k $(k = 1, \cdots, l)$ for sufficiently small ε. Let (x_0, t_0) be any point in Ω and $K^{(\varepsilon)}$ be the solution of

$$\begin{cases} K_x^{(\varepsilon)} = \displaystyle\sum_{j=0}^{m} U_j K^{(\varepsilon)}(\Gamma^{(\varepsilon)})^j + \sum_{k=1}^{l}\sum_{j=1}^{m_k} U_j^{(k)} K^{(\varepsilon)}(\Gamma^{(\varepsilon)} - \nu_k)^{-j}, \\ K_t^{(\varepsilon)} = \displaystyle\sum_{j=0}^{n} V_j K^{(\varepsilon)}(\Gamma^{(\varepsilon)})^j + \sum_{k=1}^{l}\sum_{j=1}^{n_k} V_j^{(k)} K^{(\varepsilon)}(\Gamma^{(\varepsilon)} - \nu_k)^{-j}, \\ K^{(\varepsilon)}(x_0, t_0) = K(x_0, t_0). \end{cases} \tag{20}$$

Let $K_k = K^{(1/k)}$, $\Gamma_k = \Gamma^{(1/k)}$, Ω_k be the interior of

$$\bigcap_{\varepsilon \leq 1/k} \{(x, t) \in \Omega \mid \det K^{(\varepsilon)}(x, t) \neq 0\}.$$

Then, at any given point $(x', t') \in \Omega$, $\det K^{(\varepsilon)}(x', t') \neq 0$ for sufficiently small ε since $K^{(\varepsilon)}$ depends continuously on ε in Ω and $\det K(x', t') \neq 0$.

Hence there exists k_0 such that $(x', t') \in \Omega_k$ for $k \geq k_0$. This implies

$$\bigcup_{k=1}^{\infty} \Omega_k = \Omega.$$

By (20), we have

$$K_k(x, t) \to K(x, t), \quad K_{k,x}(x, t) \to K_x(x, t), \quad K_{k,t} \to K_t(x, t)$$

as $k \to \infty$ with $\det K_k(x, t) \neq 0$ in Ω_{k_0} for $k \geq k_0$.

Since Γ_k is diagonizable, let

$$\Gamma_k = T_k \Lambda_k T_k^{-1}$$

with $\Lambda_k = \mathrm{diag}(\lambda_1^{(k)}, \cdots, \lambda_N^{(k)})$ and let

$$H_k = K_k T_k,$$

then $H_k = (h_1^{(k)}, \cdots, h_N^{(k)})$ where $h_j^{(k)}$ is a solution of (4) with $\lambda = \lambda_j^{(k)}$, and

$$S_k = H_k \Lambda_k H_k^{-1} = K_k \Gamma_k K_k^{-1} \to K \Gamma K^{-1} = S,$$
$$S_{k,x} = [H_{k,x} H_k^{-1}, S_k] = [K_{k,x} K_k^{-1}, S_k] \to [K_x K^{-1}, S] = S_x,$$
$$S_{k,t} \to S_t$$

at $(x, t) \in \Omega_k$ for $k \geq k_0$. $\square$

Theorems 1 and 2 provide all the nondegenerate DMs of first order if we do not consider the reduction. If there is reduction, DMs will be a part of that. To determine all the nondegenerate DMs of first order which keep certain reduction is usually a much more difficult problem.

We can also show that there are DMs which are not diagonizable, so in Theorem 2, (ii) is not included in (i) in general. Furthermore, The proof of Theorem 2 provides a constructive way to get nondiagonizable DMs only through the solutions of the Lax pair. We repeat it briefly here.

For given constant matrix Γ, we can construct a DM $\lambda - S$ satisfying $S(x_0, t_0) = \Gamma$ in the following way. Choose a constant matrix Θ such that $\Gamma_k = \Gamma + \Theta/k \ (k = 1, 2, \cdots)$ is diagonizable with eigenvalues different from $\nu_j \ (j = 1, \cdots, l)$ for sufficiently large k. Let $\Gamma_k = T_k \Lambda_k T_k^{-1}$ with $\Lambda_k = \mathrm{diag}(\lambda_1^{(k)}, \cdots, \lambda_N^{(k)})$. Let $H_k = (h_1^{(k)}, \cdots, h_N^{(k)})$ satisfy $H_k(x_0, t_0) = T_k$ where $h_j^{(k)}$ is a solution of (4) with $\lambda = \lambda_j^{(k)}$. Then $S = \lim_{k \to \infty} S_k$ with $S_k = H_k \Lambda_k H_k^{-1}$ gives a DM $\lambda - S$. All the DMs of form $\lambda - S$ can be constructed in this way if there is no reduction.

4   Permutability

Permutability is an important property of Bäcklund transformations. The permutability theorem for diagonizable Darboux matrices for ZS-AKNS system has already been known (cf.[3,5,8,10]). Here we make a slight generalization and show that this is true universally for any two generic nondegenerate DMs of first order of Lax pair (4).

We have

Theorem 3. *Suppose* $G_\alpha(x,t,\lambda) = \lambda - S_\alpha(x,t)$ $(\alpha = 1,2)$ *are two nondegenerate* DMs *for* (U,V), $\det(S_1 - S_2) \neq 0$. *then*

$$G_{\beta\alpha}(x,t,\lambda) = \lambda - (S_\beta - S_\alpha)S_\beta(S_\beta - S_\alpha)^{-1} \tag{21}$$

is a DM *for*

$$U_\alpha = G_\alpha U G_\alpha^{-1} + G_{\alpha,x}G_\alpha^{-1}, \quad V_\alpha = G_\alpha V G_\alpha^{-1} + G_{\alpha,t}G_\alpha^{-1}$$

$(\alpha,\beta = 1,2; \; \alpha \neq \beta)$ *and the permutability*

$$G_{21}(\lambda)G_1(\lambda) = G_{12}(\lambda)G_2(\lambda) \tag{22}$$

holds.

Proof. First suppose $S_\alpha(\alpha = 1,2)$ are diagonalizable. Let $S_\alpha = H_\alpha \Lambda_\alpha H_\alpha^{-1}$ where $\Lambda_\alpha = \mathrm{diag}(\lambda_1^{(\alpha)}, \cdots, \lambda_N^{(\alpha)})$ with $\lambda_i^{(\alpha)} \neq \nu_j$ $(i = 1, \cdots, N;$ $j = 1, \cdots, l; \; \alpha = 1,2)$, $H_\alpha = (h_1^{(\alpha)}, \cdots, h_N^{(\alpha)})$, $h_i^{(\alpha)}$ is a solution of (4) with $\lambda = \lambda_i^{(\alpha)}$. Under the action of G_α, H_β changes to

$$\begin{aligned}
H_{\beta\alpha} &= \left(G_\alpha(\lambda_1^{(\beta)})h_1^{(\beta)}, \cdots, G_\alpha(\lambda_N^{(\beta)})h_N^{(\beta)} \right) \\
&= H_\beta \Lambda_\beta - H_\alpha \Lambda_\alpha H_\alpha^{-1} H_\beta = (S_\beta - S_\alpha)H_\beta
\end{aligned}$$

$(\alpha,\beta = 1,2; \; \alpha \neq \beta)$. Therefore,

$$S_{\beta\alpha} = H_{\beta\alpha}\Lambda_\beta H_{\beta\alpha}^{-1} = (S_\beta - S_\alpha)S_\beta(S_\beta - S_\alpha)^{-1}$$

gives a DM $G_{\beta\alpha}(\lambda) = \lambda - S_{\beta\alpha}$ for U_α, V_α. Moreover,

$$\begin{aligned}
G_{\beta\alpha}(\lambda)&G_\alpha(\lambda) \\
&= \{\lambda - (S_\beta - S_\alpha)S_\beta(S_\beta - S_\alpha)^{-1}\}(\lambda - S_\alpha) \\
&= \lambda^2 - (S_\beta^2 - S_\alpha^2)(S_\beta - S_\alpha)^{-1}\lambda \\
&\qquad + (S_\beta - S_\alpha)S_\beta(S_\beta - S_\alpha)^{-1}S_\alpha.
\end{aligned} \tag{23}$$

Noticing that

$$S_2(S_2 - S_1)^{-1}S_1 + S_1(S_1 - S_2)^{-1}S_2 = 0, \tag{24}$$

we know (23) is symmetric for S_α, S_β. Hence

$$G_{21}(\lambda)G_1(\lambda) = G_{12}(\lambda)G_2(\lambda). \tag{25}$$

For nondiagonizable $G_\alpha = \lambda - S_\alpha(x,t)$, we have nondegenerate DMs $G_\alpha^{(k)}(\lambda) = \lambda - S_\alpha^{(k)}$ such that $S_\alpha^{(k)}$ is diagonizable, $\det(S_1^{(k)} - S_2^{(k)}) \neq 0$ and $G_\alpha^{(k)} \to G_\alpha$ as $k \to \infty$. From the above discussion,

$$G_{\beta\alpha}^{(k)} = \lambda - (S_\beta^{(k)} - S_\alpha^{(k)})S_\beta^{(k)}(S_\beta^{(k)} - S_\alpha^{(K)})^{-1} \tag{26}$$

is a DM for $U_\alpha^{(k)}, V_\alpha^{(k)}$, which are transformed from U, V by $G_\alpha^{(k)}$, and

$$G_{21}^{(k)}G_1^{(k)} = G_{12}^{(k)}G_2^{(k)}. \tag{27}$$

Taking that limit in (26) and (27), we derive the conclusions in the theorem. $\qquad\square$

Acknowledgment

This is a part of my Ph.D thesis. I would like to express my gratitude to Prof. Gu Chaohao and Prof. Hu Hesheng for helpful advice, assistance and encouragement. Also, I would like to thank Prof. Li Yisheng, Dr. Chenyi and Dr. Lin Junmin for helpful advice.

Part of the work in this paper was done in the international Center for Theoretical Physics, Trieste. I also would like to thank Prof. Abdus Salam, the international Atomic Energy Agency and UNESCO for the hospitality.

References

[1] Beals R. and Coifman R.R., Scattering and Inverse Scattering for First-order System II, *Inverse Problem* 3 (1987), 577.

[2] Gu C. H., On the Bäcklund Transformations for the Generalized Hierarchies of Compound MKdV-SG Equations, *Lett. Math. Phys.* 12(1986), 31.

[3] Gu C.H., On the Darboux Form of Bäcklund Transformations, *Proc. NanKai Symposium on Integrable Systems* (1987), to be published.

[4] Gu C.H. and Hu H.S., A Unified Explicit Form of Bäcklund Transformations for Generalized Hierarchies of KdV Equations. *Lett. Math. Phys.* 11 (1986),325.

[5] Gu C.H. and Zhou Z. X., On the Darboux Matrices of Bäcklund Transformations for the AKNS Systems, *Lett. Math. Phys.* 13(1987), 179.

[6] Levi D., Toward a Unification of the Various Techniques Used to Integrate Nonlinear Partial Differential Equations. Bäcklund and Darboux Transfirmations v. s. Dressing Method, *Rep. Math. Phys.* 23(1986). 41.

[7] Levi D. Ragnisco O. and Sym A., Dressing Method v. s. Classical Darboux Transformation, *I1 Nuovo Cimento* 83B (1984), 34.

[8] Li Y. S., Gu X. S. and Zou M. R. , Three Kinds of Darboux Transformations for the Evolution Equations Which Connect with A.K.N.S. Eigenvalue Problem, *Acta Math. Sinica*, New Series 3 (1987), 143.

[9] Lin J. M., Evolution of the Scattering Data under Classical Darboux Transform for su(2) Soliton System, *preprint*(1987).

[10] Neugeerbauer G. and Meinel R., General N-soliton Solution of the AKNS Class on Arbitrary Background, *Phys. Lett.* 100A(1984), 467.

[11] Rogers C. and Shadwick W. R., *Bäcklund Transformations and Their Applications*, Academic Press(1982).

[12] Scattinger D.H. and Zurkowski V.D., Gauge Theory of Bäcklund Transformations II, *Physica* 26D(1987), 325.

[13] Wadati M., Sanuki H. and Konno K., Relationships among Inverse Method, Bäcklund Transformations and Infinite Number of Conservation Laws, *Prog. Theor. Phys.* 53(1975), 418.

The First Nonzero Eigenvalue for Manifolds with Ricci Curvature Having Positive Lower Bound

Xia Changyu
Inst. of Math., Fudan Univ.
Shanghai, China

Abstract

Let M be an n-dimensional compact Riemannian manifold with nonempty connected boundary ∂M. Suppose $\mathrm{Ric}_M \geq n - 1$ and ∂M is convex. We prove that the first nonzero eigenvalue μ_1 with Neumann boundary condition of M satisfies $\mu_1 \geq n$ and $\mu_1 = n$ iff M is isometric with the standard upper half sphere $S_+^n(1)$.

1   Introduction

Let (M^n, ds^2) be a compact C^∞ Riemannian manifold with or without boundary. In local coordinates $(x^1, \cdots, x^n)$, the Riemannian metric is given by $ds^2 = \sum g_{ij} dx^i dx^j$, the Laplacian operator Δ acting on functions is defined as

$$\Delta = g^{-1/2} \sum \frac{\partial}{\partial x^i} \left(g^{1/2} g^{ij} \frac{\partial}{\partial x^j} \right).$$

Where $(g^{ij}) = (g_{ij})^{-1}$ and $g = \det(g_{ij})$.

Consider the following eigenvalue problems:

- Free membrane problem $(\partial M = \phi)$

$$(1.1) \qquad \Delta f + \lambda f = 0.$$

- Dirichlet boundary problem $(\partial M \neq \phi)$

$$(1.2) \qquad \Delta f + \eta f = 0, \quad f = 0 \text{ on } \partial M.$$

- Neumann boundary problem $(\partial M \neq \phi)$

$$(1.3) \qquad \Delta f + \mu f = 0, \quad \left.\frac{\partial f}{\partial v}\right|_{\partial M} = 0$$

where v is the outward normal vector to ∂M.

We denote by λ_1, η_1 and μ_1 the first nonzero eigenvalues of the problem (1.1), (1.2) and (1.3), respectively. It is very important to obtain optimal estimates for λ_1, η_1 and μ_1 from below in terms of geometric elements in studying analysis on a manifold. Lichnerowicz [4] initiated the task of estimating λ_1 from below for problem (1.1). He proved that: If M^n is a complete Riemannian manifold with $\mathrm{Ric}_M \geq n-1$, then $\lambda_1 \geq n$. Later, Obata [5] showed that under the same hypothesis, $\lambda_1 = n$ iff M^n is isometric to the standard unit sphere $S^n(1)$. In 1977, Reilly [6] generalized these results of Lichnerowicz and Obata to problem (1.2): If $\mathrm{Ric}_M \geq n - 1$, and if the mean curvature H of ∂M^n is nonnegative, then $\eta_1 \geq n$. Equality holds iff M^n is the standard unit upper half hemisphere $S^n_+(1)$.

In this paper, we obtain the following optimal estimate for the first nonzero eigenvalue from below to problem (1.3):

Theorem 1. *Let (M^n, g) be an n-dimensional compact Riemannian manifold with connected $\partial M \neq \phi$. If $\mathrm{Ric}_M \geq n - 1$ and ∂M is convex, then the first nonzero eigenvalue μ_1 with Neumann boundary condition of M satisfies $\mu_1 \geq n$. Equality holds iff M is isometric to the standard unit upper half hemisphere $S^n_+(1)$.*

2 A proof of Theorem

Let f be the eigenfunction corresponding to the first nonzero eigenvalue μ_1 with Neumann boundary condition of M. Start with the Bochner formula

$$(2.1) \qquad \frac{1}{2}\Delta|\nabla f|^2 = |\mathrm{Hess}f|^2 + \langle \nabla f, \nabla \Delta f\rangle + \mathrm{Ric}(\nabla f, \nabla f)$$

and integrate:

$$(2.2) \quad \int_M \left(|\mathrm{Hess}f|^2 - \mu_1|\nabla f|^2 + \mathrm{Ric}(\nabla f, \nabla f)\right) = \frac{1}{2}\int_{\partial M}\langle \nabla|\nabla f|^2, v\rangle$$

where v is the outward unit normal vector to ∂M.

Denote by II the second fundamental form of ∂M. It is easy to check from $\frac{\partial f}{\partial v}\big|_{\partial M} = 0$ that

$$(2.3) \qquad \frac{1}{2}\langle \nabla|\nabla f|^2, v\rangle = -\mathrm{II}(\tilde{\nabla}f, \tilde{\nabla}f) = -\mathrm{II}(\nabla f, \nabla f)$$

where $\tilde{\nabla}$ is the connection of ∂M induced from the connection ∇ of M.

From (2.2), (2.3) and the convexity of ∂M, we derive

$$(2.4) \qquad \int_M \left(|\mathrm{Hess}\, f|^2 - \mu_1|\nabla f|^2 + \mathrm{Ric}(\nabla f, \nabla f)\right)$$

$$= \int_{\partial M} -\mathrm{II}(\nabla f, \nabla f) \le 0.$$

Next we have
$$(2.5) \qquad |\mathrm{Hess}\, f|^2 \ge 1/n(\Delta f)^2 = \mu_1^2/n\, f^2$$

and equality holds at p if and only if $\mathrm{Hess}\, f(p) = -f(p)g$.

Also, by assumption

$$\mathrm{Ric}(\nabla f, \nabla f) \ge (n-1)|\nabla f|^2$$

so we find,

$$(2.6) \qquad 0 \;\ge\; \int_M (\mu_1^2/n\, f^2 + ((n-1) - \mu_1)|\nabla f|^2)$$

$$= \;(\mu_1^2/n + ((n-1) - \mu_1)\mu_1 \int_M f^2$$

or

$$0 \ge 1/n + (n-1)/\mu_1 - 1, \quad \mu_1 \ge n.$$

Equality means that

$$0 = \int_M \left(|\mathrm{Hess}\, f|^2 - 1/n(\Delta f)^2 + \mathrm{Ric}(\nabla f, \nabla f) - (n-1)|\nabla f|^2\right)$$

and since the integrand is pointwise ≥ 0 we conclude

$$|\mathrm{Hess}\, f|^2 = 1/n(\Delta f)^2$$

or equivalently,

$$(2.7) \qquad \mathrm{Hess}\, f(p) = -f(p)g, \quad \text{for } p \in M$$

and

$$\mathrm{Ric}(\nabla f, \nabla f) = (n-1)|\nabla f|^2$$

or equivalently,

$$(2.8) \qquad \mathrm{Ric}(\nabla f, X) = (n-1)\langle \nabla f, X\rangle = (n-1)Xf$$

for any vector field X on M, also

$$\mathrm{II}(\nabla f, \nabla f) = 0$$

or equivalently,

$$(2.9) \qquad\qquad\qquad\qquad \nabla_{\nabla f} F_0 = 0$$

for any unit normal vector field F_0 to ∂M.

This implies that H ($H := \mathrm{trII}$, the mean curvature of ∂M) satisfies a first order linear differential equation along the integral curves of $\tilde{\nabla} z = \nabla f$, where $z := f\,|_{\partial M}$:

$$(2.10) \qquad \frac{d}{dt} H \circ c(t) - z \circ c(t) H \circ c(t) = 0, \quad t \in R.$$

To prove (2.10) fix $t_0 \in R$ and determine a local orthonormal frame for $T\partial M$, $(E_i)_{i=1}^{n-1}$, such that $\tilde{\nabla}_{E_i} E_j(c(t_0)) = 0$. Denote by (F_i) the parallel frame along c obtained from $(E_i(c(t_0)))$ and by F_0 a unit normal field to ∂M.

Then one computes with $c'(t) = \tilde{\nabla} z(c(t)) = \nabla f(c(t))$

$$\frac{d}{dt} H \circ c(t) = \sum_{i=1}^{n-1} \big(R(F_i, \nabla f, F_0, F_i) + \langle \nabla_{F_1} \nabla_{\nabla f} F_0, F_i\rangle$$
$$+ \langle_{[\nabla f, F_i]} F_0, F_i\rangle \big)(c(t)).$$

Now we use (2.7) and (2.8) to find

$$(2.11) \qquad \sum_{i=1}^{n-1} R(F_i, c', F_0, F_i) = -(n-1)F_0 f = 0,$$

$$(2.12) \qquad \nabla_{F_i} \nabla f = \nabla_{F_i} \sum_k F_k f \cdot F_k$$
$$= \sum_k (\mathrm{Hess} f(F_i, F_k) F_k + (\nabla_{F_i} F_k)f \cdot F_k + F_k f \cdot \nabla_{F_i} F_k)$$
$$= -f F_i$$

hence

$$\nabla_{F_i}\nabla f(c(t_0)) = -f(c(t_0))F_i.$$

This implies, since F_i is parallel, at $c(t_0)$,

(2.13) $$[\nabla f, F_i] = \nabla_{\nabla f}F_i - \nabla_{F_1}\nabla f = fF_i$$

hence

$$\begin{aligned}
\frac{d}{dt}\Big|_{t=t_0} H\circ c(t) &= f(c(t_0))\sum_{i=1}^{n-1}\langle\nabla_{F_i}F_0, F_i\rangle(c(t_0))\\
&= f(c(t_0))H\circ c(t_0).
\end{aligned}$$

Since t_0 was arbitrary this holds for all t. It follows by integration that for $t \geq T$

(2.14) $$H\circ c(t) = H\circ c(T)\exp\left\{\int_T^t f\circ c(s)ds\right\}.$$

Since $F_0 f = 0$, we have

$$\mathrm{Hess}z = -zg_{\partial M}.$$

If $z \equiv 0$, f would be the first eigenfunction for the Dirichlet problem, and hence positive or negative off ∂M; this conflicts with f solving the Neumann problem, so $z \not\equiv 0$. When $n \geq 2$, by Obata's Theorem [5] we know that $\partial M = s^{n-1}(1)$; when $n = 2$, ∂M is a compact 1-manifold, and thus is isometric to a circle. From this, we conclude that z has exactly two critical point $p_{\max}$ and $p_{\min}$ with $z(p_{\min}) < 0$. Moreover, all integral curves $c \neq \{p_{\min/\max}\}$ of $\tilde{\nabla}z$ satisfy

$$\lim_{t\to-\infty} c(t) = p_{\min}.$$

Making $T \to -\infty$ in (2.14) we derive, since $f\circ c = z\circ c$ increases monotonically and $f(p_{\min}) < 0$,

(2.15) $$H(p) = H(p_{\min})W(p)$$

for all $p \in \partial M$ and some function W.

Finally, we choose a frame (E_i) of $T\partial M$ and compute with (2.9),

(2.16) $$\begin{aligned}
0 &= E_i\langle\nabla_{E_i}F_0, \nabla F\rangle\\
&= \langle\nabla_{E_i}\nabla_{E_i}F_0, \nabla f\rangle - f\langle\nabla_{E_i}F_0, E_i\rangle
\end{aligned}$$

since at $P_{\min}$, $0 = \tilde{\nabla} z = \nabla f$, we thus conclude from (2.16) that

$$0 = -f(p_{\min})\mathrm{II}(E_i, E_i).$$

Thus $H(p_{\min}) = 0$ hence $H(p) = 0$ and finally $\mathrm{II}(p) = 0$ for all $p \in \partial M$. i.e. *∂M is totally geodesic*. From this and the same arguments as in proof of Hopf-Rinow theorem, we conclude that every geodesic of M which does not already lie in the boundary can be extended (in both directions) either infinitely or else until it hits the boundary. Moreover, every pair of points of M can be joined by a minimizing geodesic in M.

On the other hand, since f satisfies the differential equation (2.7) and has at least one critical point (namely at the maximum of $f|_{\partial M}$ on ∂M), we can now easily carry over Obata's proof to prove that M is isometric to $S^n_+(1)$. $\square$

In [1], p.50, I. Chavel proved that the first nonzero eigenvalue μ_1 with Neumann boundary condition of the spherical cap contained in $S^n(1)$ satisfies $\mu_1 \geq n - 1$. the following corollary improves the above result in the case that the spherical cap is contained in the hemisphere.

Corollary 2. *Let $B(r) \subset S^n(1)$ be a geodesic ball of radius r and $r \leq \pi/2$. Then the first nonzero eigenvalue μ_1 with Neumann boundary condition of $B(r)$ satisfies $\mu_1 \geq n$ and $\mu_1 = n$ iff $r = \pi/2$.*

Proof. Using Theorem and the fact that $\partial B(r)$ is convex when $r \leq \pi/2$.

References

[1] I. Chavel, *Eigenvalues in Riemannian Geometry*, Academic Press, 1984.

[2] S.Gallot, Equations différentielles charactéristiques de la sphère, *Ann. Sci. Ecole Norm. Sup*(4) 12, 253–269(1978).

[3] P. Li, Poincare inequalities on Riemannian manifolds, "Seminar on Differential Geometry", *Ann. Math. Stud.* 102, 74–84(1982).

[4] A. Lichnerowicz, *Géométrie dss transformations Dunod* (1958).

[5] M. Obata, Certain conditions for a Riemannian manifold to be isometric with a sphere, *J. Math. Soc. Jap.* 14, 333–340(1962).

[6] R.C. Reilly, Applications of the Hessian Operato in a Riemannian
manifold, *Indiana Univ. Math. J.* 26, 459–472(1977).

Comparison between Maximal Function and Square Function on Positively Curved Product Manifolds

Chen Jie-cheng and *Wang Si-lei*

Department of Mathematics, Hangzhou University,
Hangzhou, P.R.China

Abstract

Comparisons between maximal functions and square functions
are basis of the theory of Hardy spaces. In this paper, we get
such a comparison relation on product manifolds by technique of
Carleson measure.

1　Introduction

Let M_i be a complete Riemannian manifold, $\triangle_i$ its Laplace-Beltrami operator, ∇_i its gradient operator, $\{P_{t_i}\}_{t_i \geq 0}$ and $\{H_{t_i}\}_{t_i \geq 0}$ its Poisson and heat diffusion semi-groups, $\{p_{t_i}\}_{t_i \geq 0}$ and $\{h_{t_i}\}_{t_i \geq 0}$ its Poisson and heat kernels, $\Delta_i = \triangle_i + (\frac{\partial}{\partial t})^2$, $\nabla_i = (\nabla_i, \frac{\partial}{\partial t})$, $d_i(x_i, y_i)$ its geodesic distance between x_i and y_i, $n_i = \dim(M_i)$, $i = 1, d$, where d is a positive integer. Set $M_i^{\perp} = M_i \times R_+^1, \overline{M} = \prod_1^d M_i, \overline{M}^{\perp} = \prod_1^d M_i^{\perp}, \overline{\nabla} = \prod_1^d \nabla_i, \nabla = \prod_1^d \nabla_i, \overline{n} = (n_1, \cdots, n_d)$. For $\overline{x} = (x_1, \cdots, x_d) \in \overline{M}, \overline{t} = (t_1, \cdots, t_d)$ and $\overline{a} = (a_1, \cdots, a_d) \in (R_+^1)^d$, set $(\overline{x}, \overline{t}) = ((x_1, t_1), \cdots, (x_d, t_d))$, $B_{\overline{x}}(\overline{t}) = \prod_1^d B_{x_i}(t_i)$, $V_{\overline{x}\overline{t}} = \prod_1^d V_{x_i}(t_i)$, $\overline{a}\overline{t} = (a_1 t_1, \cdots, a_d t_d)$, $dm_i = t_i \, dt_i \, dx_i$, $d\overline{m} = \prod_1^d dm_i$, $P_{\overline{t}} = \prod_1^d P_{t_i}$, $p_{\overline{t}}(\overline{x}, \overline{y}) = \prod_1^d p_{t_i}(x_i, y_i)$, etc.. For open set $G \subset M$, let $\hat{G}_{\overline{a}} = \{(\overline{x}, \overline{t}) : B_{\overline{x}}(\overline{a}\overline{t}) \subset G\} = (\bigcup_{\overline{w} \in G^c} \Gamma_{\overline{a}}(\overline{w}))^c$, where $\Gamma_{\overline{a}}(\overline{x}) = \{(\overline{y}, \overline{t}) \in \overline{M}^{\perp} : B_{\overline{y}}(\overline{a}\overline{t}) \ni \overline{x}\}$.

For suitable function f on $\overline{M}$, let $u(\overline{x}, \overline{t}) = P_{\overline{t}}(f)(\overline{x})$. The nontangential maximal function and the area integral function of f are defined as follows

$$N_{P,\bar{a}}(f)(\bar{x}) = \sup\{|u(\bar{y},\bar{t})| : (\bar{y},\bar{t}) \in \Gamma_{\bar{a}}(\bar{x})\}$$

$$A_{P,\bar{b}}(f)(\bar{x}) = \left(\int_{\Gamma_{\bar{b}}(\bar{x})} |\overline{\nabla} u(\bar{y},\bar{t})|^2 V_{\bar{x}}(\overline{bt})d\overline{m}\right)^{\frac{1}{2}} \tag{1}$$

In this paper, we prove

Theorem 1.   If $\mathrm{Ric}(\overline{M}) \geq 0$, $\bar{a}$ and $\bar{b} \in (R_+^1)^d, 0 < p < \infty$, then for any $f \in L_c^\infty(\overline{M})$, there holds

$$\| A_{P,\bar{b}} \|_p \leq C_{\overline{M},d,p,\bar{a},\bar{b}} \| N_{P,\bar{a}}(f) \|_p \tag{2}$$

where $C_{\overline{M},d,p,\bar{a},\bar{b}}$ is a positive number depending only on $\overline{M}, d, p, \bar{a}, \bar{b}$.

As is known, the inequality (2) is very important in H^p-theory, it is a main difficulty in extending H^p-spaces with one-parameter to H^p-spaces with multi-parameters, see[1,2,4,6,7,10]. In one parameter case, the first proof of (2) is probabilistic (see[1]), which depends on the Brown motion on R^{n+1}; the second proof of (2) is analytic, it is closely related to Green's formula on domain $\Re = \bigcup_{\bar{x} \in E} \Gamma_{\bar{a}}(\bar{x})$. In multi-parameter case, the proof given in [7] (d=2) needs treating the multi sub-linear operator[11]

$$B(u,v) = \left(\int_{\Gamma(x)} |\nabla_1 u|^2 |\nabla_2 v|^2 t_1^{-n_1} t_2^{-n_2} dm_1 dm_2\right)^{\frac{1}{2}}$$

and, the proof given in [10] depends on the dilation and translation structures of R^n. All above methods seem difficult to be applied to our case. In this paper, we generalize Chen's idea in [4] to set up Theorem 1, our main tools are Carleson measures and estimates of heat kernel [5].

In section 2, we give some basic properties of Poisson kernel and harmonic functions of one-parameter (For completeness, we give their proofs), they are important for following two sections. All of them have their multi-parameter analogues. In section 3, we prove that for any multi harmonic function u on $\overline{M}^\perp$ and a suitable choice of bounded function v_u such that $uv_u \in L^\infty(\overline{M})$, $|\overline{\nabla}(uv_u)|^2 d\overline{m}$ is a Carleson measure on $\overline{M}^\perp$. This is the heart of our proof of Theorem 1. In section 4, we give our proof of Theorem 1. The sketch is as follows. For any $\lambda > 0$, let $E_\lambda = \{N_{P,\bar{a}}(f) \leq \lambda\}$, then

$$|\{A_{P,\bar{b}}(f) \geq \lambda\}| \leq |E_\lambda^c| + \lambda^{-2} \int_{\overline{M}^\perp} |\overline{\nabla} u|^2 \tilde{\mathcal{X}}_\lambda d\overline{m}$$

where $\tilde{\mathcal{X}}_\lambda$ is a suitable smooth version of the characteristic function $\mathcal{X}_\lambda$ of $\bigcup_{\bar{x} \in E_\lambda} \Gamma_{\bar{a}}(\bar{x})$ (see(24) below). By a suitable form of Green's formula (see Lemma 5 and Lemma 6), we have

$$\int_{\overline{M}^\perp} |\overline{\nabla} u|^2 \tilde{\mathcal{X}}_\lambda d\overline{m} \leq \int_{\overline{M}^\perp} |\nabla_1 \cdots \nabla_{d-1} u|^2 |\Delta_d \tilde{\mathcal{X}}_\lambda|^2 d\overline{m} + \text{easy terms} \quad (3)$$

But, for the good choice of $\mathcal{X}_\lambda$, we have

$$|\Delta_d \tilde{\mathcal{X}}_\lambda| \leq \tilde{\mathcal{X}}_\lambda |\nabla_d P_{\bar{t}}(\mathcal{X}_{E_\lambda})|^2$$

So, if $|\nabla_1, \cdots, \nabla_{d-1} u|^2 \tilde{\mathcal{X}}_\lambda dm_1 \cdots dm_{d-1}$ is a Carleson measure on $M_1^\perp \times \cdots \times M_{d-1}^\perp$ and its C.M.norm $\leq C \cdot \lambda^2$ (technical point), then, the first term of right side of (3)

$$\leq \int_{\overline{M}^\perp} \lambda^2 \left(\int_{M_1 \times \cdots \times M_{d-1}} |\nabla_d P_{\bar{t}}(\mathcal{X}_{E_\lambda})|^2 dx_1 \cdots dx_{d-1} \right) dm_d$$

On $M_1 \times \cdots M_{d-1} \times M_d^\perp$, $\nabla_d P_{\bar{t}}(\mathcal{X}_{E_\lambda}) = \nabla_d P_{t_d}(\mathcal{X}_{E_\lambda}) = \nabla_d(1 - P_{t_d}(\mathcal{X}_{E_\lambda})) = \nabla_d P_{t_d}(\mathcal{X}_E^c)$, thus, by the L^2-boundedness of Littlewood-Paley-Stein g-function, the last term

$$= \lambda^2 \int_{M_1 \times \cdots M_{d-1}} \| g_{P_d}(\mathcal{X}_{E_\lambda^c}) \|_{L^2(M_d)}^2 \, dx_1 \cdots dx_{d-1}$$

$$\leq \lambda^2 \int_{\overline{M}} \mathcal{X}_{E_\lambda^c} d\bar{x} = \lambda^2 |E_\lambda^c|$$

The "easy terms" in (3) can be treated by induction and controlled by

$$C \cdot \lambda^2 |E_\lambda^c| + C \int_{E_\lambda} N_{P,\bar{a}}(f)^2(\bar{x}) d\bar{x}$$

Thus, we have following "good-λ inequality"

$$|\{A_{P,\bar{b}}(f) > \lambda\}| \leq C|\{N_{P,\bar{a}}(f) > \lambda\}| + \int_{N_{P,\bar{a}}(f) \leq \lambda} N_{P,\bar{a}}(f)^2(\bar{x}) d\bar{x}$$

Having this inequality, we can prove (2) easily.

In this paper, $C_{a,b,\ldots}$ denotes a positive number depending only on $a, b, \cdots$, and it may be different when it appears in different places. And, for convenience, we do not make any difference between a measure μ and its volume element $d\mu$.

Chen Jie-cheng would like to express his many thanks to Prof. Cheng Min-de for his enthusiastic encouragement.

2   Preliminary Lemmas of One Parameter

In this section, we assume M is a complete Riemannian manifold, $\mathrm{Ric}(M)$ ≥ 0. Similar to above section, we have $\triangle, \nabla, P_t, p_t, \cdots$, etc. (here, we omit the sub-index "i"). Let Φ denote the set of following function φ : (a) $\varphi \in C^2(R^1) \bigcap L^\infty(R^1)$, (b) there exists positive number C_φ such that $|\varphi'| + |\varphi''| \leq C_\varphi \cdot \varphi^{\frac{3}{4}}$, (c) $\varphi^{\frac{1}{2}} \in \Phi$. Obviously, Φ is nonempty.

Now, we give some properties of Poison Kernel.

Lemma 1.   For $k=0,1,2,...$, there hold (where $d = d(x,y)$)

$$p_t(x,y) \geq C_n t(t+d)^{-d} V_x^{-1}(t+d)$$

$$\left|\left(\frac{\partial}{\partial t}\right)^k p_t(x,y)\right| \leq C_{n,k} t^{-k+1}(t+d)^{-1} V_x^{-1}(t+d) \qquad (4)$$

$$\left|\left(\frac{\partial}{\partial t}\right)^k \nabla p_t(x,y)\right| \leq C_{n,k} t^{-k+1}(t+d)^{-1} V_x^{-1}(t+d)$$

Proof.   By lower bound estimate of heat kernel[9]

$$
\begin{aligned}
p_t(x,y) &= \int_0^\infty \frac{e^{-u}}{u^{\frac{1}{2}}} h_{t^2/4u}(x,y)\,du \\
&\geq C_n \int_0^\infty \frac{e^{-u}}{u^{\frac{1}{2}}} V_x^{-1}\left(\frac{t}{u^{\frac{1}{2}}}\right) \exp\left(-\frac{d^2}{c_n t^2/u}\right) du \\
&= C_n t \int_0^\infty \exp(-u(t^2 + d^2/c_n)) V_x^{-1}\left(\frac{1}{u^{1/2}}\right) \frac{du}{u^{1/2}} \\
&\triangleq C_n t \left(\int_0^{(t^2+d^2)^{-1}} \cdot + \int_0^{(t^2+d^2)^{-1}} \cdot\right)
\end{aligned}
\qquad (5)
$$

By [5], for $0 < t < s$

$$V_x^{-1}(s) \geq V_x^{-1}(t)\left(\frac{t}{s}\right)^n \quad \text{and} \quad V_x(s) \geq C_n\left(\frac{s}{t}\right) V_x(t) \qquad (6)$$

we have

$$
\begin{aligned}
\int_0^{(t^2+d^2)^{-1}} \cdot &\geq \int_0^{(t^2+d^2)^{-1}} C_n V_x^{-1}((t^2+d^2)^{\frac{1}{2}})\left(\frac{u}{t^2+d^2}\right)^{\frac{n}{2}} \frac{du}{u^{1/2}} \\
&\geq C_n V_x^{-1}(t+d)(t+d)^{-1}
\end{aligned}
$$

$$
\begin{aligned}
\int_{(t^2+d^2)^{-1}}^\infty \cdot &\geq \int_{(t^2+d^2)^{-1}}^\infty \exp(-u(t^2+d^2/c_n)) V_x^{-1}((t^2+d^2)^{\frac{1}{2}}) \frac{du}{u^{1/2}} \\
&\geq C_n V_x^{-1}(t+d)(t+d)^{-1}
\end{aligned}
$$

The first inequality in (4) can be obtained from above estimates and (5). To consider other two inequalities in (4), set $j = 0, 1$. By the estimates of heat kernel[5]

$$\left| \left(\frac{\partial}{\partial t} \right)^k \nabla^j p_t(x, y) \right|$$

$$\leq C_n \int_0^\infty \frac{e^{-u}}{u^{\frac{1}{2}}} V_x^{-1} \left(\frac{1}{u^{\frac{1}{2}}} \right) \exp \left(-\frac{d^2}{c_{n,k} t^2 / u} \right) \left(\frac{1}{u^{\frac{1}{2}}} \right)^{-j_t - k} du$$

$$= C_n t^{-k-j} \int_0^\infty \exp(-u(t^2 + d^2)/c_{n,k})) V_x^{-1} \left(\frac{1}{u^{\frac{1}{2}}} \right) u^{(j-1)/2} du$$

$$= C_n t^{-k-j} \left(\int_0^{(t^2+d^2)^{-1}} \cdot + \int_{(t^2+d^2)^{-1}}^\infty \cdot \right)$$

But

$$\int_0^{(t^2+d^2)^{-1}} \leq C_n \int_0^{(t^2+d^2)^{-1}} V_x^{-1}((t^2 + d^2)^{\frac{1}{2}}) u^{(j-1)/2} du$$

$$\leq C_n V_x^{-1}(t + d)(t + d)^{-j-1}$$

$$\int_{(t^2+d^2)^{-1}}^\infty \leq C_n \int_{(t^2+d^2)^{-1}}^\infty \exp(-u(t^2 + d^2)/c_{n,k}))$$

$$\cdot V_x^{-1}((t^2 + d^2)^{\frac{1}{2}} (u/(t^2 + d^2))^{\frac{n}{2}} u^{\frac{j-1}{2}} du$$

$$\leq C_n V_x^{-1}(t + d)(t + d)^{-1-j}$$

thus, (4) holds.

Lemma 2.   For $a, b \in R^1_+$, there hold

(a) $\exists \delta_0 = \delta(a, n) \in (0, 1)$, s.t. $\forall$ open set $G \subset M$

$$P_t(\chi_G)(x) \geq \delta_0 \quad (\forall (x, t) \in \hat{G}_a)$$

(b) $\forall$ closed set $E \subset M$, $\exists$ closed subset $F \subset E$, s.t.

$$|F^c| \leq C_{n,a,b} |E^c|$$

$$P_t(\chi_E)(x) \geq 1 - \frac{1}{2} \delta_0 \quad \left(\forall (x, t) \in \bigcup_{w \in F} \Gamma_b(w) \right)$$

$$P_t(\chi_E)(x) \leq 1 - \delta_0 \quad \left(\forall (x, t) \in \left(\bigcup_{w \in E} \Gamma_a(w) \right)^c \right)$$

(c) $\forall c$ and $d \in R^1$, $c < d$, $\exists \varphi_0 \in \Phi$, s.t.

$$\varphi_0(t) = 0 \ (t < c), \quad \varphi_0'(t) > 0 \ (c < t < d),$$
$$\varphi_0(t) = 1 \ (t > d)$$

Proof. $\forall (x,t) \in \widehat{G}_a$, by lemma 1 and (6)

$$
\begin{aligned}
P_t(\chi_G)(x) &= \int_M p_t(x,y)\chi_G(y)dy \geq \int_{B_x(at)} p_t(x,y)dy \\
&\geq C_n t \int_{B_x(at)} (t + d(x,y))^{-1} V_x^{-1}(t + d(x,y))dy \\
&= C_n(1+a)^{-1} V_x(at) V_x^{-1}(at + t) \geq C_n a^n (1+a)^{-n-1} > 0
\end{aligned}
$$

So, (a) holds. For (b), by Lemma 1 and (6)

$$
\begin{aligned}
N_{P,b}(f)(x) &\leq C_n \sup_{(y,t)\in\Gamma_b(x)} \int_M t(t + d(w,y))^{-1} V_w^{-1}(t \\
&\quad + d(w,y))|f(w)|dw \\
&\leq C_{n,b} \int_M t(t + d(x,w))^{-1} V_x^{-1}(t + d(x,w))|f(w)|dw \\
&= C_{n,b} \left(\sum_0^\infty \int_{2^k t < d(x,w) < 2^{1+k} t} \cdot + \int_{d(x,w) < t} \cdot \right) \\
&= C_{n,b} \sum_{-1}^\infty I_k
\end{aligned}
$$

but, by (6)

$$
\begin{aligned}
I_k &\leq (1 + 2^k)^{-1} V_x^{-1}(t + 2^k t) V_x(2^{1+k} t) M_0(f)(x) \\
&\leq (1 + 2^k)^{-1} \left(\frac{2^{1+k}}{1 + 2^k} \right)^n M_0(f)(x) \leq C_n 2^{-k} M_0(f)(x)
\end{aligned}
$$

Where $M_0(f)$ is Hardy-Littlewood maximal function of f, so

$$N_{P,b}(f)(x) \leq C_{n,b}^* M_0(f)(x) \tag{7}$$

Now, let $F = \{x \in E : M_0(\chi_{E^c})(w) \leq \delta_0/2C_{n,b}^*\}$, then $F \subset E$ and F is closed. By weak-(1,1) boundedness of M_0

$$|F^c| \leq C_n(\delta_0/2C_{n,b}^*)^{-1} \| \chi_{E^c} \|_1 = C_{n,a,b}|E^c|$$

And, for $(x,t) \in (\bigcup_{w \in E} \Gamma_a(w))^c = (\widehat{E^c})_a$,

$$P_t(\mathcal{X}_E)(x) = 1 - P_t(\mathcal{X}_{E^c})(x) \leq 1 - \delta_0 \quad \text{(by (a))}$$

So, (b) holds. Finally, the existence of φ_0 can be verified by considering $\exp(-t^{-2})$.

Lemma 3.   Suppose $v(M) \overset{\triangle}{=} \inf_{x \in M} V_x(1) > 0$. For any harmonic function u on M (i.e. $\Delta u = 0$), if $\| u \|_{h_p} \overset{\triangle}{=} \sup_{t>0} \| u(\cdot,t) \|_p < \infty$, then

$$\begin{aligned}
\| u(\cdot,t) \|_\infty &\leq C_{M,p} \| u \|_{h_p} (\inf_{x \in M} V_x(t))^{-1} \\
\| \nabla u(\cdot,t) \|_\infty &\leq C_{M,p} \| u \|_{h_p} (\inf_{x \in M} V_x(t))^{-1} t^{-1}
\end{aligned} \tag{8}$$

Where $0 < p < \infty$; furthermore, if M is noncompact, then

$$\| u(\cdot,t) \|_\infty + t \| \nabla u(\cdot,t) \|_\infty \leq C_{u,p,n,\varepsilon} t^{-1/p} \tag{9}$$

for $0 < \varepsilon < \infty$.

Proof.   From [8], we can find following fact:

For complete Riemannian manifold $\tilde{M}$ (may be $\partial \tilde{M} \neq \emptyset$), suppose $\mathrm{Ric}(\tilde{M}) \geq 0$, domain $\Omega \subset \tilde{M}$, if $r_\Omega = \sup\{r : \text{ for some } \tilde{x} \in \tilde{M}, B_{\tilde{x}}(r) \subset \Omega\}$ and $R_\Omega = \inf\{R : \text{ for some } \tilde{x} \in \tilde{M}, B_{\tilde{x}}(R) \supset \Omega\}$ satisfy $\varepsilon_1^{-1} > r/R > \varepsilon_1$ and $\Omega_{\varepsilon_2} \bigcap \partial \tilde{M} = \emptyset$ where $\Omega_{\varepsilon_2} = \{\tilde{x} \in \tilde{M} : d(\tilde{x},\Omega) < \varepsilon_2\}$, then, for any harmonic function u defined on Ω_{ε_2}, there holds

$$\sup_{\tilde{x} \in \Omega} |u(\tilde{x})|^p \leq C_{p,n,\varepsilon_1,\varepsilon_2} |\Omega_{\varepsilon_2}|^{-1} \int_{\Omega_{\varepsilon_2}} |u(\tilde{x})|^p dx \tag{10}$$

where ε_1 and ε_2 are two positive numbers. And, by gradient estimates of Yau, we have : If $B_{\tilde{x}_0}(r) \subset \tilde{M}$, then, for any positive harmonic function u on $B_{\tilde{x}_0}(r)$, there holds

$$\sup_{\text{on } B_{\tilde{x}_0}(r/2)} (|\tilde{\nabla}(\tilde{x})|/u(\tilde{x})) \leq C_{\tilde{M}} r^{-1} \tag{11}$$

where $\tilde{\nabla}$ is the gradient operator on $\tilde{M}$, $\tilde{n} = \dim(\tilde{M})$. Now, for $\tilde{M} = M^\perp$, take $\Omega = \{(w,s) : w \in B_x(t), t/2 < s < 2t\}$, by (10)

$$\begin{aligned}
|u(x,t)|^p &\leq C_{n,p} t^{-1} V_x^{-1}(2t) \int_{w \in B_x(2t), 0 < s < 3t} |u(w,s)|^p dw ds \\
&\leq C_{n,p} V_x^{-1}(t) t^{-1} \int_0^{3t} \| u \|_{h_p}^p ds \\
&\leq C_{n,p} \| u \|_{h_p}^p V_x^{-1/p}(t)
\end{aligned}$$

So, the first inequality in (8) holds. And by (10)–(11)

$$|\nabla u(x,t)| \leq C_{n,p} t^{-1} \sup_{w \in B_x(t), t/2 < s < 2t} |u(w,s)|$$
$$\leq C_{n,p} \| u \|_{h_p} V_x^{-1/p}(t)$$

So, the second inequality in (8) also holds. Finally, (9) can be obtained from (8) and (6).

Lemma 4. For any $f \in L_c^\infty(M)$, let $u = P_t(f)$, there holds

$$\left\| \left(\frac{\partial}{\partial t} \right)^k \nabla^j u(\cdot,t) \right\|_\infty \leq C_{M,k,u} t^{-k-j-1} \quad (t > 0)$$

where $k = 0, 1, \ldots, j = 0, 1, M$ is noncompact.

Proof. By Lemma 1

$$\left| \left(\frac{\partial}{\partial t} \right)^k \nabla^j u(x,t) \right|$$
$$\leq C_{M,k} \int_M t^{-k-j+1}(t + d(x,y))^{-1} V_y^{-1}(t + d(x,y)) |f(y)| dy$$

Suppose suppf$\subset B_{x_0}(R)$, by (6), we get

$$V_y(t + d) \geq C_M \inf_{y \in B_x(R)} V_y(1) \left(\frac{t+d}{1} \right) = C_{M,f}(t + d) \quad (t \geq 1)$$

because

$$\left\| \left(\frac{\partial}{\partial t} \right)^k \nabla^j u(\cdot,t) \right\|_\infty \leq C_{M,k,f} t^{-k-j-1} \| f \|_1$$

when $t \leq 1$, noticing that we have

$$\int_M (t + d(x,y))^{-1} V_x^{-1}(t + d(x,y)) dy \leq C_n t^{-1}$$

we have the left side of the inequality in Lemma 4

$$\leq C_n t^{-k-j} \| f \|_\infty$$

Therefore, Lemma 4 holds.

Secondly, we consider a kind of Green's formula on $M^\perp$. We have

Lemma 5.   Suppose M is compact, $\varphi \in \Phi, \varepsilon > 0$. For f and $g \in L^2(M)$, set $u_\varepsilon = P_{t+\varepsilon}(f)$ and $v_\varepsilon = P_{t+\varepsilon}(g)$, then

$$\int_M \int_0^\infty t\Delta(\varphi(v_\varepsilon)u_\varepsilon^2)dxdt = \int_M \varphi(v_\varepsilon(x,0)u_\varepsilon^2(x,0))dx$$

$$- \int_M \varphi(v_\varepsilon(x,\infty)u_\varepsilon^2(x,\infty))dx \qquad (12)$$

Proof.   Because M is compact, by integration by parts, so

$$\int_M \Delta\left(\varphi(v_\varepsilon)\right)u_\varepsilon^2)dx = 0$$

$$\lim_{r\to\infty}\int_M\int_0^r t \Delta\left(\varphi(v_\varepsilon)u_\varepsilon^2\right)dxdt = 0$$

On the other hand

$$\int_0^\infty t\left(\frac{\partial}{\partial t}\right)^2(\varphi(v_\varepsilon)u_\varepsilon^2)dt = r\frac{\partial}{\partial t}(\varphi(v_\varepsilon(x,r)u_\varepsilon^2(x,r))$$
$$-\varphi(v_\varepsilon(x,r))u_\varepsilon^2(x,r) + \varphi(v_\varepsilon(x,0)u_\varepsilon^2(x,0))$$

Now, when $t \longrightarrow \infty, u_\varepsilon(x,t)$, and $v_\varepsilon(x,t)$ uniformly tend to $u_\varepsilon(x,\infty) = P_\infty(f)$ and $v_\varepsilon(x,\infty) = P_\infty(g)$ respectively, where P_∞ is the orthogonal projection from $L^2(M)$ to $L_0^2(M) = \{f : f$ is constant$\}$; especially, u_ε and v_ε are bounded on $M^\perp$. And, by Lemma 1, it is not difficult to prove (Similar to [5, Lemma 2.14]) that $t\frac{\partial}{\partial t}p_t(x,y)$ uniformly tends to some constant a_M when $t \longrightarrow \infty$. But

$$\int_M \frac{\partial}{\partial t}p_t(x,y)dy = \frac{\partial}{\partial t}\int_M p_t(x,y)dy = \frac{\partial}{\partial t}(1) = 0$$

so $a_M = 0$. By these facts, we have

$$r\frac{\partial}{\partial t}v_\varepsilon(x,r) \overset{\text{uniformly}}{\longrightarrow} 0(r \longrightarrow \infty)$$

$$r\frac{\partial}{\partial t}u_\varepsilon(x,r) \overset{\text{uniformly}}{\longrightarrow} 0(r \longrightarrow \infty)$$

$$r\frac{\partial}{\partial t}(\varphi(v_\varepsilon)u_\varepsilon) \overset{\text{uniformly}}{\longrightarrow} 0(r \longrightarrow \infty)$$

Thus, Lemma 5 holds.

Now, take $\eta_r(r \geq 1)$ such that

$$\eta_r(x) = \begin{cases} 0 & d(x,x_0) \geq 2r \\ 1 & d(x,x_0) < r \end{cases} \quad \text{and} \quad \begin{array}{c} 0 \leq \eta_r \leq 1 \\ |\nabla \eta_r| \leq C_n r^{-1} \end{array} \qquad (13)$$

We have

Lemma 6. Suppose M is noncompact, $\varphi \in \Phi$, $v_\varepsilon = P_{t+\varepsilon}(g)$ and $g \in L_\infty, u \in C^2(M^\perp)$ and $\| u \|_{h_p} < \infty$ for some $0 < p \leq 1$, and

$$\left| \left(\frac{\partial}{\partial t} \right)^k \nabla^j u_\varepsilon(x,t) \right| \leq C_{u,\varepsilon} t^{-j-k-1/p} \qquad (t > 0)$$

where $u_\varepsilon(x,t) = u(x,t+\varepsilon)$. Then

$$\lim_{r \longrightarrow \infty} \int_M \int_0^{3r} \eta_r(x) \Delta(\varphi(v_\varepsilon)u_\varepsilon^2) t\, dx\, dt$$

$$= \int_M \varphi(v_\varepsilon(x,0))u_\varepsilon^2(x,0)\, dx$$

Proof. By (11), for $k \geq 0$, $j = 0$ or 1

$$\left\| \left(\frac{\partial}{\partial t} \right)^k \nabla^j v(\cdot,t) \right\|_\infty \leq C_{n,v,k} t^{-k-j} \qquad (14)$$

By integration by parts and (10)

$$\left| \int_M \eta_r(x) \,\triangle\, (\varphi(v_\varepsilon)u_\varepsilon^2) dx \right| = \left| \int_M \nabla \eta_r \,\nabla\, (\varphi(v_\varepsilon)u_\varepsilon^2) dx \right|$$

$$\leq C_{n,\varphi} r^{-1} \int_M (|u_\varepsilon^2 \,\nabla\, v_\varepsilon| + |v_\varepsilon u_\varepsilon \,\nabla\, u_\varepsilon|) dx$$

$$\leq C_{n,\varphi} r^{-1} \int_M u_\varepsilon^p C_{n,v,u,p,\varepsilon}(t+\varepsilon)^{-1}(t+\varepsilon)^{-(2-p)/p}) dx$$

so (by the conditions and (14))

$$\left| \int_M \int_0^{3r} \eta_r(x) \,\triangle\, (\varphi(v_\varepsilon)u_\varepsilon^2) t\, dx\, dt \right|$$

$$\leq C_{n,v,u,p,\varphi,\varepsilon} \int_M \int_0^{3r} u_\varepsilon^p(x,t)(t+\varepsilon)^{-2/p} t\, dx\, dt\, r^{-1}$$

$$\leq C_{n,v,u,p,\varphi,\varepsilon} \| u \|_{h_p}^p \ln r/r \overset{r \to \infty}{\longrightarrow} 0$$

On the other hand

$$\int_0^{3r} t\left(\frac{\partial}{\partial t}\right)^2 (\varphi(v_\varepsilon)u_\varepsilon^2)dt$$

$$= t\left(\frac{\partial}{\partial t}\right)(\varphi(v_\varepsilon)u_\varepsilon^2)|_0^{3r} - \varphi(v_\varepsilon)u_\varepsilon^2|_0^{3r}$$

$$= \varphi(v_\varepsilon(\cdot,0))u_\varepsilon^2(\cdot,0) + \varphi(v_\varepsilon(\cdot,3r))u_\varepsilon^2(\cdot,3r)$$

$$-r\left(\frac{\partial}{\partial t}\right)(\varphi(v_\varepsilon(\cdot,3r))u_\varepsilon^2(\cdot,3r))$$

$$= \varphi(v_\varepsilon(\cdot,0))u_\varepsilon^2(\cdot,0) + I_1 - I_2$$

Now

$$|I_1| \le C_{u,n,p,\varphi,\varepsilon}(r+\varepsilon)^{-(2-p)/p}u_\varepsilon^p(\cdot,3r) \qquad \text{(by the conditions)}$$

$$|I_2| \le C_{u,n,p,\varphi,\varepsilon}rr^{-1}r^{-(2-p)/p}u_\varepsilon^p(\cdot,3r) \qquad \text{(by the conditions and (14))}$$

so

$$\int_M\int_0^{3r} t\left(\frac{\partial}{\partial t}\right)^2 (\varphi(v_\varepsilon)u_\varepsilon^2 dtdx$$

$$= \int_M \varphi(v_\varepsilon(x,0)u_\varepsilon^2(x,0)dx + O\left(\int_M r^{1-2/p}|u(x,3r)|^p dx\right)$$

$$\longrightarrow \int_M \varphi(v_\varepsilon(x,0)u_\varepsilon^2(x,0)dx \qquad (r\longrightarrow\infty)$$

From all above discussions, we get Lemma 6.

Thirdly, we give a Poisson integral representation formula of harmonic function.

Lemma 7. Suppose $v(M) > 0$, u is a harmonic function on $M^\perp, 0 < p < \infty, \| u \|_{h_p} < \infty$, then

$$u_\varepsilon(x,t) \equiv v_\varepsilon(x,t) \stackrel{\triangle}{=} P_t(u_\varepsilon(\cdot,0))(x) \quad (\forall\varepsilon > 0)$$

where $u_\varepsilon(x,t) = u(x,t+\varepsilon)$.

Proof. When M is compact, set $f(t) = \int_M u(x,t)dt$. As a function of $t \in (0,\infty)$, $f(t)$ is constant. In fact, $f''(t) = \int_M(\frac{\partial}{\partial t})^2 u(x,t)dt =$

$-\int_M \triangle_x u(x,t)dx = 0$ (by Green's formula), so f is linear function of t; on the other hand, by (10), it is easy to see that u is bounded when $t \geq 1$, hence f is bounded. Therefore, $f(t)=$some constant C_u^*. Let $w_\varepsilon = u_\varepsilon - v_\varepsilon$, w_ε is bounded on $M^\perp$ (Similar to u). Furthermore, by (11), $|\frac{\partial}{\partial t}w_\varepsilon(x,t)| \leq C_{M,w}t^{-1}$, thus $w_\varepsilon(x,t)$ uniformly tends to some constant C_w^{**} when $t \longrightarrow \infty$. But $\int_M C_{w_\varepsilon}^{**}dx = C_{w_\varepsilon}^{**} = \int_M w_\varepsilon(x,0)dx = 0$, i.e. $C_w^{**} = 0$. Therefore $w_\varepsilon(x,t)$ uniformly tends to zero when t tends to zero or infinity. By maximal principle, $w_\varepsilon \equiv 0$, i.e. $u_\varepsilon \equiv v_\varepsilon$.

When M is noncompact, by Lemma 3, $\| u \|_{h_p}< \infty \, (q \geq p)$. So, we can assume $p \geq 1$. Take $q = 2p$, there holds

$$\lim_{r \longrightarrow \infty} \int_M \int_0^{3r} \eta_r(x)\Delta(|w_\varepsilon(x,t)|^q)tdxdt = 0. \tag{15}$$

In fact, $\| v_\varepsilon \|_{h_p}< \infty$, and by Lemma 3 and (11)

$$\left|\int_M\int_0^{3r} \eta_r \triangle (|w_\varepsilon|^q)tdxdt\right| = \left|\int_M \int_0^{3r} \nabla\eta_r \nabla (|w_\varepsilon|^q)tdxdt\right|$$

$$\leq \int_0^{3r} \int_M C_n r^{-1} C_{n,q,w_e,\varepsilon}|w_\varepsilon|^p(\varepsilon + t)^{-1-(q-p)/p}tdtdx$$

$$\leq C_{n,q,w_e,\varepsilon}\ln r/r \, \| w_\varepsilon \|_{h_p}^p \overset{r\to\infty}{\longrightarrow} 0$$

$$\left|\int_M\int_0^{3r} \eta_r \left(\frac{\partial}{\partial t}\right)^2 (|w_\varepsilon|^q)tdxdt\right|$$

$$\leq \int_M \eta_r(x)|t \left(\frac{\partial}{\partial t}\right) (|w_\varepsilon|^q)|_0^{3r} - |w_\varepsilon|^q|_0^{3r}|dx$$

$$\leq \int_M |r \frac{\partial}{\partial r}(|w_\varepsilon(x,3r)|^q) - |w_\varepsilon(x,3r)|^q|dx \quad (\because w_\varepsilon(x,0) = 0)$$

$$\leq \int_M |w_\varepsilon(x,3r)|^p C_{n,p,w_e,\varepsilon}r^{-(q-p)/p}dx \quad \text{(By Lemma 3)}$$

$$\leq C_{n,p,w_e,\varepsilon} \, \| w_\varepsilon \|_{h_p}^p \, r^{-1} \overset{r\to\infty}{\longrightarrow} 0$$

Combining all above, we get (15). Now, noticing that $|w_\varepsilon|^q \in C^2(M^\perp)$ (because $q \geq 2$) and

$$\Delta|w_\varepsilon|^q = q(q-1)|w_\varepsilon|^{q-2}|\nabla w_\varepsilon|^2 \geq 0$$

we get $|w_\varepsilon|^{q-2}|\nabla w_\varepsilon|^2 \equiv 0$ from (15), thus $w_\varepsilon \equiv$ constant. But $w_\varepsilon(x,0) \equiv 0$, so $w_\varepsilon \equiv 0$, i.e. $u_\varepsilon \equiv v_\varepsilon$.

From Lemma 7 and a limit procedure, we get

Theorem 2. Under the hypothesis of Lemma 6, If $1 < p < \infty$, then there exists $f \in L^P(M)$ such that $u = P_t(f)$; if $p = 1$, then there exists a finite Borel measure μ on M such that $u = P_t(d\mu)$.

Finally, we give some basic facts about Carleson measures. For a non-negative measure μ on M, if

$$\| \mu \|_{C.M.} \overset{\triangle}{=} \sup\{\mu(\widehat{G_1})/|G| \ : \ \text{open set } G \subset M\} \tag{16}$$

is finite, then we call μ a Carleson measure on $M^\perp$.

Following two Lemmas can be found in [5]

Lemma 8. Suppose $d\mu$ be a Carleson measure on $M^\perp, 0 < p, a < \infty, u \in C(M^\perp)$, then

$$\int_{M^\perp} |u(x,t)|^p d\mu(x,t) \leq \| d\mu \|_{C.M.} \ C_{n,p,a} \ \| N_a(u) \|_p^p$$

where $N_a(u)(x) = \sup_{(y,t) \in \Gamma_a(x)} |u(y,t)|$.

Lemma 9. For any $g \in \mathrm{BMO}(M)$, there hold: (a) $u = P_t(g) \in C^\infty(M^\perp)$, (b) if $d\mu_g = |\nabla u|^2 t dt dx$ then

$$\| d\mu_g \|_{C.M.} \leq C_n \| g \|_{\mathrm{BMO}}^2$$

But, by Lemma 1, the proofs of Lemmas 8 and 9 can be simplified now, we omit the details here.

All above lemmas have easy multiparameter generalization (except Lemmas 4 and 5). For simplicity, we only considered one parameter case in this section. For example, we have

Lemma 3$'$. Suppose $v(\overline{M}) = \inf_{\overline{x} \in \overline{M}} V_{\overline{x}}(\overline{1}) > 0 \ (\overline{1} = (1,\cdots,1)) \in (R_+^1)^d)$ and $\mathrm{Ric}(\overline{M}) \geq 0$, then form any multi-harmonic function u on $\overline{M}^\perp$ (i.e. $\Delta_i u = 0$ for $i = \overline{1,d}$), if $\| u \|_{h_p} = \sup_{\overline{t} > \overline{0}} \| u(\cdot,t) \|_p < \infty$, then

$$\left\| \prod_1^d \nabla_j^i j_{u(\cdot,t)} \right\|_\infty \leq C_{M,p} \| u \|_p \ (\inf_{\overline{x} \in \overline{M}} V_{\overline{x}}(\overline{t}))^{-1}$$

where $0 < p < \infty, i_j = 0$ or 1.

But, we should notice that $\mathrm{Ric}(\overline{M}) \geq 0$ iff $\mathrm{Ric}(M_i) \geq 0 \ (i = \overline{1,d})$.

3   Carleson Measures Produced by Harmonic Functions

Carleson measures on $M^{\perp}$ can be defined as (16), but, we should replace $\widehat{G}_1$ by $\widehat{G}_{\overline{1}}$ now. Suppose f and $g_i \in L_c^{\infty}(M)$. Let $u = P_t(f), v_i = P_t(g_i) + c$ where $c \in R^1$ and $i = \overline{1,m}$. And, for $\{\varphi_i\}_1^m \subset \Phi$, set

$$M_{\varphi}(u,v) = \sup\left\{|u(\overline{y},\overline{t})| \; : \; (\overline{y},\overline{t})' \in \bigcap_1^m \sup\varphi_j(v_j)\right\}$$

Theorem 3.   Under the above hypotheses, if $\| v_j \|_{\infty} \leq 1$, $(j = \overline{1,m})$, then

$$d\mu = \left| \overline{\nabla}\left(u \prod_1^m \varphi_j(v_j) \right) \right|^2 d\overline{m}$$

is a Carleson measure on $\overline{M}^{\perp}$ and

$$\| d\mu \|_{C.M.} \leq C_{d,\overline{n},\varphi} M_{\varphi}(u,v)^2.$$

To prove it, following lemmas are needed. We write $\{I_s\}_1^m \sim (1,\cdots,d)$ if (a). each I_s is a subset of $(1,\cdots,d)$, (b). $I_s \bigcap I_t = \emptyset$ (when $s \neq t$) and $\bigcup_1^m I_s = (1,\cdot,\cdot,\cdot,d)$. Set $\nabla_{I_s} = \prod_{i \in I_s} \nabla_i$. If I_s is empty, ∇_{I_s} denotes identity.

Lemma 10.   If $\{v_j\}_1^m$ is as in Theorem 2, then

$$d\nu(m) \triangleq \sum_{\{I_s\}_1^m \sim (1,\cdots,d)} \left(\prod_{s=1}^m |\nabla_{I_s} v_s(\overline{x},\overline{t})| \right)^2 d\overline{m}$$

is a Carleson measure on $\overline{M}^{\perp}$ and

$$\| d\nu^{(m)} \|_{C.\overline{M}.} \leq C_{\overline{n},m}$$

Proof. When $m=1$, it can be followed from the multiparameter analogue of Lemma 9. Suppose

$$\| d\nu^{(m)} \|_{C.M.} \leq C_{\overline{n},m}$$

(The first inductive assumption), we want to prove

$$\| d\nu^{(m+1)} \|_{C.M.} \leq C_{\overline{n},m+1}$$

To do so, consider a bounded open set $G \subset \overline{M}$, set $v = P_{\overline{t}}(\mathcal{X}_G)$ and

$$N_{(j)}^{(m+1)}(G) = \sum_{\{I_s\}_1^m \sim (1,\cdots,d)} \int_{\overline{M}^\perp} \varphi_0(v) |\nabla u_{m+1}|^2 \cdot \prod_{s=1}^m |\nabla_{I_s} u_s|^2 d\overline{m}$$

$$dv_{(j)}^{(m+1)} = \sum_{\{I_s\}_1^m \sim (1,\cdots,d)} |\nabla_1 \cdots \nabla_j u_{m+1}|^2 \prod_{s=1}^m |\nabla_{I_s} u_s|^2 d\overline{m}$$

where $\varphi_0 \in \Phi$, such that $\varphi_0(t) = 0$ $(\forall t < 1 - k_0)$, $\varphi_0(t) = 1$ $(\forall t > 1 - \frac{1}{2}k_0)$, $\varphi_0'(t) > 0$ $(\forall t \in (1 - k_0, 1 - \frac{1}{2}k_0))$, $k_0 = \prod_{i=1}^d \delta(1, n_i)$ and $\delta(1, n_i)$ is taken as in Lemma 2. Under the first inductive assumption, we shall prove

$$\| dv_{(j)}^{(m+1)} \|_{C.M.} \leq C_{\overline{n}, m+1} \quad (j = \overline{1, d}) \tag{17}$$

by induction. But, it is enough for (17) to prove

$$N_{(j)}^{(m+1)}(G) \leq C_{\overline{n}, m+1} |G| \tag{18}_j$$

because $\nu_{(j)}^{(m+1)}(\hat{G}_{\overline{1}}) \leq N_{(j)}^{(m+1)}(G)$ by lemma 2.

For $j = 0$, by the first inductive assumption and multiparameter analogues of Lemmas 2 and 8 and L^2-boundedness of $N_{P,\overline{a}}$, we get

$$N_{(0)}^{(m+1)}(G) \leq k_0^{-2} \int_{\overline{M}^\perp} v^2 dv_{(0)}^{(m+1)}$$

$$\leq C_{\overline{n}, k_0} \int_{\overline{M}} N_{\overline{1}}(v)^2(\overline{x}) d\overline{x} \leq C_{\overline{n}} |G|$$

Now, suppose $(18)_j$ hold for $j \leq d - 1$ (The second inductive assumption), we shall prove $(18)_{j+1}$. Simple computation gives that

$$\Delta_i(\varphi(v)\psi(u)) = \psi''(u)\varphi(v)|\nabla_i u|^2$$

$$+ \varphi''(v)\psi(u)|\nabla_i u|^2 + O(|\varphi'(v)\psi'(u)||\nabla_i u \nabla_i v|) \tag{19}$$

where u and v is harmonic on $M_i^\perp$. If $\{I_s\}_1^m \sim (j+2, \cdots, d)$, then

$$J = \int_{M^\perp} \varphi_0(v) \prod_{s=1}^m (|\nabla_{I_s} u_s|^2 |\nabla_1 \cdots \nabla_{j+1} u_{m+1}|^2) d\overline{m}$$

$$\leq C_m \left(\int_{\overline{M}^\perp} \varphi_0(v) |\Delta_{j+1} J_j(u)| |\overline{\nabla}_j u_{m+1}|^2 d\overline{m} \right.$$

$$+ \int_{\overline{M}^\perp} \varphi_0''(v) |\nabla_{j+1} v|^2 J_j(u) \overline{\nabla}_j u_{m+1}|^2 d\overline{m}$$

$$+ \int_{\overline{M}^\perp} \Delta_{j+1} (\varphi_0(v) J_j(u) |\overline{\nabla}_j u_{m+1}|^2) d\overline{m}$$

$$+ \int_{\overline{M}^\perp} |\varphi_0'(v)| |\nabla_{j+1} v| |\nabla_{j+1} J_j(u)| |\overline{\nabla}_j u_{m+1}|^2 d\overline{m}$$

$$+ \int_{\overline{M}^\perp} |\varphi_0'(v)| |\nabla_{j+1} v| J_j(u) |\overline{\nabla}_j u_{m+1}| |\nabla_j' u_{m+1}| d\overline{m}$$

$$+ \left. \int_{\overline{M}^\perp} \varphi_0(v) |\nabla_{j+1} J_j(u)| |\overline{\nabla}_j u_{m+1}| |\nabla_j' u_{m+1}| dm \right)$$

$$\triangleq C_m \sum_{i=1}^{6} K_i$$

where $\overline{\nabla}_j = \nabla_1 \cdots \nabla_j$, $\overline{\nabla}_j' = \overline{\nabla}_j \overline{\nabla}_{j+1}$, and $J_j(u) = \prod_{s=1}^{m} |\nabla_{I_s} u_s|^2$. By the first inductive assumption

$$K_1 \leq C_m N_{(j)}^{(m+1)}(G) \leq C_{\overline{n},m} |G|;$$

by the second inductive assumption and the multiparameter analogue of Lemma 8

$$K_2 \leq C_{\overline{n}} \int_{\overline{M}_{j+1}^\perp} \left(\int_{\prod_{i=1,i\neq j+1}^{d} \overline{M}_i^\perp} |\nabla_{j+1} v|^2 \prod_{i=1,i\neq j+1}^{d} dx_i \right) dm_{j+1}$$

$$= \frac{1}{2} C_n \int_{\overline{M}} v^2(\overline{x}, \overline{0}) d\overline{x} = \frac{1}{2} C_n |G|;$$

applying Green's formula on $M_j^\perp$ (Similar to lemmas 5 and 6)

$$K_3 = \int_{\prod_{i=1,i\neq j+1}^{d} M_i^\perp} \left(\int_{M_{j+1}} \varphi_0(v) J_j(u) |\overline{\nabla}_j u_{m+1}|^2 dx_{j+1} \right) \prod_{i=1,i\neq j+1}^{d} dm_i$$

$$\leq C_{\overline{n},m} \int_{M_{j+1}} |G_{x_{j+1}}| dx_{j+1} = C_{\overline{n},m} |G|$$

where $G_{x_{j+1}} = \{ \overline{x} \in \overline{M} : x_{j+1} \text{ fixed} \}$. For K_4, writing $|\overline{\nabla}_j u_{m+1}|^2$ as $|\overline{\nabla}_j u_{m+1}|^{1+1}$ and noticing that $|\varphi_0'| \leq C_{\varphi_0} \cdot \varphi_0^{\frac{1}{2}+\frac{1}{4}}$, then applying Hölder inequality, we get (where $\sum$ is a finite sum)

$$K_4 \le C_m \sum ((\text{term like } K_1)^{\frac{1}{2}} (\text{term like } K_2)^{\frac{1}{2}}) \le C_{\bar{n},m} |G|$$

similarly

$$K_5 \ \le \ C_m \sum ((\text{term like } K_2)^{\frac{1}{2}} (N_{(j+1)}^{(m+1)}(G))^{\frac{1}{2}})$$

$$\le \ C_{\bar{n},m} |G|^{\frac{1}{2}} (N_{(j+1)}^{(m+1)}(G))^{\frac{1}{2}}$$

$$K_6 \ \le \ C_m \sum ((\text{term like } K_1)^{\frac{1}{2}} (N_{(j+1)}^{(m+1)}(G))^{\frac{1}{2}})$$

$$\le \ C_{\bar{n},m} |G|^{\frac{1}{2}} (N_{(j+1)}^{(m+1)}(G))^{\frac{1}{2}}$$

Combining all above, we get

$$N_{(j+1)}^{(m+1)}(G) \le C_{\bar{n},m}(|G| + (|G| \cdot N_{(j+1)}^{(m+1)}(G))^{\frac{1}{2}})$$

Thus, $(18)_{j+1}$ holds. By inductive principle, $\{d\nu_{(j)}^{(m+1)}\}_{j=0}^d$ are Carleson measures and $(18)_j$ $(j = \overline{0,d})$ hold. By symmetry, Lemma 10 holds.

Lemma 11. Under the hypotheses of Theorem 2 (but, $d = 1, m = 1$, and the subindex "1" of φ_0 and v_1 is omitted),

$$d\mu = |\nabla_1 u|^2 \varphi(v) dm_1$$

is a Carleson measure on $M_1^{\perp}$ and

$$\| d\mu \|_{C.M.} \le C_{n_1,\varphi} M_{\varphi}(u,v)^2$$

Proof. For any bounded open set $G \subset M_1$, let $w = P_t(\mathcal{X}_G)$ and take $\varphi_0 \in \Psi$ as in proof of Lemma 10. Then, set

$$A(G) = \int_{M_1^{\perp}} |\nabla_1 u|^2 \varphi(v) \varphi_0(w) dm_1$$

By lemma 2, $\mu(\hat{G}_1) \le C_{n_1} A(G)$. Now, by an analogue of (19)

$$A(G) \ = \ \frac{1}{2} \int_{M_1^{\perp}} \Delta_1(\varphi(v)\varphi_0(w)u^2) dm_1$$

$$-\frac{1}{2}\int_{M_{\bar{1}}^{\perp}} u^2 \varphi''(v)\varphi_0(w)|\nabla_1 v|^2 dm_1$$

$$-\frac{1}{2}\int_{M_{\bar{1}}^{\perp}} u^2 \varphi_0''(w)\varphi(v)|\nabla_1 w|^2 dm_1$$

$$+O(1)\int_{M_{\bar{1}}^{\perp}} |\varphi'(v)\varphi_0(w)u||\nabla_1 u||\nabla_1 v| dm_1$$

$$+O(1)\int_{M_{\bar{1}}^{\perp}} |\varphi(v)\varphi_0'(w)u||\nabla_1 u||\nabla_1 w| dm_1$$

$$+O(1)\int_{M_{\bar{1}}^{\perp}} |u^2 \varphi'(v)\varphi_0'(w)||\nabla_1 v||\nabla_1 w| dm_1$$

$$\triangleq \sum_{1}^{6} I_j$$

By Lemmas 4, 5 and 6

$$|I_1| = \frac{1}{2}\int_{M_1} u^2(x_1,\varepsilon)\varphi(v(x_1,\varepsilon))\varphi_0(w(x_1,\varepsilon))dx_1$$

$$\leq \frac{1}{2}\|\varphi\|_\infty M_\varphi(u,v)^2 \int_{M_1} \varphi_0(w(x_1,\varepsilon))dx_1$$

$$= C_\varphi M_\varphi(u,v)^2 |G|$$

by lemmas 8,9 and L^2-boundedness of $N_{P,\bar{1}}$

$$|I_2| \leq \frac{1}{2}\|\varphi''\|_\infty M_\varphi(u,v)^2 \int_{M_{\bar{1}}^{\perp}} C_{\varphi_0} w^2 |\nabla_1 v|^2 dm_1$$

$$\leq C_{n,\varphi,\varphi_0} M_\varphi(u,v)^2 |G|$$

by L^2-boundedness of Littlewood-Paley-Stein g-function

$$|I_3| \leq \frac{1}{2}\|\varphi_0''\|_\infty M_\varphi(u,v)^2 \int_{M_{\bar{1}}^{\perp}} |\nabla_1 w|^2 dm_1$$

$$= C_{\varphi_0} M_\varphi(u,v)^2 \|g_\Gamma(\mathcal{X}_G)\|_2^2$$

$$\leq C_{n_1} M_\varphi(u,v)^2 |G|$$

by Hölder inequality

$$|I_4| \leq C_{n_1,\varphi} A(G)^{\frac{1}{2}} M_\varphi(u,v)\left(\int_{M_{\bar{1}}^{\perp}} |\nabla_1 v|^2 \varphi_0(w) dm_1\right)^{\frac{1}{2}}$$

$$\leq C_{n_1,\varphi} M_\varphi(u,v)|G|^{\frac{1}{2}} A(G)^{\frac{1}{2}}$$

$$|I_5| \leq C_{n_1,\varphi} M_\varphi(u,v)|G|^{\frac{1}{2}} A(G)^{\frac{1}{2}}$$

$$|I_6| \leq C_{n_1,\varphi} M_\varphi(u,v)^2 \left(\int_{M_1^\perp} |\nabla_2 w|^2 dm_1\right)^{\frac{1}{2}} \left(\int_{M_1^\perp} |\varphi_0'(w)|^2 |\nabla_1 v|^2 dm_1\right)^{\frac{1}{2}}$$

$$\leq C_{n_1,\varphi} M_\varphi(u,v)^2 |G|$$

Combining all above, we get

$$A(G) \leq C_{n_1,\varphi}(M_\varphi(u,v)^2 |G| + A(G)^{\frac{1}{2}} |G|^{\frac{1}{2}} M_\varphi(u,v))$$
$$A(G) \leq C_{n_1,\varphi} M_\varphi(u,v)^2 |G|$$

Lemma 11 is proved.

Now, we shall prove Theorem 2 by induction. It is obvious that Theorem 3 is equivalent to following

Theorem 3'. Under the hypotheses of Theorem 3,

$$d\mu^{(d)} = |\overline{\nabla} u|^2 \prod_1^m \varphi_j(v_j) d\overline{m}$$

is a Carleson measure on $\overline{M}^\perp$, and

$$\| d\mu^{(d)} \|_{C.M.} \leq C_{d,\bar{n}} M_\varphi(u,v)^2 \tag{20}$$

For simplicity, we only consider the case $m = 1$ and omit the subindex "1" of v_1 and φ_1.

When $d = 1$, Theorem 3' holds by Lemma 11. Suppose Theorem 3' holds for some $d \geq 1$ (The first inductive assumption), we shall prove that Theorem 3' holds for $d + 1$. To do so, set $\overline{\nabla}_j = \nabla_1 \cdots \nabla_j, \overline{\nabla}_j' = \nabla_j \nabla_{j+1}$ and

$$E^{(k,j)} = \varphi(v) \sum_{\{I_*\}_1^k \sim (1,\cdots,k)} \prod_1^k |\nabla_{I_*}|^2$$

$$d\nu^{(k,j)} = |\overline{\nabla}_j u|^2 E^{(k,j)} dm_1 \cdots dm_k$$

$$K_G^{(k,j)} = \int_{\prod_1^k M_i^\perp} \varphi_0(w) d\nu^{(k,j)}$$

where $j = \overline{0,k}$, G is a bounded open set of $\prod_1^k M_i$. Under the first inductive assumption, we shall prove that every $d\nu^{(d+1,j)}$ is a Carleson measure on $\prod_1^{d+1} M_i^\perp$ and

$$\| \, d\nu^{(d+1,j)} \, \|_{C.M.} \leq C_{d,\overline{n},\varphi} M_\varphi(u,v)^2 \qquad (21)_j$$

When $j = 0$, $(21)_0$ can be followed from Lemma 10 and the fact $u^2\varphi(v) \leq M_\varphi(u,v)^2$. Suppose $(21)_j$ holds for some $j \leq d$ (The second inductive assumption), we shall prove $(21)_{j+1}$. But, it is enough to prove

$$K_G^{(d+1,j+1)} \leq C_{\overline{n},d,\varphi} |G| M_\varphi(u,v)^2 \qquad (22)$$

for any bounded open set $G \subset \prod_1^{d+1} M_i$. By an analogue of (19), we get

$$
\begin{aligned}
K_G^{(d+1,j+1)} \; = \; & \frac{1}{2} \int_{\prod_{i=1}^{d+1} M_i^\perp} \Big(\Delta_{j+1}(|\overline{\nabla}_j u|^2 E^{(d+1,j+1)} \varphi_0(w)) \\[6pt]
& - |\overline{\nabla}_j u|^2 E^{(d+1,j+1)} \varphi_0''(w)|\nabla_{j+1} w|^2 \\[6pt]
& - |\overline{\nabla}_j u|^2 \varphi_0(w) \Delta_{j+1} E^{(d+1,j+1)} \\[6pt]
& + O(1)|\overline{\nabla}_j u||\overline{\nabla}_j' u|\nabla_{j+1} E^{(d+1,j+1)}|\varphi_0(w) \\[6pt]
& + O(1)|\overline{\nabla}_j u||\overline{\nabla}_j' u||\nabla_{j+1}\varphi_0(w)|E^{(d+1,j+1)} \\[6pt]
& + O(1)|\overline{\nabla}_j u|^2|\nabla_{j+1} E^{(d+1,j+1)}||\nabla_{j+1}\varphi_0(w)|\Big) \prod_{i=1}^{d+1} dm_i \\[6pt]
\triangleq \; & \sum_1^6 \text{II}_i
\end{aligned}
$$

By Lemma 5 (when M_{j+1} is compact) or Lemma 6 and Lemma 4 (when M_{j+1} is noncompact), and the second inductive assumption (when $j < d$) or the first inductive assumption (when $j = d$), we get

$$
\begin{aligned}
\text{II}_1 \; \leq \; & C_{\overline{n},d,\varphi} \int_{\prod_{i=1,i\neq j+1}^{d+1} M_i^\perp} \left(\int_{M_{j+1}} |\overline{\nabla}_j u|^2 E^{(d+1,j+1)} w^2 dx_{j+1} \right) \\[6pt]
& \cdot \prod_{i=1,i\neq j+1}^{d+1} dm_i
\end{aligned}
$$

$$\leq \ C_{\overline{n},d,\varphi} M_\varphi(u,v)^2 \int_{\prod_{i=1}^{d+1} M_i} w^2(\overline{x},\overline{0})\mathrm{d}\overline{x}$$

$$= \ C_{\overline{n},d,\varphi}|G| M_\varphi(u,v)^2$$

by the inductive assumptions. Lemma 8 and the L^2-boundedness of $N_{P,\overline{a}}$

$$|\mathrm{II}_2| \ \leq \ C_{\overline{n},d,\varphi} M_\varphi(u,v)^2 \int_{M_{j+1}^\perp} \left(\int_{\prod_{i=1,i\neq j+1}^{d+1} M_i} |\nabla_{j+1} w|^2 \right.$$

$$\left. \cdot \prod_{i=1,i\neq j+1}^{d+1} dx_i \right) dm_{j+1}$$

$$\leq \ C_{\overline{n},d,\varphi} M_\varphi(u,v)^2 \int_{\prod_{i=1}^{d+1} M_i} w^2(\overline{x},\overline{0})\mathrm{d}\overline{x}$$

$$= \ C_{\overline{n},d,\varphi}|G| M_\varphi(u,v)^2$$

For II_3, similar to (19), it is not difficult to prove that

$$|\Delta_{j+1} E^{(d+1,j+1)}| \leq C_d E^{(d+1,j+1)}$$

so, by the second inductive assumption

$$|\mathrm{II}_3| \ \leq \ C_{\overline{n},d,\varphi} M_\varphi(u,v)^2 \int_{\prod_{i=1}^{d+1} M_i^\perp} w^2 d\nu^{(d+1,j+1)}$$

$$\leq \ C_{\overline{n},d,\varphi} M_\varphi(u,v)^2 |G|$$

For $\mathrm{II}_4 - \mathrm{II}_6$, noticing that

$$|\nabla_{j+1} E^{(d+1,j+1)}| \leq C_d \big(E^{(d+1,j+1)} E^{(d+1,j)} \big)^{\frac{1}{2}}$$

by Hölder inequality, we get

$$|\mathrm{II}_4| \ \leq \ C_{\overline{n},d,\varphi} \big(K_G^{(d+1,j+1)} |\text{a term like } \mathrm{II}_3| \big)^{\frac{1}{2}}$$

$$\leq \ C_{\overline{n},d,\varphi} |G|^{\frac{1}{2}} M_\varphi(u,v) \big(K_G^{(d+1,j+1)} \big)^{\frac{1}{2}}$$

$$|\mathrm{II}_5| \ \leq \ C_{\overline{n},d,\varphi} \big(|\text{a term like } \mathrm{II}_2| K_G^{(d+1,j+1)} \big)^{\frac{1}{2}}$$

$$\leq \ C_{\overline{n},d,\varphi} M_\varphi(u,v) |G|^{\frac{1}{2}} \big(K_G^{(d+1,j+1)} \big)^{\frac{1}{2}}$$

$$|\mathrm{II}_6| \ \leq \ C_{\overline{n},d,\varphi} \big(|\text{a term like } \mathrm{II}_2| |\text{a term like } \mathrm{II}_2| \big)^{\frac{1}{2}}$$

$$\leq \ C_{\overline{n},d,\varphi} |G| M_\varphi(u,v)^2$$

Combining all above, we get

$$K_G^{(d+1,j+1)} \leq C_{\overline{n},d,\varphi}(M_\varphi(u,v)^2|G| + |G|^{\frac{1}{2}} M_\varphi(u,v)(K_G^{(d+1,j+1)})^{\frac{1}{2}})$$

Therefore, (22) holds. By inductive principle, $(21)_j$ holds for $j = \overline{1, d+1}$. Thus, (20) holds.

4   Proof of Theorem 1

When $2 \leq p < \infty$, the inequality (2) is a corollary of the fact $|f(\overline{x})| \leq N_{P,\overline{a}}(f)(\overline{x})$ and following

Lemma 12. For $\overline{b} \in (R_+^1)^d, f \in L^p(\overline{M}), 2 \leq p < \infty$, there holds

$$\| A_{P,\overline{b}}(f) \|_p \leq C_{\overline{M},\overline{b},p} \| f \|_p$$

Proof. When $p = 2$, by (6), $V_{\overline{w}}(\overline{bt})^{-1} \leq C_{\overline{M},\overline{b}} V_{\overline{x}}(t)^{-1}$ for $\overline{x} \in \Gamma_{\overline{b}}(\overline{w})$, so

$$\begin{aligned}
\| A_{P,\overline{b}}(f) \|_2^2 &= \int_{\overline{M}} \left(\int_{\Gamma_{\overline{b}}(\overline{w})} |\overline{\nabla} P_{\overline{t}}(f)(\overline{x})|^2 V_{\overline{w}}^{-1}(\overline{bt}) d\overline{m} \right) d\overline{w} \\
&\leq C_{\overline{n},\overline{b}} \int_{\overline{M}} \overline{\nabla} P_{\overline{t}}(f)(\overline{x})|^2 V_{\overline{x}}(t)^{-1} |\{\overline{w} : \overline{x} \in \Gamma_{\overline{b}}(\overline{w})\}| d\overline{m} \\
&\leq C_{\overline{n},\overline{b}} \int_{\overline{M}} \overline{\nabla} P_{\overline{t}}(f)(\overline{x})|^2 d\overline{m} \\
&= C_{\overline{n},\overline{b}} \| g_P(f) \|_2^2 \leq C_{\overline{n},\overline{b}} \| f \|_2^2
\end{aligned}$$

by L^2-boundedness of Littlewood-Paley-Stein g-function.

When $\infty > p > 2$, for any nonnegative function $\varphi \in L^{p'} (p^{-1} + p'^{-1} = 1)$, by Lemma 1, there holds

$$\begin{aligned}
\int_{\overline{M}} A_{P,\overline{b}}^2(f)(\overline{w})\varphi(\overline{w}) d\overline{w} &= \int_{\overline{M}} \left(\int_{\Gamma_{\overline{b}}(\overline{w})} |\overline{\nabla} P_{\overline{t}}(f)(\overline{x})|^2 V_{\overline{w}}^{-1}(\overline{b}\,\overline{t}) d\overline{m}\varphi(\overline{w}) \right) d\overline{w} \\
&\leq C_{\overline{n}} \int_{\overline{M}} \left(\int_{\overline{M}^\perp} |\overline{\nabla} P_{\overline{t}}(f)(\overline{x})|^2 p_{\overline{bt}}(\overline{x}, \overline{w}) d\overline{m} \right) d\overline{w} \\
&= C_{\overline{n}} \int_{\overline{M}^\perp} |\overline{\nabla} P_{\overline{t}}(f)(\overline{x})|^2 P_{\overline{bt}}(\varphi)(\overline{w}) d\overline{m} \\
&\leq C_{\overline{n}} \int_{\overline{M}} g_P^2(f)(\overline{w}) N_{P,\overline{1}}(\varphi)(\overline{w}) d\overline{w}
\end{aligned}$$

$$\leq C_{\overline{n}} \int_{\overline{M}} \| \, g_P(f) \, \|_p^2 \| \, N_{P,\overline{1}} \, \|_{p'}$$

$$\leq C_{\overline{n},p} \| \, f \, \|_p^2 \| \, \varphi \, \|_{p'}$$

by boundedness of g_P and $N_{P,\overline{1}}$. So

$$\| \, A_{P,\overline{b}}(f) \, \|_p = \sup_{\|\varphi\|_{p'} \leq 1} \left(\int_{\overline{M}} A^2_{P,\overline{b}}(f)\varphi \right)^{\frac{1}{2}} \leq C_{\overline{n},p} \| \, f \, \|_p$$

Note that Lemma 12 also holds for $1 < p < 2$, we do not discuss this in detail here (Refers to [5] for $d = 1$).

Now we consider the case $0 < p \leq 2$. At first, we shall prove that

$$L_u^{(d)} \triangleq t^{-2} \int_{\overline{M}^{\perp}} \varphi_0(v)|\overline{\nabla}u|^2 d\overline{m}$$

$$\leq C_{\overline{n},d,\overline{a}} \left(|E_t^c| + \int_{E_t} t^{-2} N_{P,\overline{a}}(u)(\overline{x})^2 d\overline{x} \right) \tag{23}$$

where $E_t = \{\overline{x} \in \overline{M} : N_{P,\overline{a}}(f)(\overline{x}) \leq t\}$, $v = P_{\overline{t}}(\mathcal{X}_G)$, φ_0 as in proof of Lemma 10 (but, $k_0 = \prod_{i=1}^d \delta(a_i, n_i)$ now).

For fixed d, let $\overline{M}_j = \prod_1^j M_i$, $\overline{M}_j' = \prod_{j+1}^d M_i$, $\overline{M}_j^{\perp} = \prod_1^j M_i^{\perp}$, $\overline{M}'^{\perp}_j = \prod_{j+1}^d M_i^{\perp}$, $\overline{t}_j = (t_1,\cdots,t_j)$, $\overline{t}'_j = (t_{j+1},\cdots,t_d)$, $\overline{w}_j = (w_1,\cdots,w_j)$, $\overline{w}'_j = (w_{j+1},\cdots,w_d)$, $\overline{\nabla}_j = \nabla_1\cdots\nabla_j$, $d\overline{m}_j = \prod_1^j dm_i$, and

$$L_u^{(d,j)} = t^{-2} \int_{\overline{M}_j'} \left(\int_{\overline{M}_j^{\perp}} |\overline{\nabla}_j u|^2 \varphi_0(v)d\overline{m}_j \right) d\overline{w}_j'$$

Then, $L_u^{(d,d)} = L_u^{(d)}$, and

$$L_u^{(d,0)} = t^{-2} \int_{\overline{M}_d} u^2 \varphi(v)d\overline{w}_d \leq \int_{E_t} t^{-2} N_{P,\overline{a}}(f)(\overline{x})^2 d\overline{x}$$

So, it is enough for (23) to prove

$$L_u^{(d,j)} \leq C_{\overline{n},d,\overline{a}}(|E_t^c| + L_u^{(d,j+1)}) \quad (j = \overline{1,d})$$

Similar to (19), we have

$$L_u^{(d,j)} = \frac{1}{2}t^{-2} \int_{\overline{M}_j'} \left(\int_{\overline{M}_j^{\perp}} \Delta_j(|\overline{\nabla}_{j-1}u|^2 \varphi_0(v))d\overline{m}_j \right) d\overline{w}_j'$$

$$-\frac{1}{2}t^{-2} \int_{\overline{M}_j'} \left(\int_{\overline{M}_j^{\perp}} |\overline{\nabla}_{j-1}u|^2 \varphi_0''(v)|\nabla_j v|^2 d\overline{m}_j \right) d\overline{w}_j'$$

$$+O(1)t^{-2}\int_{\overline{M}_j'}\left(\int_{\overline{M}_j^{\perp}}|\overline{\nabla}_ju|\cdot|\overline{\nabla}_{j-1}u||\varphi_0'(v)\nabla_jv|d\overline{m}_j\right)d\overline{w}_j'$$

$$\stackrel{\triangle}{=}\ \sum_1^3 \text{III}_i$$

By Lemma 5 (when M_j is compact) or Lemmas 4 and 6 (when M_j is noncompact)

$$\text{III}_1\ =\ \frac{1}{2}t^{-2}\int_{\overline{M}_{j-1}'}\left(\int_{\overline{M}_{j-1}^{\perp}}|\overline{\nabla}_{j-1}u|^2\varphi_0(v)d\overline{m}_{j-1}\right)d\overline{w}_{j-1}'$$

$$=\ \frac{1}{2}L_u^{(d,j-1)}$$

By Lemma 2

$$d\mu_j=|\nabla_{j-1}u|^2\varphi_0^{\frac{1}{2}}(v)d\overline{m}_{j-1}$$

is a Carleson measure on $\overline{M}_{j-1}$, and its $C.M.$ norm is less than or equal to $t^2C_{\overline{n},j-1}$, thus (because $|\varphi_0''|\leq C_{\varphi_0}\cdot\varphi_0^{\frac{3}{4}}$)

$$|\text{III}_2|\ \leq\ C_{\overline{n},d,\overline{a}}t^{-2}\int_{\overline{M}_j'}\left(\int_{\overline{M}_j^{\perp}}\left(\int_{\overline{M}_{j-1}^{\perp}}|\nabla_jv|^2d\mu_j\right)dm_j\right)d\overline{w}_j'$$

$$\leq\ C_{\overline{n},d,\overline{a}}\int_{\overline{M}_j'}\left(\int_{\overline{M}_j^{\perp}}\left(\int_{\overline{M}_{j-1}^{\perp}}|\nabla_j\tilde{v}|^2d\overline{w}_{j-1}\right)dm_j\right)d\overline{w}_j'$$

(by Lemma 8)

where $\tilde{v}=1-v$, $\nabla_jv=\nabla_j\tilde{v}$, by Fubini theorem, L^2-boundedness of Littlewood-Paley-Stein g-function

$$|\text{III}_2|\leq C_{\overline{n},d,\overline{a}}\int_{\overline{M}_j'}\left(\int_{\overline{M}_{j-1}}\left(\int_{\overline{M}_j}\tilde{v}^2dw_j\right)d\overline{w}_{j-1}\right)d\overline{w}_j'=C_{\overline{n},d,\overline{a}}|E_t^c|$$

For III_3, by Hölder inequality and $|\varphi_0'|\leq C_{\varphi_0}\cdot\varphi_0^{\frac{3}{4}}$

$$|\text{III}_3|\ \leq\ C_{\overline{n},d,\overline{a}}\left(L_u^{(d,j)}\int_{\overline{M}_j'}\left(\int_{\overline{M}_j^{\perp}}|\nabla_{j-1}u|^2\cdot\varphi_0^{\frac{1}{2}}(v)|\nabla_jv|^2d\overline{m}_j\right)d\overline{w}_j'\right)^{\frac{1}{2}}$$

$$\leq\ C_{\overline{n},d,\overline{a}}(L_u^{(d,j)}|E_t^c|)^{\frac{1}{2}}$$

Thus

$$L_u^{(d,j)}\leq C_{\overline{n},d,\overline{a}}(L_u^{(d,j-1)}+|E_t^c|+(L_u^{(d,j)}|E_t^c|^{\frac{1}{2}}))$$

and

$$L_u^{(d,j)} \leq C_{\overline{n},d,\overline{a}}(|E_t^c| + L_u^{(d,j-1)})$$

because $L_u^{(d,j)} \leq C \cdot \parallel u(\cdot,0) \parallel_2^2 < \infty$. (23) is proved.

Now, by a multiparameter analogue of Lemma 2, there exists a closed set $F_t \subset E_t$ such that $|F_t^c| \leq C_{\overline{n},\overline{a},\overline{b}}|E_t^c|$, and a $\varphi_0 \in \Phi$ such that

$$|\{A_{P,b}(f) > t\}| \leq |F_t^c| + |\{A_{P,b}(f) \leq t\} \bigcap F_t|$$
$$\leq C_{\overline{M},\overline{a},\overline{b}} \left(|E_t^c| + t^{-2} \int_{\overline{M}^\perp} |\overline{\nabla} u|^2 \varphi_0(v)d\overline{m} \right)$$
$$\leq C_{\overline{M},\overline{a},\overline{b}} \left(|E_t^c| + t^{-2} \int_{E_t} N_{P,\overline{a}}(f)^2(\overline{x})\mathrm{d}\overline{x} \right) \tag{24}$$

by (23). Combining (24) with the following identity

$$\parallel f \parallel_p^p = p \int_0^\infty t^{p-1}|\{|f| > t\}|dt$$

we get

$$\parallel A_{P,\overline{b}}(f) \parallel_p \leq C_{\overline{M},p,\overline{a},\overline{b}} \parallel N_{P,\overline{a}}(f) \parallel_p$$

easily. Theorem 1 is proved now.

As a corollary of Theorem 1, we get

Theorem 4. Under the hypotheses of Theorem 1, if $v(\overline{M}) = \inf V_{\overline{x}}(\overline{1}) > 0$, then for any multiharmonic function w on $\overline{M}^\perp$, there holds

$$\parallel A_{\overline{b}}(w) \parallel_p \leq C_{\overline{M},\overline{a},\overline{b},p} \parallel N_{\overline{a}}(w) \parallel_p \tag{25}$$

where the definition of $A_{\overline{b}}$ and $N_{\overline{a}}$ is similar to $A_{P,\overline{b}}$ and $N_{P,\overline{a}}$ but, $u(\text{in}(1))$ is replaced by w.

Proof. When $2 \leq p < \infty$, by Theorem 2, there exists $f \in L^p$, such that $w = P_{\overline{t}}(f)$. By Lemma 12

$$\parallel A_{\overline{b}}(w) \parallel_p \leq C_{\overline{M},\overline{b},p} \parallel f \parallel_p \leq C_{\overline{M},\overline{b},p} \parallel N_{\overline{a}}(w) \parallel_p$$

When $0 < p < 2$, by Lemma 4, $w_\varepsilon(\cdot,\overline{t}) = w(\cdot,\overline{t} + \varepsilon\overline{1}) \in L^2(\overline{M})$. But, $L_c^\infty(\overline{M})$ is dense in $L^2(\overline{M})$, so (24) holds when we replace $A_{P,\overline{b}}(f)$ and $N_{P,\overline{a}}(f)$ by $A_{\overline{b}}(w)$ and $N_{\overline{a}}(w)$ respectively. Therefore, (25) holds for w_ε. Let $\varepsilon \longrightarrow 0$. we get (25).

Acknowledgement. The first author would like to express his many thanks to Prof. Cheng Min-de for his enthusistic support.

References

[1] D.L.Burkholder and R. F. Gundy and M. L. Silverstein, A maximal function characterization of the class H^p, *Trans. Amer. Math. Soc.*, 157(1971), 137–153.

[2] A.P.Calderon and A.Torchinsky, Parabolic maximal functions associated with a distribution, *Adv. in Math.*, 16(1975), 1–64.

[3] S.-Y. Chang and R. Fefferman, A continuous version of duality of H^1 and BMO on the bi-disc, *Ann. of Math.*, 112(1980), 179–201.

[4] Jie-cheng Chen, Thesis for Master's Degree, Hangzhou University, Hangzhou, 1984.

[5] Jie-cheng Chen, Thesis for Ph.D., Hangzhou University, Hangzhou, 1987.

[6] C. Fefferman and E.M. Stein, H^p spaces of several variables, *Acta Math.*, 129(1972), 137–193.

[7] R.F.Gundy and E.M.Stein, H^p theory for the poly-disc, *Proc. Natl. Acad. Sci.U.S.A.*, 76 (1979), 1026–1029.

[8] P.Li and R.Schoen, L^p and mean value properties of subharmonic functions on Riemannian manifolds, *Acta Math.*, 153(1984), 280–301.

[9] P.Li and S. -T. Yau, On the Parabolic kernel of the Schordinger operator, *Acta Math.*, 156(1986), 153–201.

[10] K.G.Merryfield, Doctoral dissertation, The University of Chicago, 1980.

[11] E.M.Stein, A variant of the area integral, *Bul. Sc. Math.*, 103(1979), 449–461.

Surfaces of General Type with $\chi(O_S) = 1$ and Fibrations of Genus Two (VII)

Weng Lin
Mathematics Institute, Fudan University, P. R. China

Abstract

In this series papers, we will classify surfaces of general type with $\chi(O_S) = 1$ and fibrations of genus two completely, and construct some new surfaces with $p_g = 0$.

Following the last paper, this paper will show the classification of fibrations of genus 2 with $p_g = q = g(C) = 1, K^2 = 5$ or 6 and $\sum_F (t_F - 1) = s$.

1   The Main Theorem

In this paper, we will use the same notations as [3].

In [4], [5] and [6], we have shown the classification of fibrations of genus 2 with $p_g = q = g(C) = 1$ and $K^2 = 2, 3, 4$. In [4], we have pointed out that $f : S \to C$ is a fibration of genus two with $p_g = q = g(C) = 1$, $K^2 = 2, 3, 4, 5, 6$. So we must classify such fibrations with $K^2 = 5, 6$. Also from[4], we know such fibrations must satisfy $\sum_F (t_F - 1) - s \leq 2$. In[3], we have given some properties of these fibrations. Now, we will use them to classify what are in our problem.

Theorem: *Suppose $f : S \to C$ is a fibration of genus 2 with $K^2 = 5, 6$ and $\sum_F (t_F - 1) = s$. Let $\Phi : S \to P$ be a double covering determined by f, R the main part of branch locus of Φ. Then the classification of R must be as follows:*

(1) $e(P) = 1$.

Case 1: $K^2 = 5$.

type of C_i's	$p_y(C_i)$	(a_{ij})			(b_{ij})		
$(6,7)$	3	(3	3	3)	(3	3	3)
$(3,3)$	1	1	1	1	2	2	2
$(3,4)$	3	2	2	2	1	1	1
$(2,3)$	3	1	1	1	1	1	1
$(4,4)$	1	2	2	2	2	2	2
$(2,2)$	1	2	1	1	0	1	1
$(4,5)$	3	1	2	2	3	2	2
$(1,1)$	1	1	1	1	0	0	0
$(5,6)$	3	2	2	2	3	3	3
$(1,1)$	1	1	1	0	0	0	1
$(1,1)$	1	1	0	1	0	1	0
$(4,5)$	3	1	2	2	3	2	2
$(1,1)$	1	1	0	0	0	1	1
$(2,2)$	1	0	1	1	2	1	1
$(3,4)$	3	2	2	2	1	1	1
$(1,1)$	1	1	1	1	0	0	0
$(3,3)$	1	1	1	1	2	2	2
$(2,3)$	3	1	1	1	1	1	1
$(2,2)$	1	2	1	1	0	1	1
$(2,2)$	1	0	1	1	2	1	1
$(2,3)$	3	1	1	1	1	1	1
$(1,1)$	1	0	1	1	1	0	0
$(1,1)$	1	0	0	0	1	1	1
$(2,2)$	1	2	1	1	0	1	1
$(2,3)$	3	1	1	1	1	1	1
$(1,1)$	1	1	0	0	0	1	1
$(1,1)$	1	0	1	0	1	0	1
$(1,1)$	1	0	0	1	1	1	0
$(3,4)$	3	2	2	2	1	1	1
$(1,1)$	1	1	1	1	0	0	0
$(1,1)$	1	1	0	0	0	1	1
$(1,1)$	1	0	1	0	1	0	1
$(1,1)$	1	0	0	1	1	1	0
$(2,3)$	3	1	1	1	1	1	1

Case 2: $K^2 = 6$.

type of C_i's	$p_g(C_i)$	(a_{ij})	(b_{ij})
$(6,8)$	2	$\begin{pmatrix} 3 & 3 & 3 & 3 \end{pmatrix}$	$\begin{pmatrix} 3 & 3 & 3 & 3 \end{pmatrix}$
$(2,3)$	2	$\begin{pmatrix} 0 & 1 & 1 & 1 \\ 3 & 2 & 2 & 2 \end{pmatrix}$	$\begin{pmatrix} 2 & 1 & 1 & 1 \\ 1 & 2 & 2 & 2 \end{pmatrix}$
$(4,5)$	2		

(2) $e(P) = -1$.

Case 1: $K^2 = 5$.

> *In this case, it is enough to change coordinates of C_i's into* $(6, 1)$; $(3, 0)$, $(3, 1)$; $(2, 1)$, $(4, 0)$; $(2, 0)$, $(4, 1)$; $(1, 0)$, $(1, 0)$, $(4, 1)$; $(1, 0)$, $(2, 0)$, $(3, 1)$; $(1, 0)$, $(3, 0)$, $(2, 1)$; $(2, 0)$, $(2, 0)$, $(2, 1)$; $(1, 0)$, $(1, 0)$, $(2, 0)$, $(2, 1)$; $(1, 0)$, $(1, 0)$, $(1, 0)$, $(3, 1)$; $(1, 0)$, $(1, 0)$, $(1, 0)$, $(1, 0)$, $(2, 1)$.

Case 2: $K^2 = 6$. *In this case, it is enough to change coordinates of* C_i*'s into* $(6, 2)$; $(2, 1)$, $(4, 1)$.

Remark: *In the theorem, we assume every $t_j = 1$. Surely, from this, we can obtain other cases.*

In the next section, we only prove the theorem for $e(P) = 1$, as the proof of $e(P) = -1$ is very similar to the case of $e(P) = 1$.

2   The Proof of The Theorem

In [2], we have shown the singular types of fibrations of genus 2 with $\sum_F (t_F - 1) = s$. Now we use it to prove our theorem.

Obviously, in this case, $R \sim (6, K^2 + 2)$. If (α, β) is an irreducible curve on P and $\alpha \geq 0$, we have $\beta \geq \alpha$. Moreover, we have $\sum^h p_g(C_i) = -1 + p_a(R) - 2 \times 3 \times 2 \times s/2 = h + 7 - K^2$.

Remarks: $p_a((1,\beta)) = 1$. $p_a((2,\beta)) = \beta$. $p_a((3,\beta)) = 2\beta - 2$. $p_a((4,\beta)) = 3\beta - 5$. $p_a((5,\beta)) = 4\beta - 9$. $p_a((6,\beta)) = 5\beta - 14$.

Next we denote the number of irreducible components of R as h.

(1) $h = 6$. It does not occur;

(2) $h = 5$.

Case 1: $K^2 = 5$. From above we have $p_g((2, x)) = 3$, so $x = 3$ and this curve is smooth. Thus $C_i \sim (1, 1), C_5 \sim (2, 3)$ and

$$
(a_{ij}) = \begin{pmatrix} 1 & 1 & 1 \\ 1 & 0 & 0 \\ 0 & 1 & 0 \\ 0 & 0 & 1 \\ 1 & 1 & 1 \end{pmatrix}, \qquad (b_{ij}) = \begin{pmatrix} 0 & 0 & 0 \\ 0 & 1 & 1 \\ 1 & 0 & 1 \\ 1 & 1 & 0 \\ 1 & 1 & 1 \end{pmatrix};
$$

Case 2: $K^2 = 6$. If C_i's are with types $(2, 3)$, $(1, 2)$, $(1, 1)$, $(1, 1)$, $(1, 1)$. As $p_g((2, 3)) = 2$, this curve has a double point. But $(2, 3)(1, 1) = 3$, it is impossible;

If C_i's are with types $(2, 4)$, $(1, 1)$, $(1, 1)$, $(1, 1)$, $(1, 1)$, with the same reason as above, we have $(2, 4)$ has two double points. By intersection theory, it is impossible;

(3) $h = 4$.

Case 1: C_i's are with types $(3, x), (1, y_j), j = 1, 2, 3$.

If $K^2 = 5$, as $s = 3$ and $p_g((3, x)) = 3$, we have $2x - 2 \geq 3 + 3$. Thus $x \geq 4$. Therefore $x = 4$ and curve $(3, 4)$ has three double points,

$$
(a_{ij}) = \begin{pmatrix} 1 & 0 & 0 \\ 0 & 1 & 0 \\ 0 & 0 & 1 \\ 2 & 2 & 2 \end{pmatrix}, \qquad (b_{ij}) = \begin{pmatrix} 0 & 1 & 1 \\ 1 & 0 & 1 \\ 1 & 1 & 0 \\ 1 & 1 & 1 \end{pmatrix};
$$

If $K^2 = 6$, we also have $x \geq 4$, and curve $(3, 4)$ must have 2 double points, curve $(3, 5)$ must have three double points. From this, by intersection theory, we deduce that this case does not happen;

Case 2: C_i's are with types as follows: $(1, u), (1, v), (2, x), (2, y)$.

    (a) $K^2 = 5$. If $x = y = 2$, we have $p_g((2, 2)) = 2$. So curve $(2, 2)$ has no singular point. Thus $(1, 1)(2, 2) = 3$, contradiction;

    If $x = 2, y = 3$. Obviously, curve $(2, 3)$ is smooth. Thus

$$
(a_{ij}) = \begin{pmatrix} 0 & 1 & 1 \\ 0 & 0 & 0 \\ 2 & 1 & 1 \\ 1 & 1 & 1 \end{pmatrix}, \qquad (b_{ij}) = \begin{pmatrix} 1 & 0 & 0 \\ 1 & 1 & 1 \\ 0 & 1 & 1 \\ 1 & 1 & 1 \end{pmatrix};
$$

(b) $K^2 = 6$. as $s = 4$, $(1, 1)$ and $(2, 2)$ could not occur at the same time. In fact, otherwise, we have $(1, 1)(2, 2) = 3$, it is impossible. Thus C_i's must be with types as $(2, 3)$, $(2, 3)$, $(1, 1)$, $(1, 1)$ or $(2, 2)$, $(2, 2)$, $(1, 2)$, $(1, 2)$. Surely, $p_g((2, 3)) = 1$ or 2, and curve $(2, 3)$ must have a double point or two double points. From intersection theorem, it is impossible. In fact, we must have curve $(2, 2)$ be smooth.

(4) $h = 3$.

Case 1: C_i's are with types $(2, x), (2, y), (2, z)$.

If $K^2 = 5$, we have $x = y = 2$, $z = 3$ and curve $(2, 3)$ is smooth, curve $(2, 3)$ has a double point. Others are trivial;

If $K^2 = 6$, as $(2, 2)(2, 2) = 4 \neq 5$, we have $x = 2$ and $y = z = 3$. But $(2, 3)(2, 3) = 9$ is an odd number, contradiction;

Case 2: C_i's are with types $(3, x), (2, y), (1, z)$.

If $K^2 = 5$, we have $(x, y, z) = (4, 2, 1), (3, 3, 1), (3, 2, 2)$.

(a) If $(x, y, z) = (4, 2, 1)$, we have curve $(2, 2)$ has a double point. Otherwise, $(1, 1)(2, 2) = 3$. Thus

$$(a_{ij}) = \begin{pmatrix} 1 & 0 & 0 \\ 0 & 1 & 1 \\ 2 & 2 & 2 \end{pmatrix}, \qquad (b_{ij}) = \begin{pmatrix} 0 & 1 & 1 \\ 2 & 1 & 1 \\ 1 & 1 & 1 \end{pmatrix};$$

(b) If $(x, y, z) = (3, 3, 1)$, we have $(2, 3)$ is smooth. In fact, otherwise, $(2, 3)(3, 3) \neq 9$. Thus we have

$$(a_{ij}) = \begin{pmatrix} 1 & 1 & 1 \\ 1 & 1 & 1 \\ 1 & 1 & 1 \end{pmatrix}, \qquad (b_{ij}) = \begin{pmatrix} 0 & 0 & 0 \\ 1 & 1 & 1 \\ 2 & 2 & 2 \end{pmatrix};$$

(c) $(x, y, z) = (3, 2, 2)$. Obviously, curve $(2, 2)$ is singular. Otherwise $(2, 2)(1, 2) = 3$. But in this case, $(1, 2)(3, 3) = 6$, contradiction;

If $K^2 = 6$, we have $x \geq 4$. In fact $p_g((3, x)) \geq 1$. But in this case, if $(2, 2)$ and $(1, 1)$ occur at the same time, we have $(2, 2)(1, 1) = 3$, it is impossible. Thus C_i's must be with types $(3, 4)$, $(2, 3)$, $(1, 1)$ or $(3, 4)$, $(2, 2)$, $(1, 2)$. With the same reason as above, we have $(2, 3)$ and $(2, 2)$ are singular. So $p_g((3,4)) = 1$ and $3 = (2,2)(1,2)$, contradiction;

Case 3: C_i's are with types $(1, w), (1, y), (4, x)$.

If $K^2 = 5$, we have $w = y = 1, x = 5$ and

$$(a_{ij}) = \begin{pmatrix} 0 & 0 & 1 \\ 0 & 1 & 0 \\ 3 & 2 & 2 \end{pmatrix}, \qquad (b_{ij}) = \begin{pmatrix} 1 & 1 & 0 \\ 1 & 0 & 1 \\ 1 & 2 & 2 \end{pmatrix};$$

If $K^2 = 6$, $x \neq 5$. In fact, $p_g((4, x)) = 2$, so $3x - 5 \geq 2 + 8$, i.e. $x \geq 5$. But if $(4, 5)$ and $(1, 1)$ occur at the same time, we would have $(4,5)(1,1) = 8$, contradiction. Thus $x = 6, w = y = 1$. In this case, $p_a((4,6)) = 13$, so $(4, 6)$ has three triple points and a double point. From this, $(1,1)(1,1) \geq 2$, contradiction;

(5) $h = 2$.

Case 1: C_i's are with types $(5, x), (1, y)$. As $5 = 2 + 3$, if $K^2 = 5$, we have $4x - 9 \geq 3 + 12$, i.e. $x \geq 6$. Thus $x = 6$ and $y = 1$. Furthermore

$$(a_{ij}) = \begin{pmatrix} 2 & 2 & 2 \\ 1 & 1 & 1 \end{pmatrix}, \qquad (b_{ij}) = \begin{pmatrix} 3 & 3 & 3 \\ 0 & 0 & 0 \end{pmatrix};$$

If $K^2 = 6$, we would have $4x - 9 = 2 + 16$. But it is impossible;

Case 2: C_i's are with types $(2, y), (4, x)$.

If $K^2 = 5$, we have $x = 4$ or 5.

If $x = 4$, we have

$$(a_{ij}) = \begin{pmatrix} 1 & 1 & 1 \\ 2 & 2 & 2 \end{pmatrix}, \qquad (b_{ij}) = \begin{pmatrix} 1 & 1 & 1 \\ 2 & 2 & 2 \end{pmatrix};$$

If $x = 5$, we have

$$(a_{ij}) = \begin{pmatrix} 0 & 1 & 1 \\ 3 & 2 & 2 \end{pmatrix}, \qquad (b_{ij}) = \begin{pmatrix} 2 & 1 & 1 \\ 1 & 2 & 2 \end{pmatrix};$$

If $K^2 = 6$, as $3x - 5 \geq 1 + 8$, we have $x \geq 5$. If $C_1 \sim (4,6)$, we would have $(4,6)(2,2) = 12 \neq 14$ or 16, it is impossible; If $x = 5$, we have curve $(2, 3)$ has a singular point, thus

$$(a_{ij}) = \begin{pmatrix} 0 & 1 & 1 & 1 \\ 3 & 2 & 2 & 2 \end{pmatrix}, \qquad (b_{ij}) = \begin{pmatrix} 2 & 1 & 1 & 1 \\ 1 & 2 & 2 & 2 \end{pmatrix};$$

Case 3: C_i's are with types $(3, x), (3, y)$.

If $K^2 = 5$, we have $x = 4, y = 3$ and

$$(a_{ij}) = \begin{pmatrix} 1 & 1 & 1 \\ 2 & 2 & 2 \end{pmatrix} , \qquad (b_{ij}) = \begin{pmatrix} 2 & 2 & 2 \\ 1 & 1 & 1 \end{pmatrix} ;$$

If $K^2 = 6$, by a simple discussion, we know it does not occur;

(6) $h = 1$. Trivial. $\qquad\qquad\qquad\qquad\qquad\qquad\qquad\qquad\qquad\qquad$ $\square$

Acknowledgment: I would thank Professor G. Xiao, Professor K. Feng and Professor M. Chern for their help.

References

[1] Barth, W. -Peters, C. -Van de Ven, A., Compact Complex Surfaces, *Springer-Verlag*, 1984.

[2] Weng, L., Surfaces of general type with $\chi(O_S) = 1$ and fibrations of genus two(I), *preprint*, Shanghai, 1986.

[3] Weng, L., Surfaces of general type with $\chi(O_S) = 1$ and fibrations of genus two(II), *preprint*, Shanghai, 1986.

[4] Weng, L., Surfaces of general type with $\chi(O_S) = 1$ and fibrations of genus two(IV), *preprint*, Hefei, 1987.

[5] Weng, L., Surfaces of general type with $\chi(O_S) = 1$ and fibrations of genus two(V), *preprint*, Hefei, 1987.

[6] Weng, L., Surfaces of general type with $\chi(O_S) = 1$ and fibrations of genus two(VI), *preprint*, Hefei, 1987.

[7] Xiao, G., Surfaces Fibrees en Courbes de Genre Deux, *Springer LNM*. n. 1137, 1985.

The Tail Probability for the Kolmogorov Distance in the Worst Direction

Zhu Li-xing
Institute of Systems Science
Academia Sinica, Beijing

Abstract

Consider D_n the maximum Kolmogorov distance between P_n and P among all possible one-dimensional projections, where P_n is an empirical measure based on d-dimensional i.i.d vectors with elliptically symmetric probability measure P. For the supreme of Gaussian process, $\sup_{\mathcal{F}} | W_P(f) |$,indexed by the class of half-spaces of $\mathcal{F}$, which is the weak limit of $\sqrt{n}D_n$, it is shown that

$$\sum_{m=1}^{\infty} \left(\frac{1}{2}m^2\lambda^2 - \frac{1}{8} \right) \exp\left(-2m^2\lambda^2\right)$$
$$\leq P(\sup_{\mathcal{F}} | W_P(f) | > \lambda)$$
$$\leq c\lambda^{2(d-1)}\exp(-2\lambda^2)$$

for large $\lambda, d \geq 2$ and an appropriate constant c. From this, when dimension d is fixed, I give a negative answer for Huber's conjecture $P(D_n > \varepsilon) \leq N\exp(-2n\varepsilon^2)$, where N is a constant depend only on dimension d.

1   Introduction

Suppose that P is an elliptical symmetric non-atomic probability measure in R^d $(d \geq 1)$. Let $X_1, \cdots, X_n$ be a sample from P, and let P_n be the empirical measure determined by this sample.

Put

$$S_d = \{a : A \in R^d, \|a\| = 1\},$$

and

$$D_n = \sup_{a \in S_d} \sup_{b \in R^1} | P_n I(a^T X > b) - PI(a^T X > b) | \qquad (1.1)$$

where

$$PI(a^T X > b) = \int I_{[a^T X > b]}(x)\mathrm{d}P(x),$$

$$P_n I(a^T X > b) = \int I_{[a^T X > b]}(x)\mathrm{d}P_n(x),$$

$I_{[a^T X > b]}$ is the indicate function of set $\{a^T X > b\}$.

D_n is the maximum Kolmogorov distance between P_n and P among all possible one-dimensional projections. Let

$$Q(n, d, \varepsilon) = P(D_n > \varepsilon). \tag{1.2}$$

For $d = 1$, from the continuity of P, we easily know that

$$\sup_{b \in R'} \mid P_n I(X > b) - PI(X > b) \mid = \sup_{b \in R'} \mid P_n I(-X > b) - PI(-X > b) \mid$$

then

$$D_n = \sup_{b \in R'} \mid P_n I(X > b) - PI(X > b) \mid$$

so we have the well-known asymptotic approximation to $Q(n, 1, \varepsilon)$

$$Q(n, 1, \varepsilon) \sim 2 \sum_{j=1}^{\infty} (-1)^{j-1} \exp(-2nj^2 \varepsilon^2). \tag{1.3}$$

As a special case of Devroye(1982) [1], who refined an earlier result of Vapnik and Červonenkis(1971)[2], we know that

$$Q(n, d, \varepsilon) \leq 4e^8 (n^2 e/d)^d \exp(-2n\varepsilon^2). \tag{1.4}$$

Huber(1988)[3], under the condition that P is spherically symmetric non-atomic probability measure in R^d, showed that for $\varepsilon > d/n$

$$Q(n, d, \varepsilon) \leq 2(en/d)^d \exp(-2n(\varepsilon - d/n)^2). \tag{1.5}$$

Öhrvik(1988)[4], for d large, presented an empirical formula on the basis of simulations

$$Q(n, d, \varepsilon) \sim 2 \exp(-2n\varepsilon^2 + 2.464(d - 1)). \tag{1.6}$$

Huber(in[3], section 3) put forward the conjecture based on Öhrvik's empirical formula

$$Q(n, d, \varepsilon) \leq N \exp(-2n\varepsilon^2). \tag{1.7}$$

where N is a constant independent of size of sample n.

In this paper, when dimension d is fixed and size of sample is large enough, we show that the formula(1.7) does not hold for the elliptically symmetric probability measure.

The proving procedure is that we initially find the probability bounds for the supreme of some Gaussian process W_P, then apply the result of Dudley(1978)[5], Dudley and Philipp(1983)[6] (Theorem7.1) to empirical process.

Let

$$\mathcal{F} = \{f = I(\{a^T X > b\}) : \quad a \in S_d, b \in R'\}. \tag{1.8}$$

let $\{W_P(f) : f \in \mathcal{F}\}$ be zero means a.s continuous Gaussian process indexed by $\mathcal{F}$ satisfying

$$EW_P(f_1)W_P(f_2) = Pf_1f_2 - Pf_1Pf_2, \quad \forall f_1, f_2 \in \mathcal{F}. \tag{1.9}$$

This process is also called as P-bridge (cf. Pollard 1984, chapter7).

We derive the following result:

Theorem 1. *Suppose that P is the elliptically symmetric, non-atomic probability measure in $R^d(d > 1)$, then for $\lambda > 0$*

$$P\{\sup_{\mathcal{F}} | W_P(f) | > \lambda\} \geq \frac{1}{2} \sum_{m=1}^{\infty} (m^2\lambda^2 - 1/4)\exp(-2m^2\lambda^2). \tag{1.10}$$

If $d > 3$, we have for $\lambda > 1$

$$P\{\sup_{\mathcal{F}} | W_P(f) | > \lambda\} \leq c\lambda^{2(d-1)}\exp(-2\lambda^2) \tag{1.11}$$

where c depends only on dimension d.

For $d = 2$, (1.11) still holds when P possesses density function.

Theorem 2. *Suppose that P is the elliptically symmetric, non-atomic probability measure, then*

$$Q(n, d, \varepsilon/\sqrt{n}) \geq \frac{1}{2}\varepsilon^2\exp(-2\varepsilon^2)(1 + o(1)) \tag{1.12}$$

for n large enough, where $o(1)$ depends only on n.

2 The Proof of Theorem 1

The basic idea for the proof of the theorem is to use Slepian's inequality.

SLEPIAN INEQUALITY (see[7] lemma2.1).   Let X and Y be two zero means a.s continuous Gaussian processes defined on some set T. If

$$Var(X(t)) = Var(Y(t)), \quad \forall t \in T$$

and

$$Cov(X(t), X(s)) \leq Cov(Y(t), Y(s)), \quad \forall s, t \in T \tag{2.1}$$

then for any λ

$$P\{\sup_T X(t) > \lambda\} \geq P\{\sup_T Y(t) > \lambda\}. \tag{2.2}$$

Note that Slepian's inequality does not extend to comparison of $\sup|X|$ and $\sup|Y|$, nevertheless, we can always use the fact that for symmetric process

$$P\{\sup X > \lambda\} \leq P\{\sup | X |> \lambda\} \leq 2P\{\sup X > \lambda\}. \tag{2.3}$$

In order to simplify the proof of the theorem, we can consider the spherically symmetric non-atomic probability measure. In fact, by the properties of elliptically symmetric probability measure P we know that if X is distributed with P, then there are a nonsigular matrix A, a constant vector μ, and a random vector Y with spheri-cal symmetric distribution P' such that $X = AY + \mu$ in distribution. Without loss of generality, we assume that μ is zero vector. Since A is nonsigular, we derive that $\|a^T A\| = \sqrt{a^T A A^T a} > 0$ holds for every $a \in S_d$. And by spherical symmetry, $a^T Y$ is distributed with identical probability measure for every $a \in S_d$. Consequently, using continuity of sup norm, we easily obtain that in distribution

$$
\begin{aligned}
D_n &= \sup_{a \in S_d} \sup_{b \in R'} | P_n I(\{a^T X > b\}) - P I(\{a^T X > b\}) | \\
&= \sup_{a \in S_d} \sup_{b \in R'} | P_n I(\{a^T AY/\|a^T A\| > b/\|a^T A\|\}) \\
&\qquad - P' I(\{a^T AY/\|a^T A\| > b/\|a^T A\|\}) | \\
&= \sup_{a \in S_d} \sup_{b \in R'} | P_n I(\{a^T Y > b\}) - P' I(\{a^T Y > b\}) | \overset{\triangle}{=} D'_n.
\end{aligned}
$$

˙Without confusion, D'_n is written simply as D_n.

So in the following procedure, we consider that P is spherical symmetrical.

The inequality (1.11) is just as corollary 3 in our recent work[8], here we only prove the inequality (1.10).

We consider the case of $d = 2$ first.

For any $a \in S_d$, it can be written as $a = (\cos\varphi, \sin\varphi), 0 \le \varphi \le 2\pi$ Let $H(\varphi, b) = \{(\cos\varphi, \sin\varphi)X > b\}$, so $\mathcal{F} = \{f = I(H(\varphi, b)) \quad 0 \le \varphi \le 2\pi, b \in R'\}$

Define $f(\varphi, b) = P(H(0,0)\triangle H(\varphi, b))$, where $\triangle$ denotes the symmetric difference of two sets. By spherical symmetry, we easily know that

Proposition 1. *$PI(H(\varphi, b))$ is independent of φ and is a strict decreasing, continuous function of b. We write it simply as $\varphi_2(b)$*

For $f(\varphi, b)$, we have obtained the following properties in [11].

Proposition 2. *For any fixed b, $f(\varphi, b)$ is a nondecreasing, continuous function on $\varphi \in [0, \pi]$.*

Furthermore, for b which makes $0 < \varphi_2(b) < 1$, there is a $\varphi_b \in [0, \pi/2]$ such that in the interval $[\varphi_b, \pi - \varphi_b], f(\varphi, b)$ increases strictly with respect to φ.

From simple calculation, one can verify that

$$\left.\begin{array}{l} \text{for} \quad \varphi_2(b) \le 1/2, \quad \{f(\varphi, b) : \varphi_b \le \varphi \le \pi - \varphi_b\} \\ \qquad\qquad\qquad = [\frac{1}{2} - \varphi_2(b), \frac{1}{2} + \varphi_2(b)], \\ \text{for} \quad \varphi_2(b) > 1/2, \quad \{f(\varphi, b) : \varphi_b \le \varphi \le \pi - \varphi_b\} \\ \qquad\qquad\qquad = [\varphi_2(b) - \frac{1}{2}, \frac{3}{2} - \varphi_2(b)], \end{array}\right\} \qquad (2.4)$$

In order to use Slepian's inequality to verify theorem 1, we shall define two zero means a.s. continuous Gaussian processes.

Assume that P_1 is probability measure uniformly on the line segment $\{(s, t) : 0 \le s \le 1, 0 \le t \le 1, s + t = 1\}$, let $\{Y_{P_1}(g), g \in \mathcal{G}\}$ be a zero means a.s. continuous Gaussian process satisfying

$$EY_{P_1}(g_1)Y_{P_1}(g_2) = P_1 g_1 g_2 - P_1 g_1 P_1 g_2 \quad \forall g_1, g_2 \in \mathcal{G} \qquad (2.5)$$

where $\mathcal{G} = \{g = I(\{X^T \le (s, t)\}) : \quad 0 \le s \le 1, \quad 0 \le t \le 1, \quad s + t - 1 \ge 0\}$. Indeed, $\{Y_{P_1}(g) : g \in \mathcal{G}\}$ is weak limit process of an empirical process indexed by $\mathcal{G}$, $\{\sqrt{n}(\tilde{P}_n g - P_1 g) : g \in \mathcal{G}\}$, where $\tilde{P}_n$ is an empirical measure based on a sample of size n with distribution P_1.

Let $\mathcal{G} = \{g \in \mathcal{G} : 1/4 \le s \le 3/4, 3/4 \le t \le 1\}$

$$\mathcal{F}_1 = \left\{ I(H(\psi_b, b)) : \begin{array}{l} b \in R', \psi_b \in [\varphi_b, \pi - \varphi_b], \\ 1 + \varphi_2(b) + f(\psi_b, b) \le 2, \\ \text{and} \, 1 + \varphi_2(b) - f(\psi_b, b) \le 2, \end{array} \right\}.$$

Define a mapping $T_1 : \mathcal{F}_1 \to \mathcal{G}_1$ by

$$T_1(I(H(\psi_b, b)))$$
$$= I(\{X^T \le (1 + \varphi_2(b) - f(\psi_b, b))/2, (1 + \varphi_2(b) + f(\psi_b, b))\}),$$
$$\tag{2.6}$$

and define a process on $\mathcal{G}_1$ by $X_P(g) = W_P(T_1^{-1} g)$. We shall compare probability of $\sup_{\mathcal{G}_1} | Y_{P_1}(g) |$ with that of $\sup_{\mathcal{G}_1} | X_P(g) |$. First we show that $X_P(g)$ is well-defined, it is enough to show that T_1 is one-to-one and onto.

If for (ψ_b, b) and $(\psi_{b'}, b')$ such that

$$T_1(I(H(\psi_b, b))) = T_1(I(H(\psi_{b'}, b'))), \tag{2.7}$$

then

$$\left\{ \begin{array}{l} (1 + \varphi_2(b) - f(\psi_b, b)) = (1 + \varphi_2(b') - f(\psi_{b'}, b')), \\ (1 + \varphi_2(b) + f(\psi_b, b)) = (1 + \varphi_2(b') + f(\psi_{b'}, b')). \end{array} \right. \tag{2.8}$$

So we have $b = b'$ and $\varphi_b = \varphi_{b'}$ by propositions 1 and 2. Consequently, T_1 is one-to-one. On the other hand, for any $g = I(\{X^T \le (s, t)\}) \in \mathcal{G}_1$, we shall take $\varphi_2(b) = (s + t - 1)$ and $f(\psi_b, b) = (t - s)$ to obtain a $f \in \mathcal{F}_1$ corresponding to g, that is,

$$s = (1 + \varphi_2(b) - f(\psi_b, b))/2, \quad t = (1 + \varphi_2(b) + f(\psi_b, b))/2. \tag{2.9}$$

(1) If $0 \le s + t - 1 \le 1/2$, one takes b such that $s + t - 1 = \varphi_2(b) \le 1/2$, meanwhile, as the result of $t \ge 3/4$ and $s \ge 1/4$, we have

$$t - s \ge -t - s + 3/2 = 1/2 - \varphi_2(b),$$

and

$$t - s \le t + s - 1/2 = 1/2 + \varphi_2(b).$$

So by the continuity of $f(\varphi, b)$ and (2.4), we can take a ψ_b such that $t - s = f(\psi_b, b)$, it implies (2.9).

(2) In the case $1/2 < t + s - 1 \leq 1$. As above, one first take b such that $s + t - 1 = \varphi_2(b)$. Moreover, the inequalities

$$t - s \geq t + s - 3/2 = \varphi_2(b) - 1/2,$$

and

$$t - s \leq 3/2 - t - s + 1 = 3/2 - \varphi_2(b),$$

follow from $t \leq 1$ and $s \leq 3/4$. Then from (2.4) and the continuity of $f(\varphi, b)$ there exists a $\psi_b \in [\varphi_b, \pi - \varphi_b]$ such that $f(\psi_b, b) = t - s$, it also implies(2.9). Hence T_1 is onto.

Clearly, $\{X_P(g), g \in \mathcal{G}\}$ is Gaussian with zero means, and it follows that

$$
\begin{aligned}
EX_P^2(g) &= EW_P^2(T_1^{-1}g) = \varphi_2(b) - (\psi_2(b))^2 \\
&= (((1 + \varphi_2(b) + f(\psi_b, b))/2) - (1 + f(\psi_b, b) - \varphi_2(b))/2) \\
&\quad -\{((1 + \varphi_2(b) + f(\psi_b, b))/2) - (1 + f(\psi_b, b) - \varphi_2(b))/2)\}^2 \\
&= (s + t - 1) - (s + t - 1)^2 = EY_{P_1}^2(g) \qquad (2.10)
\end{aligned}
$$

the last equation following from which $P_1 g = (s + t - 1)$. For any two $g = I(\{X^T \leq (s, t)\})$ and $g' = I(\{X^T \leq (s', t')\})$, their correspondences in $\mathcal{F}_1$ are $T_1^{-1}g = I(H(\psi_b, b))$ and $T_1^{-1}g' = I(H(\psi_{b'}, b'))$. From Slepian's inequality, we know that if the following inequality

$$
\begin{aligned}
EY_{P_1}(g)Y_{P_1}(g') &= P_1 g g' - P_1 g P_1 g' \\
&\geq PT_1^{-1}g T_1^{-1}g' - PT_1^{-1}g PT_1^{-1}g' \\
&= EX_P(g)X_P(g') \qquad (2.11)
\end{aligned}
$$

holds, then

$$P\{\sup_{\mathcal{G}_1} Y_{P_1}(g) > \lambda\} \leq P\{\sup_{\mathcal{G}_1} X_P(g) > \lambda\}. \qquad (2.12)$$

Since $P_1 g P_1 g' = PT_1^{-1}g PT_1^{-1}g'$, hence we only need to prove that

$$P_1 g g' \geq PT_1^{-1}g T_1^{-1}g',$$

so as to obtain (2.11).

Now prove $P_1gg' \geq PT_1^{-1}gT_1^{-1}g',$. By simple calculation, we have

$$
\begin{aligned}
P_1gg' &= (t \wedge t' + s \wedge s' - 1) \\
&= ((1 + \varphi_2(b) + f(\psi_b, b))/2 \wedge ((1 + \varphi_2(b') + f(\psi_{b'}, b'))/2) \\
&\quad + ((1 + \varphi_2(b) - f(\psi_b, b))/2 \wedge ((1 + \varphi_2(b') - f(\psi_{b'}, b'))/2) - 1 \\
&= (((f(\psi_b, b) + \varphi_2(b))/2) \wedge ((f(\psi_{b'}(b') + \varphi_2(b'))/2) \\
&\quad - ((f(\psi_b, b) - \varphi_2(b))/2) \vee ((f(\psi_{b'}(b') - \varphi_2(b'))/2). \quad (2.13)
\end{aligned}
$$

Without loss of generality, assume that $f(\psi_b, b) + \varphi_2(b) \geq \varphi_2(b') + f(\psi_{b'}, b')$. If $f(\psi_b, b) - \varphi_2(b) \leq f(\psi_{b'}, b') - \varphi_2(b')$, it follows that

$$
\begin{aligned}
P_1gg' &= ((f(\psi_{b'}, b') + \varphi_2(b'))/2) - ((f(\psi_{b'}, b') - \varphi_2(b'))/2) \\
&= \varphi_2(b') = (PI(H(\psi_{b'}, b')) \\
&= PT_1^{-1}g' \geq PT_1^{-1}g'T_1^{-1}g. \quad (2.14)
\end{aligned}
$$

If $f(\psi_b, b) - \varphi_2(b) \geq f(\psi_{b'}, b') - \varphi_2(b')$, we obtain

$$
\begin{aligned}
P_1gg' &= ((f(\psi_{b'}, b') + \varphi_2(b'))/2) - (f(\psi_b, b) - \varphi_2(b))/2 \\
&= ((f(\psi_{b'}, b') - f(\psi_b, b))/2) + (\psi_2(b') + \varphi_2(b))/2 \\
&= ((f(\psi_{b'}, b') - f(\psi_b, b))/2) + P\{H(\psi_{b'}, b') \triangle H(\psi_b, b)\}/2 \\
&\quad + PI(H(\psi_b, b))I(H(\psi_{b'}, b')). \quad (2.15)
\end{aligned}
$$

Hence to prove the above inequality, we need only to show

$$
-(f(\psi_{b'}, b') - f(\psi_b, b)) \leq P\{H(\psi_{b'}, b') \triangle H(\psi_b, b)\}. \quad (2.16)
$$

when $f(\psi_{b'}, b') - f(\psi_b, b) \geq 0$, the inequality (2.16) is obvious.

If $f(\psi_b, b) - f(\psi_{b'}, b') > 0$, by the definition of $f(\psi_b, b)$, we have

$$
\begin{aligned}
&f(\psi_b, b) - f(\psi_{b'}, b') \\
&= P\{H(0,0) \triangle H(\psi_b, b)\} - P\{H(0,0) \triangle H(\psi_{b'}, b') \\
&= P\{\{X' \leq 0\} \cap \{(\cos\psi_b, \sin\psi_b) > b\}\} \\
&\quad - P\{\{X' \leq 0\} \cap \{(\cos\psi_{b'}, \sin\psi_{b'})X > b'\}\} \\
&\quad + P\{\{X' > 0\} \cap \{(\cos\psi_b, \sin\psi_b)X \leq b\}\} \\
&\quad - P\{\{X' > 0\} \cap \{(\cos\psi_{b'}, \sin\psi_{b'})X \leq b'\}\} \\
&\leq P\{\{X' < 0\} \cap \{(\cos\psi_b, \sin\psi_b)X > b\} \cap \{(\cos\psi_{b'}, \sin\psi_{b'})X \leq b'\}\} \\
&\quad + P\{\{X' > 0\} \cap \{(\cos\psi_b, \sin\psi_b)X \leq b\} \cap \{(\cos\psi_{b'}, \sin\psi_{b'})X > b'\}\} \\
&\leq P\{H(\psi_b, b) \triangle H(\psi_{b'}, b')\}, \quad (2.17)
\end{aligned}
$$

where X' is the first component of X, the inequality (2.16) holds. So $P_1 g g' \geq P T_1^{-1} g T_1^{-1} g'$. Then it follows that from the definition of X

$$P\{\sup_{\mathcal{F}} | W_P(f) | > \lambda\} \geq P\{\sup_{\mathcal{F}_1} | W_P(f) | > \lambda\}$$

$$= P\{\sup_{\mathcal{G}_1} | X_P(g) | > \lambda\} \geq (1/2) P\{\sup_{\mathcal{G}_1} | Y_{P_1}(g) | > \lambda\}. \qquad (2.18)$$

Let

$$\mathcal{G}_2 = \left\{ g' = I\left(\left\{ X^T \leq (s',t'), \frac{1}{2} \leq s' \leq 1, \frac{1}{2} \leq t' \leq \frac{3}{4}, s' \leq t' \right\}\right) \right\}.$$

Define a mapping $T_2 : \mathcal{G}_1 \to \mathcal{G}_2$ by

$$T_2 I(\{ X^T \leq (s,t)\}) = I\left(\left\{ X^T \leq \left(s - \frac{1}{4}, t + \frac{1}{4}\right) \right\}\right) = I(\{ X^T \leq (s',t')\}),$$

T_2 is obviously one-to-one and onto. Furthermore, the identical relation $s \wedge s_1, +t \wedge t_1, -1 = s' \wedge s_1' + t' \wedge t_1' - 1$ yields that for $g, g_1 \in \mathcal{G}_1$ and $g', g_1' \in \mathcal{G}_2$

$$EY_{P_1}(g)Y_{P_1}(g_1) = EY_P(g')Y_{P_1}(g_1'). \qquad (2.19)$$

Following from this formula and normality, we obtain that

$$\{Y_{P_1}(g) : g \in \mathcal{G}_1\} = \{Y_{P_1}(g') : g' \in \mathcal{G}_2\} \qquad (2.20)$$

in distribution, and then for any $\lambda > 0$

$$P\{\sup_{\mathcal{G}_1} | Y_{P_1}(g) | > \lambda\} = P\{\sup_{\mathcal{G}_2} | Y_{P_1}(g') | > \lambda\}. \qquad (2.21)$$

Furthermore, let

$$\mathcal{G}_3 = \mathcal{G}_1 \cup \mathcal{G}_2 = \left\{ g \in \mathcal{G} : \frac{1}{4} \leq s \leq 1, \frac{1}{2} \leq t \leq 1, s \leq t \right\}.$$

We have from (2.21)

$$P\{\sup_{\mathcal{G}_3} | Y_{P_1}(g) | > \lambda\} \leq 2P\{\sup_{\mathcal{G}_1} | Y_{P_1}(g) | > \lambda\}. \qquad (2.22)$$

Put

$$\mathcal{G}_4 = \left\{ g \in \mathcal{G} : 0 \leq s \leq \frac{1}{4}, 0 \leq t \leq 1, s \leq t \right\}.$$

Define a mapping T_3 such that

$$T_3 I(\{X^T \le (s,t)\}) = I\left(\left\{X^T \le \left(s + \frac{1}{4}, t - \frac{1}{4}\right)\right\}\right), \qquad (2.23)$$

so $T_3 \mathcal{G}_4 = \{g \in \mathcal{G} : \frac{1}{4} \le s' \le \frac{1}{2}, \frac{1}{2} \le t' \le \frac{3}{4}, s' \le t'\} \subset \mathcal{G}_3$. In the same procedure as above, we have in distribution

$$\{Y_{P_1}(g) : g \in \mathcal{G}_4\} = \{Y_{P_1}(g') : g' \in T_3 \mathcal{G}_4\}. \qquad (2.24)$$

Put $\overline{\mathcal{G}} = \{g \in \mathcal{G} : 0 \le s \le 1, 0 \le t \le 1, s \le t\} = \mathcal{G}_3 \cup \mathcal{G}_4$, we have the following inequality

$$P\{\sup_{\overline{\mathcal{G}}} \mid Y_{P_1}(g) \mid > \lambda\}$$
$$\le P\{\sup_{\mathcal{G}_3} \mid Y_{P_1}(g) \mid > \lambda\} + P\{\sup_{\mathcal{G}_4} \mid Y_{P_1}(g) \mid > \lambda\}$$
$$\le 2P\{\sup_{\mathcal{G}_3} \mid Y_{P_1}(g) \mid > \lambda\}$$
$$\le 4P\{\sup_{\mathcal{G}_1} \mid Y_{P_1}(g) \mid > \lambda\}. \qquad (2.25)$$

Denote that

$$\tilde{D}_n(g) = (\tilde{P}_n g - Pg), \quad D_n^{(1)} = \sup_{\mathcal{G} - \overline{\mathcal{G}}} \mid \tilde{D}_n(g) \mid,$$

and

$$D_n^{(2)} = \sup_{\mathcal{G}} \mid \tilde{D}_n \mid,$$

using the symmetry of $\tilde{D}_n(g)$ with respect to s and t, we easily know that $\sqrt{n} D_n^{(1)} = \sqrt{n} D_n^{(2)}$ in distribution, then their weak limit $\sup_{\mathcal{G} - \bar{\mathcal{G}}} \mid Y_{P_1}(g) \mid$ and $\sup_{\bar{\mathcal{G}}} \mid Y_{P_1}(g) \mid$ have relation as follows

$$\sup_{\mathcal{G} - \bar{\mathcal{G}}} \mid Y_{P_1}(g) \mid = \sup_{\bar{\mathcal{G}}} \mid Y_{P_1}(g) \mid \quad \text{in} \quad \text{distribution.} \qquad (2.26)$$

But the $P\{\sup_{\mathcal{G}} \mid Y_{P_1}(g) \mid > \lambda\}$ is know exactly, having been determined by Kac, Kiefer and Wolfowitz[9](1955, equation(4.6), or see[10],(1.3)). So from (2.18), (2.25) and (2.26)

$$P\{\sup_{\mathcal{F}} \mid W_P(f) \mid > \lambda\}$$

$$\geq \frac{1}{2} P\{\sup_{\mathcal{G}_1} \mid Y_{P_1}(g) \mid > \lambda\}$$

$$\geq \frac{1}{8} P\{\sup_{\tilde{\mathcal{G}}} \mid Y_{P_1}(g) \mid > \lambda \geq\}$$

$$\geq \frac{1}{16} P\{\sup_{\mathcal{G}} \mid Y_{P_1}(g) \mid > \lambda \geq\}$$

$$= \frac{1}{2} \sum_{m=1}^{\infty} (m^2\lambda^2 - \frac{1}{4}) \exp(-2m^2\lambda^2). \tag{2.27}$$

For the case $d > 2$, let

$$\mathcal{F}' = \{f = I(\{a^T X > b\}), \quad a = (\cos\varphi, \sin\varphi, 0, \cdots, 0), \quad 0 \leq \varphi \leq 2\pi\},$$
$$S'_d = \{a \in S_d, \quad a = (\cos\varphi, \sin\varphi, 0, \cdots, 0), \quad 0 \leq \varphi \leq 2\pi\}.$$

Since two-dimensional marginal probability measure P' of P is still spherically symmetric non-atomic, we can obtain that

$$P\{\sup_{\mathcal{F}} \mid W_P(f) \mid > \lambda\}$$

$$\geq P\{\sup_{\mathcal{F}'} \mid W_P(f') \mid > \lambda\}$$

$$= P\{\sup_{\mathcal{F}'} \mid W_{P'}(f) \mid > \lambda\}. \tag{2.28}$$

Now, the proof of Theorem 1 is completed.

3   The Proof of Theorem 2

In order to use Theorem 1 to prove Theorem 2, we shall need proceed Via the embedding argument. Theorem 7.1 of Dudley and Philipp(1983) implies the following embedding theorem.

Theorem A (Dudley and Philipp). *Let $X_1, X_2, \cdots$ be an infinite sequence of i.i.d r.v.s with common probability measure P. Let P_n be the empirical measure based on $X_1, \cdots, X_n$. Then, enlarging the probability space if necessary, for every $\theta > 0$ and $H < \infty$ there exist $Y_1, Y_2, \cdots$ be an infinite sequence of i.i.d. Gaussian processes with the same distribution law as that of W_P such that*

$$P\left\{\sqrt{n} \max_{1\leq k\leq n} \sup_{\mathcal{F}} \left| P_k f - Pf - (1/k)\sum_{j\leq k} Y_j(f) \right| > (\log n)^{-\theta}\right\}$$

$$\leq c(\theta, H)(\log n)^{-H}, \tag{3.1}$$

where $c(\theta, H)$ is the constant depending only on θ and H.

By the properties of $\{Y_1, Y_2, \cdots\}$ we have

$$\frac{1}{n} \sum_{j=1}^{n} Y_j = W_P \quad \text{in} \quad \text{distribution}, \tag{3.2}$$

consequently,

$$\sup_{\mathcal{F}} \left| \frac{1}{n} \sum_{j=1}^{n} Y_j(f) \right| = \sup_{\mathcal{F}} | W_P(f) | \quad \text{in} \quad \text{distribution}, \tag{3.3}$$

Both (3.1) and (3.3) imply (1.12). The proof of Theorem 2 is complete.

References

[1] Devroye,L.,(1982) Bounds for the uniform deviation of empirical measure, *J. Multivariate Anal.*, 12, 72–79.

[2] Vapnik, V.N. and Cervonenkis, A, YA. (1971) on the uniform convergence of relative frequencies of events to their probabilities, *Theory Prob. Appl.* 16, 264–280. (Teor. Verojatnost. i Ptimenen. 16, 264-297, in Russian).

[3] Huber, P.J. (1988) Spurious structure? or improved bounds for the Kolmogorov index. *Research Report, Harvard Univ.* Dept of Statistics.

[4] Ohrvik, J. (1988) On the distribution of the Kolmogorov distance in the worst direction. *Research Report. Univ. of Stockholm*, Dept. of Statistics.

[5] Dudley. R.M. (1978) Central limit theorems for empirical measure, *Ann. Probab.* 6, 899–929.

[6] Dudley, R.M. and Philipp, W.(1983) Invariance principle for sums of Banach space valued random elements and empirical processes, *Z.Wahrsch. Verw. Gebiete*, 62, 509–552.

[7] Adler,R.J. and Brown, L.D.(1986) Tail behaviour for supreme of empirical processes, *Ann. Probab.*, 14, 1–30.

[8] Zhu L.X.(1989) The tail probability for the suprema of Gaussian processes indexed by half-spaces with applications (in Chinese).

[9] Kac, M, Kiefer, J. and Wolfowitz,J.(1955) On tests of normality and other tests of goodness fit based on distance methods, *Ann Math. statist,* 26, 189–211.

[10] Kiefer, J.,(1961) On large deviations of the empiric D.F. of vector chance variables and a law of the iterated logarithm, *Pacific,J.Math.,* 11, 649–660.

[11] Zhu Li-Xing (1989). The tail behaviour for the distribution of Kolmogorov distance in the worst direction.